Encyclopedia of Mathematics and Society

Mathematics and the Social Sciences

Encyclopedia of Mathematics and Society

Mathematics and the Social Sciences

Editors
Sarah J. Greenwald
and
Jill E. Thomley
Applachian State University

SALEM PRESS
A Division of EBSCO Publishing
Ipswich, Massachusetts

Cover Photo: © Justin Lane/epa/Corbis

Copyright © 2013, by Salem Press, A Division of EBSCO Publishing, Inc.
All rights reserved. No part of this work may be used or reproduced in any manner whatsoever or transmitted in any form or by any means, electronic or mechanical, including photocopy, recording, or any information storage and retrieval system, without written permission from the copyright owner. For permissions requests, contact proprietarypublishing@ebscohost.com.

ISBN 978-1-4298-3752-1

Printed in the United States of America

Contents

List of Contributors   ix

Accounting	1
Advertising	4
Africa, Central	7
Africa, Eastern	9
Africa, Northern	11
Africa, Southern	12
Africa, Western	14
America, Caribbean	15
America, Central	17
America, North	18
America, South	20
Arabic/Islamic Mathematics	22
Asia, Central and Northern (Russia)	24
Asia, Eastern	26
Asia, Southeastern	28
Asia, Southern	30
Asia, Western	32
Atomic Bomb (Manhattan Project)	35
Babylonian Mathematics	37
Bankruptcy, Business	39
Bankruptcy, Personal	41
Bar Codes	43
Budgeting	44

Census	46
Chinese Mathematics	48
Civil War, U.S.	52
Cold War	55
Comparison Shopping	59
Coupons and Rebates	61
Credit Cards	63
Currency Exchange	65
Data Mining	67
Egyptian Mathematics	69
Elections	72
Europe, Eastern	77
Europe, Northern	81
Europe, Southern	82
Europe, Western	84
Gerrymandering	86
Greek Mathematics	88
Gross Domestic Product (GDP)	91
Home Buying	93
Incan and Mayan Mathematics	96
Income Tax	99
Industrial Revolution	102
Infantry (Aerial and Ground Movements)	104
Insurance	106
Intelligence and Counterintelligence	108
Inventory Models	111
Loans	112
Market Research	114
Mathematics: Discovery or Invention	116
Middle Ages	118
Military Draft	120
Missiles	122
Mutual Funds	123
National Debt	125
Native American Mathematics	127
Oceania, Australia and New Zealand	129

Oceania, Pacific Islands	131
Payroll	132
Pearl Harbor, Attack on	134
Pensions, IRAs, and Social Security	135
Predicting Attacks	138
Predicting Preferences	139
Prehistory	142
Quality Control	144
Renaissance	146
Revolutionary War, U.S.	148
Roman Mathematics	149
Sales Tax and Shipping Fees	151
Scheduling	153
Stock Market Indices	154
Strategy and Tactics	156
Unemployment, Estimating	159
Vedic Mathematics	161
Vietnam War	163
Voting Methods	164
World War I	167
World War II	170
Resource Guide	**175**
Index	**179**

List of Contributors

Micah Altman
  Harvard University
Mohamed Amezziane
  DePaul University
Judith E. Beauford
  University of the Incarnate Word
Bonnie Ellen Blustein
  West Los Angeles College
Casey Borch
  University of Alabama at Birmingham
Sarah Boslaugh
  Washington University School of Medicine
David Brink
  University College Dublin, Ireland
Patrick L. Brockett
  University of Texas at Austin
Darrah Chavey
  Beloit College
John T. Chen
  Bowling Green State University
Loren Cobb
  University of Colorado, Denver
Dogan Comez
  North Dakota State University
Justin Corfield
  Geelong Grammar School
Richard De Veaux
  Williams College
Maria Droujkova
  Natural Math
Gisela Ernst-Slavit
  Washington State University, Vancouver
Amy Everton
  Independent Scholar

Jonathan David Farley
  University of Oxford
Daniel J. Galiffa
  Penn State Erie, The Behrend College
Rick Gorvett
  University of Illinois at Urbana-Champaign
Judith V. Grabiner
  Pitzer College
Sarah J. Greenwald
  Appalachian State University
Juan B. Gutierrez
  University of Miami
Simone Gyorfi
  O. Goga High School, Jibou, Romania
Thomas W. Hair
  Florida Gulf Coast University
Holly Hirst
  Appalachian State University
Calli A. Holaway
  University of Alabama
Liang Hong
  Bradley University
Jerry Johnson
  Western Washington University
Pete Johnson
  Eastern Connecticut State University
Ugur Kaplan
  Kadir Has University, Istanbul
Matt Kretchmar
  Denison University
Bill Kte'pi
  Independent Scholar
Konnie G. Kustron
  Eastern Michigan University

List of Contributors

Alistair Kwan
Yale University

James Landau
Independent Scholar

Carmen M. Latterell
University of Minnesota, Duluth

Michael G. Lovorn
University of Alabama

Chad T. Lower
Pennsylvania College of Technology

Yiu-Kwong Man
The Hong Kong Institute of Education

Mariana Montiel
Georgia State University

David C. Royster
University of Kentucky

Alun Salt
University of Leicester

Shahriar Shahriari
Pomona College

Barbara A. Shipman
University of Texas at Arlington

Lawrence H. Shirley
Towson University

Daniel Showalter
Ohio University

David Slavit
Washington State University

Ravi Sreenivasan
University of Mysore

Kristi L. Stringer
University of Alabama at Birmingham

Stephen Szydlik
University of Wisconsin, Oshkosh

Jill E. Thomley
Appalachian State University

Christopher J. Weinmann
Independent Scholar

Accounting

Category: Business, Economics, and Marketing.
Fields of Study: Algebra; Number and Operations.
Summary: Accounting applies mathematics to the recording and analysis of a business's financial status.

Accounting is the recording, interpretation, and presentation of financial information about a business entity, typically with the goal of producing financial statements that describe the business's economic resources in standardized terms. Formal accounting began with the work of Franciscan friar Luca Pacioli, who introduced accounting techniques in his 1494 mathematical work *Summa de Arithmetica, Geometria, Proportioni et Proportionalita*. During the Industrial Revolution, Josiah Wedgwood introduced cost accounting, a technique to ensure a profit margin by calculating the costs of materials and labor at every stage of production and setting the price accordingly. The needs of stockholders and other interested parties within the business, and an increasingly complex business environment, have increased the need for financial record-keeping techniques that are thorough and produce useful financial statements. Modern accounting is assisted by a variety of software packages, but the accountant must still be well-versed in mathematics in order to interpret the information. The fundamental accounting equation can be stated as the following:

Assets = Liabilities + Owners' Equity.

For any given company, assets can be thought of as what the company owns. This includes cash (actual cash and bank accounts), money that is owed to the business (called accounts receivables), inventory, buildings, land, equipment, and intangibles like patents and goodwill. Liabilities are what the company owes. This includes money owed to a bank (notes payable), suppliers (accounts payable), or the government (taxes payable). Owners' equity can take several forms depending on who the owners are: a single person (sole proprietor), a few people (partnership), or shareholders (corporation). Each method of ownership has advantages and disadvantages, but regardless of the method, the owners' equity can be thought of as a net asset since it can be found by subtracting liabilities from assets.

Accounting is the process of keeping track of the operations and financial status of a business. (Photos.com)

Accounting as Record Keeping

Whenever a financial transaction takes place, it must be recorded in at least three locations. First, it will be recorded in the general ledger (a book of entry summarizing a company's financial transactions). When recorded, the entry should contain the date of the transaction, a brief description of the transaction, and the monetary changes to all accounts affected (which will be at least two).

From there, the transaction gets recorded a second time in a secondary (or subsidiary) ledger for each of the accounts affected. When the amounts are recorded, they are put into the left (debit) column or the right (credit) column of the ledger. (In bookkeeping, "debit" and "credit" mean left and right, respectively; they are not related to debit or credit cards in this situation.) The total of each column of the general ledger record must add to the same sum. In that manner, all money can be accounted for as going into or out of an account.

In order to determine whether to credit or debit an account, a general rule that works for most accounts is to first look at the fundamental accounting equation. Since assets are listed on the left, to increase assets, the transaction is recorded in the left column (debiting the account) and to decrease assets, the transaction is recorded in the right column (crediting the account). Similarly, since liabilities and owners' equity are listed on the right-hand side of the equation, to increase liabilities and owners' equity, the transaction is recorded in the right column (crediting the account) and to decrease liabilities and owners' equity, the transaction is recorded in the left column (debiting the account). For example, suppose a company needed to purchase $100 worth of office supplies. Furthermore, suppose the company pays $40 with cash and puts the remaining $60 on account (store credit). The general ledger may look like the following:

Figure 1. Purchased office supplies.

04-31-2017	Office supplies	$100	
	Cash		$40
	Accounts payable		$60

In Figure 1, notice that both the right and left columns add up to $100; this shows that no money was lost in the process. Office supplies are considered an asset, so since the company increased the amount of office supplies, that account was recorded on the left—in other words, it debited office supplies for $100. Cash is also an asset, but the company decreased the amount of cash it had. As a result, cash was credited (the transaction was recorded on the right for that account). Accounts payable is a liability the company owes to the retailer it purchased the products from. Since the company increased the amount it owed the retailer, that account was recorded on the right as an increase to the company's liabilities—accounts payable was credited.

Once this transaction was recorded in the general ledger, the company would also need to record this transaction in the Office Supplies ledger, the Cash ledger, and the Accounts Payable ledger. Accounts are debited or credited in their specific ledgers in the exact same manner that they are debited or credited in the general ledger. In a similar manner, the retailer who sold the office supplies would need to record this same transaction into his or her general and secondary ledgers. However, the retailer's transaction would use the opposite side to denote the sale as follows:

Figure 2. Sold office supplies.

04-31-2017	Cash	$40	
	Account receivable	$60	
	Inventory		$100

Again, the right and left columns add up to the same amount. Contrary to the purchasing company, the receiving company lists three assets to record the transaction. Cash and accounts receivable are both being increased, so debited. The asset "inventory" is being decreased and results in a credit to inventory. If this were a large company, rather than record each individual transaction, the retailer would most likely record an entire day's transactions as a single entry at the end of each business day. Once the general ledger has been recorded, the secondary ledgers need adjusting entries as well to denote the transaction(s).

Accounting as Record Sharing

In addition to keeping records of transactions for a business, accounting is responsible for creating reports that summarize the journals to share with others. To learn about the reports and how to create reports intended for people outside the business (such as shareholders, creditors, or government agencies), a person can take a

Benford's Law

Benford's law, named after physicist Frank Benford, gives the probability with which the numbers 1 through 9 will occur as the first digit in many types of real-life data. For example, in a list of actual bank account deposits in a given day, about 30% of the time the first digit of the deposit amount will be a 1. Fraudulent data that has been created by people often does not match the expected probabilities.

In very large modern data sets, highly focused tests use this principle to find deviations in selected subsets; for example, the occurrence of a suspiciously large frequency of $24 receipts submitted in a company that has a $25 maximum meal allowance.

---

class in financial accounting. To learn about the reports and how to create reports intended for people inside the business (such as managers), a person can take a class in managerial accounting.

The most common reports created for people outside the business are balance sheets, income statements, cash flow statements, and retained earnings statements. Of the four statement types, the balance sheet is written as a snapshot of the company at a point in time. In contrast, the other three statements are created to show what happened over a period of time such as a month, quarter, or year. When creating these reports, the income statement is usually completed first. As its name implies, the income statement is created to determine the company's income during a specific time period. The income statement is also known as a profit and loss statement (P&L) or earnings statement. Information from the income statement is then used to create the retained earnings statement. Finally, the information from the retained earnings statement is used on the balance sheet.

The balance sheet first lists all of the company's assets in order of liquidity (the ability to turn the asset into cash easily) from the most liquid to the least liquid. The assets are then added together to find the total assets of the company. The balance sheet next lists all of the company's liabilities in order of due date from the soonest due to the latest due. Below the liabilities is listed the owners' equity (which includes retained earnings from the retained earnings statement). The liabilities and owners' equity are added together. Referring back to the fundamental accounting equation, both of these amounts (the total assets and the sum of the liabilities and owners' equity) should equal one another.

Reports created for internal users vary widely depending on the reasoning and the need for the report. Internal reports are usually created and specifically designed for making decisions within the company. For example, manufacturers could use internal reports to determine the optimal price of their product.

Manufacturers may also use internal reports to determine if it is more cost effective to create a needed part or to purchase the part from another company. They may need to consider continuing or eliminating a division of their company. Managerial accounting is also responsible for budgeting and forecasting.

Mathematical Models

Many areas in financial accounting rely on mathematical models for explanation and prediction. For example, models have played important roles in applications such as understanding the consequences of public disclosure, formalizing market efficiency or competition, measuring income, and evaluating equilibrium pricing for goods and services. Some important mathematical techniques used in accounting models include linear regression, systems of simultaneous equations, equilibrium notions, and stochastic analysis. In the latter, random rather than constant inputs are used to model scenarios where decisions must be made under realistic conditions of uncertainty. The data used in these models may be cross-sectional (representing a single snapshot in time) or longitudinal (one or more variables are measured repeatedly to detect trends and patterns). Probability theory is also used to detect instances of accounting fraud.

Further Reading

Davis, Morton D. *The Math of Money: Making Mathematical Sense of Your Personal Finances.* New York: Copernicus, 2001.

Hoyle, Joe Ben, Thomas F. Schaefer, and Timothy S. Doupnik. *Fundamentals of Advanced Accounting.* New York: McGraw-Hill, 2010.

Kimmel, Paul D., Jerry J. Weygandt, and Donald E. Keiso. *Financial Accounting: Tools for Business Decision Making*. 5th ed. Hoboken, NJ: Wiley, 2009.

Mullis, Darrell, and Judith Handler Orloff. *The Accounting Game: Basic Accounting Fresh From the Lemonade Stand*. Naperville, IL: Sourcebooks, 2008.

Verrecchia, Robert. "The Use of Mathematical Models in Financial Accounting." *Journal of Accounting Research* 20 (1982).

Weygandt, Jerry J., Paul D. Kimmel, and Donald E. Keiso. *Managerial Accounting: Tools for Business Decision Making*. 4th ed. Hoboken, NJ: Wiley, 2008.

Chad T. Lower

Advertising

Category: Business, Economics, and Marketing.
Fields of Study: Algebra; Data Analysis and Probability; Number and Operations.
Summary: Mathematics is used to weigh the costs and gains of advertising and to profile and target consumers.

Advertising delivers product information from suppliers to consumers—suppliers may be manufacturers, hospitals, software developers, educators—and is critical to the success of a business in marketing development. Advertising media may be traditional (such as television, newspapers, and posters) or technological (via Internet and e-mail), as well as commercial (to sell products for profit) or noncommercial (in political campaigns or for religious purposes). The annual advertising cost in the United States amounts to more than $100 billion.

Advertising includes two stages: the planning stage for marketing strategies, whose goal is business development, and the analysis stage of cost analysis involved with the forms and the contents of communication between suppliers and potential customers. Mathematics and statistics play critical roles in both stages of advertising.

Market Shares

In the planning stage, the analysis of market shares for advertising necessitates matrix operations and multivariate probability inequalities to portray the dynamics of market shares over time. The following is an example of matrix operations, which bridge advertising with market shares. Consider the market shares of General Motors (GM) and Ford in the U.S. automobile industry. Assume that the current market shares distribute as follows:

General Motors:	21%
Ford:	17%
Other Manufacturers:	62%

If GM starts an advertising campaign with the goal of increasing the market share to 29% in three years, GM may count on customers to switch from Ford or other manufacturers to GM. However, in reality, some of the GM customers may switch to Ford or to other manufacturers.

Let a_1, a_2, a_3 be the percentages of original GM users who, at the end of the advertising campaign, remain with GM, who switch to Ford, and who switch to other manufacturers, respectively. Let b_1, b_2, b_3 be the percentages of original Ford users who switch to GM, who remain with Ford, and who switch to other manufacturers, respectively. Let c_1, c_2, c_3 be the percentages of the other customers who switch to GM, who switch to Ford, and who remain with their manufacturers, respectively. Then, the market shares x_{GM}, x_{Ford}, and x_{Others} at the end of the three years are determined by the following simple matrix equation:

$$\begin{bmatrix} x_{GM} \\ x_{Ford} \\ x_{Others} \end{bmatrix} = \begin{bmatrix} a_1 & b_1 & c_1 \\ a_2 & b_2 & c_2 \\ a_3 & b_3 & c_3 \end{bmatrix} \begin{bmatrix} 21\% \\ 17\% \\ 62\% \end{bmatrix}.$$

If GM intends to increase x_{GM} to 29%, GM should advertise specifically to different groups of customers. This is mathematically equivalent to manipulating the elements in the 3×3 matrix above within plausible ranges of the elements.

The foregoing scenario is a simplified example to illustrate the role of matrix operations in advertising. In reality, the story is more complex. For example, the 3×3 matrix above will become an $n \times n$ matrix, where n is the number of competing suppliers in the market. Also, the stochastic feature of the supply-demand market, the market shares, and the corresponding elements for the $n \times n$ matrix change constantly under the influence of the advertising campaign.

Thus, it is more appropriate to treat the market shares as a vector consisting of random variables. In this case, one of the convenient approaches to evaluating the market shares is the method of multivariate probability inequalities in conjunction with the construction of Hamilton-type circuits.

Advertising Costs and Effects

The analysis stage examines costs and effects associated with various communication channels and advertising media. For instance, in Internet advertising, typical cost considerations are cost per mile (CPM), cost per click (CPC), and conversion rate. These terms have strong connections with mathematics and statistics.

For Web advertising, CPM usually refers to the cost for every thousand visits to the publisher's Web site. For example, assume that an ad network offers a $5 CPM for a banner, which was put on three Web sites for three months. If the total page views for the three Web sites are 80,000, 110,000, and 140,000 during the three-month period, the total cost of Web advertising for the ad network is

$$\$5\frac{80{,}000}{1000}+\$5\frac{110{,}000}{1000}+\$5\frac{140{,}000}{1000}=\$1{,}650$$

In general, if an ad is posted in n Web sites, the total cost is

$$\sum_{i=1}^{n} \text{CPM} \times (W_i/1000)$$

where W_i is the number of Web impressions (visits) to the ith publisher's Web site for the same period of time.

Consider that the number of Web impressions on each publisher's Web site depends on many continuously changing factors; then W_i is a random number. Let $E(W_i)$ be the expected value of W_i, which measures the long-term average of the number of Web impressions of the banner on the ith publisher's Web site. The long-term average cost is

$$\sum_{i=1}^{n} \text{CPM} \times (E(W_i)/1000).$$

CPC refers to the amount that the advertiser pays for each click generated from the Web publisher. For example, if the cost per click is $0.04, and three Web publishers generate 1700, 1600, and 900 clicks in three months, the cost of Web advertising is

$$\$0.04(1700)+\$0.04(1600)+\$0.04(900)=\$168$$

In general, if a Web ad is posted in m Web sites, the total cost is

$$\sum_{i=1}^{m} \text{CPC} \times C_i$$

where C_i is the number of clicks generated on the ith publisher's Web site for a given period of time.

Consider the fact that the number of clicks on each publisher's Web site depends on various unexpected factors: C_i is actually a random variable. Let $E(C_i)$ be the expected value of C_i, which measures the long-term average of the number of clicks generated from the ith publisher's Web site over a given period of time. The long-term average ad cost is then

$$\sum_{i=1}^{m} \text{CPC} \times E(C_i).$$

The foregoing two concepts, CPM and CPC, measure the potential impact of the internet ad only in terms of clicks or Web visits. However, these two concepts are unable to provide the advertiser with information regarding whether the Web impression has been transferred into the desired action (such as buying the advertised product). A useful measurement in Web advertising to help account for the advertising effect is the "conversion rate" (or CR, the average number of people taking the action encouraged by the ad per 100 visits to the publisher's Web site). For example, if out of 2000 clicks on an ad posted on a publisher's Web site, 12 people end up buying the product, the conversion rate of the ad for this Web site is then

$$\left(\frac{12}{2000}\right) \times 100 = 0.6\%.$$

Being highly associated with key factors such as the design of the publisher's Web site, the conversion rate is an index that directly measures the final impact of the ad for the Web site.

Since the conversion rate directly reflects the performance of the Web site, it can be used to compare advertising effects of two or more Web sites. However, it is risky to compare conversion rates directly. The example in Figure 1 helps illustrate this point. Consider two Web sites: Google AdSense and Chitika. If the conversion rates of the two Web sites are as follows

Figure 1.

	May	June	July	August
Google AdSense	5%	6.1%	4.3%	7.5%
Chitika	7.3%	5.2%	5.7%	6.4%

in the past four months, it is impossible to claim which site has better performance on Web advertising.

In fact, the raw values shown in Figure 1 include the stochastic influence of many online factors. In this case, to evaluate the monthly advertising effect of different Web sites accurately, statistical data analysis is needed.

Because of random effects, the expected value of the conversion rate of each Web site should be considered when comparing two or more publishers' Web sites in terms of the conversion rates. Given a set of historical data involving all the Web sites of interest, one of the statistical estimation approaches is the method of "simultaneous confidence intervals," which compares the ranges of expected conversion rates with a pre-specified confidence level. For example, with a set of data for the conversion rates of three Web sites over a period of time, if a 95% simultaneous confidence interval reads

$$0.5\% < CR_{Google} - CR_{Chitika} < 2\%$$

and

$$1.3\% < CR_{Google} - CR_{Yahoo} < 3.4\%$$

it means that at 95% confidence level, the advertising performance (in terms of conversion rate) of Google is better than that of Chitika and Yahoo.

To enhance the accuracy of the simultaneous confidence ranges, or to improve the power of testing multiple advertising effects, the two-stage estimation procedure can be considered. When the underlying distribution of the monthly conversion rates is skewed, the two-stage estimation procedure can be used with nonparametric tests to make inferences on the performance of multiple Web sites.

Data Mining and Advertisements

Masses of personal data being collected every day about consumers, via mechanisms like credit card applications, consumer discount cards, and product views and ratings on shopping Web sites are poised to revolutionize the field of advertising. Data mining is the mathematical and statistical method for sifting through large volumes of data to find patterns and create prediction models, in this case of consumer behavior. In 2009, the online video rental company Netflix awarded a $1 million prize to the winners of its three-year contest to develop a better algorithm to predict what movies its users would prefer, based on ratings data provided by the company.

Finally, mathematics is used not only to decide when, where, and how to advertise products and services but also to determine what to emphasize within the advertisements themselves: discounts on pricing or the number of calories per serving, just to name two. However, it is often difficult to verify those numbers. Many will remember Trident Gum's 1960s slogan, "Four out of five dentists surveyed would recommend sugarless gum to their patients who chew gum." Although the statement was popular at the time, its legitimacy was later questioned, since it came from a survey whose details have never been released.

IBM has initiated a Smarter Planet campaign focused on dispersed or cloud computing (Internet-based computing). Its "Smarter Math Builds Equations for a Smarter Planet" commercial cites mathematics as the universal language and gives a number of ways in which mathematics will be used to create a "smarter planet."

Further Reading

Baines, Paul. "A Pie in the Face." *Alternatives Journal* 27, no. 2 (2001).

Graydon, Shari. *Made You Look—How Advertising Works and Why You Should Know*. Toronto: Annick Press, 2003.

Kotabe, Masaki, and Kristiaan Helsen. *Global Marketing Management*. Hoboken, NJ: Wiley, 2004.

Laermer, Richard, and Mark Simmons. *Punk Marketing*. New York: HarperCollins, 2007.

Murray, David, Joel Schwartz, and S. Robert Lichter. *Ain't Necessarily So: How the Media Remake Our Picture of Reality*. New York: Penguin, 2002.

Russell, J. Thomas, and W. Ronald Lane. *Kleppner's Advertising Procedure*. Upper Saddle River, NJ: Prentice Hall, 1999.

John T. Chen

Africa, Central

Category: Mathematics Around the World.
Fields of Study: All.
Summary: Central African contributions include counting games and decorative geometric patterns.

Central Africa comprises Angola, the Central African Republic, Chad, Congo, the Democratic Republic of the Congo, Equatorial Guinea, Gabon, and Sao Tome and Principe. Mathematical concepts developed in central Africa include variations of the counting game Mancala and the sophisticated geometric patterns used in traditional art. These patterns, in sand art and pottery, woven into mats and baskets, and displayed in tattoos, include complex symmetries and fractals. Some educators have advocated incorporating these indigenous African manifestations of mathematics into school curriculums.

Mancala

As with much of Africa, variations of the mathematical counting game Mancala were played throughout the region. The mathematics of Mancala games are discussed in more detail in the entry "Africa, East," but some description here is warranted. *The Complete Mancala Games Book* gives rules for 28 different versions of this game played in central Africa. These variations arise throughout much of central Africa but especially in Cameroon and the Congo. While the version of Mancala best known in the United States is a two-row version (also called Wari or Oware), many of the variations played in the Congo have four rows, which adds substantially to the complexity of the game, as well as the complexity of the arithmetic calculations and logical thinking required to play them well. Even with the two-row version, the Congolese variation Mbele uses a complicated game board (a two-row version with many holes in each row, with the rows pinched together near the ends). Again, this adds mathematical complexity to the game.

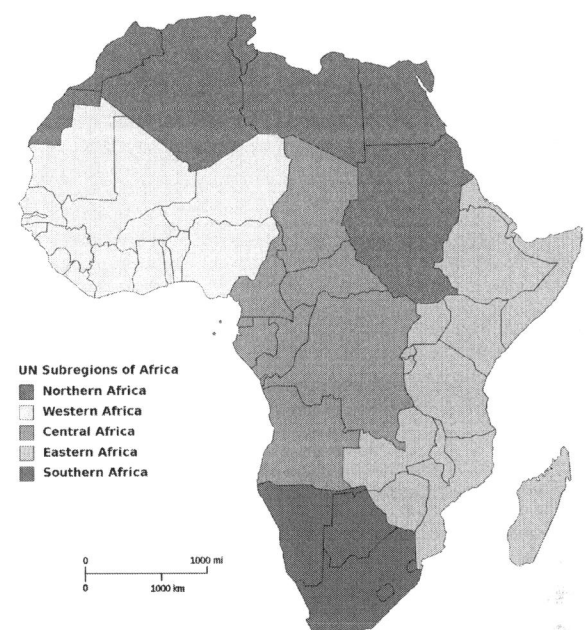

Central Africa is composed of eight countries and is shown in the medium gray shaded area. (Darrah Chavey)

Geometric Patterns

Many of the most interesting mathematics developed by the peoples of central Africa have been geometric in nature. A significant part of African art traditions include quite complex—and mathematically sophisticated—geometric patterns. These patterns include symmetries in various combinations, between different elements, and between various colors. Claudia Zaslavsky writes: "If one wanted to survey the whole field of geometric design in Africa, one would have to catalogue almost every aspect of life." In central Africa, such geometric patterns are found on pottery, cloths, mats, carvings, baskets, bowls, tattoos, and other objects of daily use.

The Kuba people of the Congo are particularly famous for such art, especially their raffia embroidered cloth. Both *Africa Counts* and *Geometry From Africa* show many examples of Kuba artwork, along with artwork of other African peoples. The woven mats of the Yombe women of the Congo are another example of complex geometric design. Paulus Gerdes has studied these mat designs as an interplay between cultural values and mathematics.

The art of the Chokwe people of the Congo and Angola includes a mathematically challenging art form called "sona," usually drawn in the sand. These drawings are made with a single line continuously weaving through an arrangement of dots, such as the "Lion With Cubs" drawing of the accompanying figure. The heads and tails of the animals are added after the principal line is drawn. These drawings represent stories, morals, or values of the Chokwe, or just an animal or object from their environment. The techniques for determining which dot arrangements will generate such one-line drawings are fundamentally mathematical in nature. Drawings that can be done in a single line, without retracing, are a mathematics topic known as Eulerian Graphs. This artwork of the Chokwe is strongly connected to this mathematical idea, and was being investigated by the Chokwe artists about the same time that the idea was first studied by European mathematicians in the mid-eighteenth century.

The geometric patterns of central Africa extend to include fractal designs. Fractals are a mathematical structure that can be viewed as a repetition of the same shapes at many different sizes or scales. For example, trees have branches, each with smaller branches, and then even smaller branches. Western architecture often has rectangular blocks with rectangular houses, but rarely are such shapes repeated at more than two scales, and rarely is this a conscious shape imitation. African fractals often use circular, oval, or diamond shapes at several scales, with smaller shapes inside or around the larger shapes. There is substantial evidence that at least some of these fractal designs are a conscious choice of the artists and builders, and not accidental. *African Fractals* shows several Cameroonian examples of fractal designs in cities and villages, and even in hair braiding. This book also shows a similar style of pattern, using increasingly smaller but otherwise identical shapes in the art of the Mangbetu people of the Congo.

Education

Several African educators have suggested incorporating these traditional mathematical elements into their schools. The Cameroonian educator A. N. Boma writes: "In African traditional education, the curriculum was organized holistically rather than in discipline areas such as mathematics, history. . . .Education for all cannot afford the luxury of isolating education in terms of disciplines, rather it should take the holistic approach in developing a total person. . . ." The ideas described here integrate mathematics with cultural, artistic, and other elements to achieve this holistic approach. Unfortunately, the schools in Central Africa cannot easily incorporate such ideas. The 2009 *Mathematics in Africa* report describes low percentages of the population attending schools, high student-to-teacher ratios, heavy use of recycled European mathematics textbooks, and few prepared teachers in most of central Africa outside of Cameroon. All of these facts make it difficult to customize mathematics education for African students. Cameroon has a more developed education system, but at the college level it is struggling with filling the mathematics faculty positions that have been approved, and most mathematics teaching there is done in large classes by low-level staff. Nevertheless, with more than half of the central African Ph.D.s

"Lion With Cubs" drawing made with a single line weaving through an arrangement of dots. (Darrah Chavey)

in mathematics, Cameroon may become a leader in mathematics education for the region.

Further Reading

Boma, A. N. "Some Lessons From Traditional Practices for Present-Day Education in Africa." In *African Thoughts on the Prospects of Education for All*. Dakar, Senegal: United Nations Educational, Scientific and Cultural Organization (UNESCO)-United Nations Children's Fund (UNICEF), 1990.

Eglash, Ron. *African Fractals: Modern Computing and Indigenous Design*. Piscataway, NJ: Rutgers University Press, 1999.

Gerdes, Paulus. *African Doctorates in Mathematics: A Catalogue*. Maputo, Mozambique: Research Centre for Mathematics, Culture and Education, 2007.

———. *Geometry From Africa*. Washington, DC: Mathematical Association of America, 1999.

Gerdes, Paulus, and Ahmed Djebbar. *Mathematics in African History and Cultures: An Annotated Bibliography*. Cape Town, South Africa: African Mathematical Union, 2004.

International Mathematical Union. "Mathematics in Africa: Challenges and Opportunities." 2009. http://www.mathunion.org/publications/reports-recommendations.

Russ, Laurence. *The Complete Mancala Games Book*. New York: Marlowe Co., 1999.

Zaslavsky, Claudia. *Africa Counts: Number and Pattern in African Cultures*. 3rd ed. Chicago: Chicago Review Press, 1999.

Darrah Chavey

Africa, Eastern

Category: Mathematics Around the World.
Fields of Study: All.
Summary: East African contributions include Mancala, logic games, and games similar to Tic-Tac-Toe.

Eastern Africa is the birthplace of the human species, and includes Burundi, Comoros, Djibouti, Eritrea, Ethiopia, Kenya, Madagascar, Malawi, Mauritius, Mayotte, Mozambique, Reunion, Rwanda, Seychelles,

The game of Mancala has many variations, one of which is played on this board in Zanzibar. (iStockphoto)

Somalia, Tanzania, Uganda, Zambia, and Zimbabwe. Mancala, an ancient counting game with many variations throughout the continent, originates in East Africa, which is also home to complicated geometric patterns in woven art and a number of logic puzzles and other mathematical games. The quality of mathematics education continues to be a serious concern.

Mancala

Eastern Africa is home to an impressive variety of mathematically based games. The most well known are the many variations of Mancala, often called the "African national game." Although there are hundreds of variations, the general idea is: (1) stones or seeds are placed in pits laid out with two to four rows and several pits per row; (2) players collect the seeds from one pit and "sow" them one at a time into other pits around the board; (3) under some circumstances, the player picks up the seeds from the final pit and continues sowing those seeds; (4) when the move ends, the player will, in some cases, capture seeds from his or her opponent. These games generally involve a substantial amount of counting, adding, and subtracting (for example, to determine where the final seed will land), as well as consideration of multiple possibilities, analysis to calculate where an opponent can move afterward, strategy, and logic. It is no wonder that some leaders (including Tanzanian president Julius Nyerere) were first noticed as good Mancala players. It is uncertain where the game originated, but the oldest dated game boards come from Ethiopia and Eritrea about 1300 years ago. The game is surely older than that, possibly as much as 3300 years old. *The Complete Mancala Games Book* includes 61 different variations of this game played in eastern Africa, including variations specific to every country except Burundi.

Other Puzzles and Games

Logic puzzles come in many forms. One puzzle type common to eastern Africa is the river-crossing puzzle. For example, a man with a wolf, a goat, and a cabbage must use a boat to cross a river except that (1) he can take only one item across at a time; and (2) the goat cannot be left alone with the wolf (who would eat it) or the cabbage (which it would eat). These kinds of puzzles are mathematical because, as Marcia Ascher writes, "A stated goal must be achieved under a given set of logical constraints." Variations of this puzzle, with different logical constraints, appear in Ethiopia, Zambia, and Mozambique.

Several "three-in-a-row" games, related to Tic-Tac-Toe, are played in eastern Africa. In Shisima, from Kenya, players start with an octagonal board, the eight corners, a center point, and lines connecting opposite corners through that center. Players start with three stones each, on the corners closest to them. During a turn, players move one of the stones to one of the nine points (eight corners and the center) connected to it, if it is empty. The goal is to get three stones in a row (a straight line), which must include the center and two corners opposite each other. *Africa Counts* describes two other three-in-a-row games from Zimbabwe, each of which begins like Tic-Tac-Toe where players place stones on points on the board, then continues like Shisima with players moving their stones to get a triple. In Tsoro Yematatu, the board has seven spots, each player has three stones, and one spot is always empty. In African Morris, there are 24 spots, and each player has 12 stones. Here, it could happen that the board becomes filled, but if there is a three-in-a-row during that stage, the player does not win; instead, the player captures an opponent's stone. Hence, the game usually continues into the second phase. These three-in-a-row games are logic puzzles and are examples of games of position, which have been widely studied in mathematics.

Geometric Patterns

The geometric patterns of art from eastern Africa contain a great deal of mathematical and geometric structure and symmetry. Some of the most well known of such crafts are the woven *sipatsi* baskets of Mozambique, and other types of woven baskets and mats from Mozambique, Kenya, Tanzania, Uganda, and Madagascar. This artwork contains varied types of symmetries and dramatic patterns. Paulus Gerdes writes that this art "reveals the force of the imagination and the artistic and geometric creativity of the women and men who weave [these baskets]." Examples exist in the Ba-ila settlement in Zambia and in Ethiopian processional crosses.

Mathematical Education

Mathematical education in eastern Africa shares many of the challenges that exist throughout the continent, especially the lack of prepared teachers at the secondary level. As the South African mathematics educator Jan Persons writes, "At the departure of the Portuguese from Mozambique in the early 1970s, there were only a handful of qualified secondary mathematics teachers. In general, starving the local population of decent and effective education was used as a weapon to halt or, at least, retard development."

This issue has been a major problem in eastern and central Africa, which combined have 48% of Africa's population but have produced less than 8% of Africa's mathematics Ph.D.s. Kenya has a strong college-level mathematics program, having produced nearly half of all Ph.D.s in eastern Africa. Unfortunately, as also happens in central Africa, most of the mathematics students are attracted into professions other than teaching because of the low salaries for teachers. There are several efforts in place to improve mathematics education in these countries, but much work on the educational structures remains to be done throughout this region.

Further Reading

Ascher, Marcia. "A River-Crossing Problem in Cross-Cultural Perspective," *Mathematics Magazine* 63, no. 1 (1990).

Eglash, Ron. *African Fractals: Modern Computing and Indigenous Design*. Piscataway, NJ: Rutgers University Press, 1999.

Gerdes, Paulus. *African Doctorates in Mathematics: A Catalogue*. Maputo, Mozambique: Research Centre for Mathematics, Culture and Education, 2007.

———. *Geometry From Africa*. Washington, DC: Mathematical Association of America, 1999.

Gerdes, Paulus, and Ahmed Djebbar. *Mathematics in African History and Cultures: An Annotated Bibliography*. Cape Town, South Africa: African Mathematical Union, 2004.

International Mathematical Union. "Mathematics in Africa: Challenges and Opportunities." 2009. http://www.mathunion.org/publications/reports-recommendations.

Russ, Laurence. *The Complete Mancala Games Book*. New York: Marlowe Co., 1999.

Zaslavsky, Claudia. *Africa Counts: Number and Pattern in African Cultures*. 3rd ed. Chicago: Chicago Review Press, 1999.

Darrah Chavey

Africa, Northern

Category: Mathematics Around the World.
Fields of Study: All.
Summary: North Africa has been a major contributor to mathematics, particularly in ancient Egypt and the Islamic Golden Age.

North Africa, comprised of Algeria, Egypt, Libya, Morocco, Sudan, Tunisia, and Western Sahara, has long been geographically and culturally distinct from the rest of the continent because of the Sahara desert (which includes most of the region) and the proximity to southern Europe and the Middle East. The mathematics of ancient Egypt is among the oldest known mathematics traditions, and the Egyptian city of Alexandria was an important center of learning in the ancient world. Centuries later, Egyptian mathematicians were among the contributors to the Islamic Golden Age, translating classical works, which also helped bring about the Renaissance and Age of Enlightenment.

Mathematics historians and teachers have explored a variety of historical mathematics in the area, such as string figures and precolonial mathematics in Sudan, or the work of Gaston Julia, who was born in Algeria at the end of the nineteenth century and is known for his investigations on dynamical systems. The Julia set is named for him. Modern mathematicians and scholars in North Africa continue to take part in mathematics research and teaching.

Ancient Developments

Papyrus scrolls predating 1500 B.C.E. have been found in Egypt that discuss mathematical topics. One of the more famous is the Ahmes scroll (after the name of the scribe to whom it is attributed), currently held in the British Museum, which describes many problems in algebra and geometry and demonstrates their solutions. It is of particular interest for its use of unit fractions (fractions with a numerator of 1, such as 1/8) and for demonstrating a method of calculating circular areas.

In the Hellenistic period (c. 323–146 B.C.E.), and in the Roman period that followed, the city of Alexandria in Egypt was a center of learning, and the Great Library of Alexandria was the most important library in the ancient world. Euclid (c. 300 B.C.E.), a Greek mathematician who worked in Alexandria, is best known for his treatise *Elements*, which formed the basis for how geometry has been understood and taught for more than 2000 years. Eratosthenes of Cyrene (276–194 B.C.E.) was born in what is now Libya. He estimated the circumference of Earth and is known for the Sieve of Eratosthenes, which is useful in number theory.

One of the best-known Egyptian mathematicians from the Roman period was Ptolemy (c. 90–168 C.E.), a Roman citizen who lived in Egypt. One of his well-known works is the *Almagest*, the most comprehensive surviving ancient treatise on astronomy. Hypatia (c. 350–415), a Greek who lived in Alexandria, was a female mathematician who wrote commentaries and was also known as a teacher of astronomy and philosophy.

Islamic Period

Mathematics flourished during the Islamic Golden Age (c. mid-eighth to mid-thirteenth century). One impetus to this development was the translation of classical

Greek works, such as Ptolemy's *Almagest* and Euclid's *Elements*. These translations were often the only surviving copies and their preservation by Islamic scholars allowed them to be reintroduced into Western thought. Besides the appreciation of knowledge for its own sake, the development of mathematical sciences had practical uses in the Islamic world; for instance, knowledge of astronomy was required to understand the phases of the moon and thus correctly observe Islamic holy days, while algebraic notation was developed in part to solve problems relating to the laws of inheritance. Geometric motifs are very common in Islamic art and design, in part because, for religious reasons, Islamic artists did not create representational art, such as portraits. Instead, complex patterns such as tessellation figures (tilings) were developed for artistic use.

Many mathematicians worked in Egypt during the Islamic Golden Age. Ahmed ibn Yusuf (c. 835–912) was born in what is now Iraq but moved to Egypt and died in Cairo. He worked with his father, Yusuf ibn Ibrahim, on mathematics and wrote a book on ratio and proportion, which commented on Euclid's *Elements* and was translated into Latin in the twelfth century. Abu Kamil Shuja ibn Aslam (c. 850–930) was a mathematician who made important contributions to the study of real numbers, irrational numbers, and combinatorics, and some of whose techniques were adopted by the thirteenth-century Italian mathematician Fibonacci. Ibn Yunus (c. 950–1009) was an Egyptian astronomer and mathematician whose most famous work is a handbook of astronomical tables, which is notable for the accuracy of his observations and for his meticulous description of numerous planetary conjunctions and lunar eclipses. Abu Ali al-Hasan ibn al-Hasan ibn al-Haytham (c. 965–1039) was born in Persia but lived primarily in Egypt and died in Cairo. He worked as an engineer, reportedly attempting to develop a method to dam the Nile River, and made important contributions to optics and to the development of the scientific method. Al-Marrakushi ibn Al-Banna (c. 1256–1321) lived in Morocco and may have been born there. He worked on Euclid's *Elements* and texts on algebra and arithmetic operations.

Modern Developments

In the early twenty-first century, mathematical study and research continues in North Africa. Mathematicians belong to professional organizations like the Association Mathématique Algérienne, the Egyptian Mathematical Society, the Tunisian Mathematical Society, and the Société des Sciences Naturelles et Physiques du Maroc. Egypt and Tunisia are members of the International Mathematical Union, which is a worldwide organization designed to promote mathematics. North African countries have participated in the International Mathematical Olympiad, an annual competition held since 1959 for high school students. Algeria first participated in 1977, Morocco in 1983, and Tunisia in 1981.

Further Reading

Gerdes, Paulus. *African Doctorates in Mathematics: A Catalogue*. Maputo, Mozambique: Research Centre for Mathematics, Culture and Education, 2007.

Gerdes, Paulus, and Ahmed Djebbar. *Mathematics in African History and Cultures: An Annotated Bibliography*. 2nd ed. Cape Town, South Africa: African Mathematical Union, 2007.

Joseph, George Gheverghese. *The Crest of the Peacock: Non-European Roots of Mathematics*. Revised ed. Princeton, NJ: Princeton University Press, 2000.

Kani, Ahmad. "Arithmetic in the Pre-Colonial Central Sudan." In *Science and Technology in African History With Case Studies From Nigeria, Sierra Leone, Zimbabwe, and Zambia*. G. Thomas-Emeagwali, ed. Lewiston: Edwin Mellin Press, 1992.

Sarah Boslaugh

Africa, Southern

Category: Mathematics Around the World.
Fields of Study: All.
Summary: Southern Africa is the home of ancient mathematical artifacts and modern mathematical innovations.

Southern Africa comprises the five nations of the Southern African Customs Union: Botswana, Lesotho, Namibia, South Africa, and Swaziland. Colonization led to significant European populations, especially in South Africa and Namibia.

The oldest known mathematical artifact is the Lebombo bone, discovered in a rock shelter in the Lebombo Mountains near the South Africa/Swaziland border. There is evidence of the cave having been inhabited continuously beginning some 200,000 years ago, and the bone itself is estimated to be 35,000 years old. The Lebombo bone is a fragment of baboon fibula with 29 notches, most likely used as a tally stick—a notched object used to keep track of quantities. In this case it may have been a menstrual calendar.

Historically, the Dutch and British were particularly influential in this region. For example, the nineteenth-century Boer (also known as Afrikaner) community established the Boer States, including Transvaal and the Orange Free State. It has been documented that the Boer farmers, who were largely descendants of Dutch and some other European settlers, relied heavily on education at home. The migration of large numbers of predominantly British settlers into South Africa in the nineteenth century saw the establishment of more schools and later, universities in the European style. The mathematics heritage of southern Africa reflects both the diversity of the native cultures and the effects of this European colonialism.

South African Mathematicians

One early South African mathematician was Francis Guthrie (1831–1899), who proposed the Four Color Problem. It stemmed from a problem he first explored as a student in which only four colors could be used to denote the counties of England, and no two counties sharing a border could have the same color. Guthrie was born in London but immigrated to South Africa, where he worked as both a mathematician and a botanist. Mathematician Stanley Skewes (1899–1988), who was a faculty member at the University of South Africa and grew up near Johannesburg, postulated his Skewes number, which is an important concept in number theory.

Within South Africa, one well-known mathematician is Chris Brink, who grew up in a town on the edge of the Kalahari Desert and studied at Johannesburg. He earned a degree in mathematics before earning a scholarship to Cambridge University in England, where he completed his doctoral thesis on algebraic logic.

Returning to South Africa, he worked on Boolean modules and was vice-chancellor of the University of Stellenbosch from 2002 until 2007. Outside of the country of South Africa, another early mathematics Ph.D. from the southern Africa region is Abraham Busa Xaba. He was born in Swaziland in 1938 and earned his Ph.D. in 1984. His doctoral dissertation was titled "Maintaining an optimal steady state in the disturbances."

During the latter years of the twentieth century, some South African mathematicians also became known for their work overseas. For example, Lionel Cooper (1915–1979) left the country for political reasons. He grew up in Cape Town and won a Rhodes scholarship to study mathematics at Oxford University. Afterward, he served as a lecturer at Birkbeck College, London, and at Cardiff University, then became head of the Mathematics Department at Chelsea College, London. Abraham Manie Adelstein (1916–1992) was born in South Africa but left to live in England in 1961, where he became a leading medical statistician.

Organizations

As well as these important role models, there have been many attempts to encourage collaboration and development of mathematics in the southern African region. The Southern Africa Mathematical Sciences Association was founded in 1981 and is headquartered in Botswana. Its serves as a forum for the sharing of mathematical ideas for the countries in southern Africa as well as some neighboring countries that may be more broadly defined as being in the southern portion of the African continent.

The African Institute for Mathematical Sciences was founded in 2003 as a partnership of six universities: Cambridge University (England), University of Cape Town (South Africa), Oxford University (England), Université Paris-Sud XI (France), Stellenbosch University (South Africa), and University of the Western Cape (South Africa). Its three primary goals are: promoting mathematics and science in Africa; recruiting and training talented students and teachers of science and mathematics; and building capacity for educational, research, and technological initiatives in Africa. The South African Mathematics Olympiad is held each year for high school students, and teams from southern Africa have participated in the International Mathematical Olympiad since 1992.

Further Reading

Gerdes, Paulus. *African Doctorates in Mathematics: A Catalogue*. Maputo, Mozambique: Research Centre for Mathematics, Culture and Education, 2007.

———. "On Mathematics in the History of Sub-Saharan Africa." *Historia Mathematica* 21, no. 3 (1994).

Gerdes, Paulus, and Ahmed Djebbar. *Mathematics in African History and Cultures: An Annotated Bibliography*. Cape Town, South Africa: African Mathematical Union, 2004.

Simkins, C. E. W., with Andrew Paterson. *Learner Performance in South Africa: Social and Economic Determinants of Success in Language and Mathematics*. Cape Town, South Africa: HSRC Press, 2005.

Vithal, Renuka, Jill Adler, and Christine Keitel. *Researching Mathematics Education in South Africa: Perspectives, Practices and Possibilities*. Cape Town, South Africa: HSRC Press, 2005.

Justin Corfield

Africa, Western

Category: Mathematics Around the World.
Fields of Study: All.
Summary: Mathematics has long been used in west African art, architecture, industry, and music.

The peoples of west Africa have a long history of using mathematics. Everyday uses were similar to mathematics in other traditional societies around the world. Farmers measured their fields and counted their crops, anticipating the production figures. Fishers designed boats to carry them off the coast and prepared nets for catching fish. For both, there were processes to handle their products, either for immediate consumption or—with additional mathematics—for sale in local or distant markets. Markets served as centers of trade and also as centers of mathematical calculations of quantities and sizes, profits and losses. Everyone designed and built houses, often round in shape, which calculus shows to provide the maximum area for a given perimeter. As larger societies and governing units grew beyond the villages, mathematics played a role in governments, from taxation and salaries to the design of palaces and warehouses.

Mathematics in West African Art

West Africa has long been known for its art, textiles, music, and dance. Mathematics is central to the creative and performing arts. Some particular west African examples include carved sculptures, wall paintings, tie-dyed textiles, and woven cloth. Sculptures often show symmetries, not only of human features but also of geometrical designs and proportions of animals, village scenes and daily life, and abstractions of circles, rhombi, stars, and repeating patterns. Often, the palaces of chiefs or emirs became sites of art, especially with designs on the walls or in the architecture of the structure—all incorporating geometrical designs.

Throughout west Africa, textiles have been a central part of culture. From the multicolored patterns in Sierra Leone to the deep blues and indigos of the Hausas, the techniques of dyeing cloth have been popular, especially with tied or sewn folds of the cloth to yield intricate patterns of dyed and nondyed areas of the material. Often, the use of symmetries and Euclidean geometric constructions is necessary to produce the desired circular, radial, rhombic, and zigzag patterns. Woven cloth includes the brightly colored *kente* of Ghana, the metallic shine of the Okenne cloth of western Nigeria, and others. Weaving requires engineering mathematics to design and build looms, and then careful planning so that the strips of material that come off the looms will fit together in two-dimensional symmetrical arrangements.

These traditional artistic products have been carried into the present day. Traditional designs are now seen in modern buildings throughout the region. Fashionable textiles sometimes use new materials or printed cloth but continue the geometric traditions. *Kente* has become a popular material not only in Ghana but also in the United States, especially the symmetrical strips used as wraps and ties. Recent studies by ethnomathematician Ron Eglash have demonstrated a variety of uses of fractal patterns in traditional west African arts, ranging from repeating smaller patterns in the geographical arrangement of savannah villages, to necklaces and bracelets, carvings from Mali of increasingly small antelopes, and even corn-row hair braids that repeat smaller shapes as the pattern goes from the forehead and temples to the rear of the head.

Music and dance from west Africa are famous to both ethnomusicologists and jazz aficionados—and to ethnomathematicians. The rhythm patterns, especially from complex drumming structures, often involve unusual time signatures and alternations of loud and soft sounds. The three-dimensional movements of dance, like the carvings and textiles, show complex symmetries and geometrical arrangements of the dancers.

Early in the second millennium, Islam was introduced in west Africa, along with Islamic mathematical studies. This introduction added to the original practical base of west African mathematics, as west Africans adapted Islamic counting methods, reflected not only in the languages of west Africa but also in theoretical mathematics studied at scholarly centers such as Timbuktu (in modern Mali) and Katsina (in modern Nigeria).

Mathematics and West African Development

Since gaining independence, mostly in the 1960s, west African countries have moved rapidly to modernize. In the process, they have shown a dynamic use of mathematics—on a smaller scale than but similar to the technical mathematics of the developed world. Oil production in Nigeria, gold mining in Ghana, and diamond mining in Sierra Leone all use modern mathematical techniques, including those employed by geological surveys, sophisticated industrial equipment design, accounting, marketing, and business management. New businesses are being established to work with cell phones, the Internet, automatic teller machines, television and film production, and other industries that rely on technical mathematics and engineering. Modern freeways connect the larger cities and are designed by civil engineers and urban planners.

Education and West African Mathematics

Education throughout west Africa has grown dramatically since independence—universal primary education remains elusive, but the percentage of children attending school is approaching that goal in several countries. Political independence also brought educational independence, including national curricula offered by the Ministries of Education, the West African Examinations Council's system of standardized examinations, and locally produced textbooks and teaching materials, using familiar names, places, and situations in examples. Local researchers are studying their own cultures, seeking examples of ethnomathematics in traditional life, often with the goal of using these findings to strengthen the content of school mathematics curricula. With only a few universities in existence at the time of independence, west African countries now have numerous universities. These are often managed by the national governments—though some states of Nigeria operate their own universities and research centers, and the number of private universities is growing. These universities offer degree programs in mathematics, the sciences, engineering, and computer science, all with curricula based on the accepted world standards of these fields. Most countries have professional and scholarly organizations of mathematicians and mathematics educators, and periodically there are regional and continent-wide conferences, such as the meetings of the African Mathematical Union (AMU). The AMU's activities include the Commissions on Mathematics Education in Africa, Women in Mathematics in Africa, the African Mathematics Olympiads, and publishing the journal *Afrika Matematica*. Thus, even as west Africa maintains its traditional uses of mathematics in the arts and music, it has also become a part of the modern world mathematics community.

Further Reading

Eglash, Ron. *African Fractals: Modern Computing and Indigenous Design*. New Brunswick, NJ: Rutgers University Press, 1999.

LaGamma, Alisa, and Christine Giuntini. *The Essential Art of African Textiles: Design Without End*. New York: Metropolitan Museum of Art, 2008.

Mendonsa, Eugene L. *West Africa: An Introduction to Its History, Civilization and Contemporary Situation*. Durham, NC: Carolina Academic Press, 2002.

Zaslavsky, Claudia. *Africa Counts: Number and Pattern in African Culture*. 3rd ed. Chicago: Lawrence Hill Books, 1999.

Lawrence H. Shirley

America, Caribbean

Category: Mathematics Around the World.
Fields of Study: All.
Summary: The diverse islands of Caribbean America have produced notable mathematicians.

The Arawaks, Caribs, and other pre-Columbian peoples lived in the area of the Caribbean Sea before Spanish, French, English, or Scottish sea traders settled there. Linguists explore the different languages that were spoken in the Caribbean, traces of which can be found in the twenty-first century. Along with these languages, there were possibly different numerical systems. Sea merchants needed bookkeepers and accountants to keep track of their business, and although Port Royal and other places in the seventeenth-century Caribbean were notorious for piracy and lawlessness, there were also many counting houses and legitimate business operations.

The development of schools and universities led to more mathematical opportunities. According to the United Nations, the Caribbean America region encompasses Anguilla, Antigua and Barbuda, Aruba, the Bahamas, Barbados, the British Virgin Islands, Cayman Islands, Cuba, Dominica, the Dominican Republic, Grenada, Guadeloupe, Haiti, Jamaica, Martinique, Montserrat, Netherlands Antilles, Puerto Rico, Saint-Barthélemy, Saint Kitts and Nevis, Saint Lucia, Saint Martin (French part), Saint Vincent and the Grenadines, Trinidad and Tobago, Turks and Caicos Islands, and the U.S. Virgin Islands. By the end of the twentieth century, there were numerous Caribbean mathematicians, and *The Caribbean Journal of Mathematical and Computing Sciences* has published volumes of research articles.

Mathematicians in the Caribbean, and around the world, have also worked on mathematics history and research that is specifically related to the Caribbean area, like C. Allen Butler, who investigated optimal search techniques for smugglers in the Caribbean. Mathematicians have also discussed the high numbers of university graduates who have left the Caribbean, and they have created educational initiatives and mathematical texts designed for Caribbean children. The Caribbean and Central America areas combine for a joint Mathematical Olympiad. The most well-known mathematician in the region is perhaps Keith Michell from Grenada who, after completing his doctoral thesis from the American University, was a professor at Howard University, and then returned to Grenada, becoming prime minister in 1995, a position he held until 2008.

Barbados

On the island of Barbados, although education was an important facet of colonial life from the late nineteenth century on, few students were able to continue with mathematics. One exception was Merville O'Neale Campbell, who had become fascinated with mathematics at an early age and won a scholarship to study at Cambridge University in England. He then went to teach at the Gold Coast (now Ghana), completing his doctoral thesis, "Classification of Countable Torsion-Free Abelian Groups," from the University of London, and is noted as the first Barbadian to have a Ph.D. in mathematics. His daughter, Lucy Jean Campbell, also completed her doctoral thesis in mathematics, and specializes in geophysical fluid dynamics, nonlinear waves, and a variety of numerical and analytical methods at Carleton University in Ottawa. Other prominent Barbadian mathematicians include Charles C. Cadogan, who has edited the *Caribbean Journal of Mathematical and Computing Sciences* and has contributed papers in journals around the world; and Hugh G. R. Millington, who completed his doctorate, "Cylinder Measures," from the University of British Columbia and then worked at the University of the West Indies, Cave Hill, Barbados.

British Caribbean

Well-known mathematicians from the British Caribbean include those from Jamaica. Earl Brown, who was the head of the Department of Science & Mathematics at University of Technology (Jamaica) from 1997 to 2000, completed his doctoral thesis at the Massachusetts Institute of Technology. Joshua Leslie completed his doctoral thesis from the Sorbonne in Paris, and was the chair of the Mathematics Department at Howard University; and Kweku-Muata Agyei Osei-Bryson from Kingston completed his doctoral thesis, "Multi-objective and Large-Scale Linear Programming," at the University of Maryland—College Park in 1988, and from 1993 until 1997 was the Faculty Fellow (Information Systems) for the U.S. Army, The Pentagon. Other prominent mathematicians from Jamaica, or whose ancestors were from Jamaica, include Garth A. Baker, Charles Gladstone Costley, Leighton Henry, Fern Hunt, Lancelot F. James, Clement McCalla, Bernard Mair, Claude Packer, Paul Peart, Donald St. P. Richards, and Karl Robinson.

Elsewhere in the British Caribbean, there have also been a number of mathematicians who held senior positions in the region and in the United States including Ron Buckmire from Grenada, who has specialized in computational aerodynamics; Edward Farrell from Trinidad, who has published extensively on polynomials; and Velmer Headley from Barbados, who has concentrated on the study of differential equations.

Cuba

One notable Cuban mathematician is Argelia Velez-Rodriguez, who was born in Havana and won her first mathematics competition when she was 9. She was the first Afro-Cuban to complete a doctorate from the University of Havana but left Cuba two years later to live in the United States. Since the 1959 Revolution, there has been an increased emphasis on the education system in Cuba, and Cuban students have long shown a high aptitude for mathematics.

French Caribbean

French Caribbean mathematicians include those from Haiti, with a desperately poor education system, and Guadeloupe. Louis Beaugris completed his doctoral thesis, "Some Results Related to the Generators of Cyclic Codes Over Zm," at the University of Iowa. Serge A. Bernard completed his doctoral thesis, "A Multivariate EWMA Approach to Monitor Process Dispersion," at the University of Maryland—College Park; and Jean-Michelet Jean-Michel completed his doctorate at Brown University. Alex Meril from Guadeloupe completed his thesis at the University of Bordeaux and worked at the University of Guadeloupe.

Further Reading

Nieto Said, José, and Rafael Sánchez Lamoneda. "Ten Years of the Mathematical Olympiad of Central America and the Caribbean." *World Federation of National Mathematics Competitions* 22, no. 1 (2009).

University of the West Indies. "Caribbean Journal of Mathematical and Computing Sciences." http://www.cavehill.uwi.edu/fpas/cmp/journal/cjmcs.htm.

Williams, Scott. "Mathematics Today in the Caribbean." http://www.math.buffalo.edu/mad/Caribbean/Caribbean.html.

Justin Corfield

America, Central

Category: Mathematics Around the World.
Fields of Study: All.
Summary: Mesoamericans were sophisticated mathematicians, and mathematics continues to be important in the area.

Central America is defined as the southern part of the North American continent, reaching from Mexico to Panama. The portion of the region in which corn, beans, and squash were reliable crops during the pre-Columbian era is referred to as "Mesoamerica," reaching from the mountains of Mexico to Guatemala and down the Pacific coast into Nicaragua. Teotihuacan, Olmec, Maya, and Aztec were among the many cultures sharing the same prehistoric land and cultural legacy.

The development of the area and its perspective on mathematics were shaped in part by the origins of civilization isolated from the other large centers of civilization in the Eastern Hemisphere. Spanish colonization in the sixteenth century brought the first introduction to European cultures. Efforts at spreading Christianity resulted in the loss of much of their rich, ancient heritage. The area had gained independence by the mid-nineteenth century, variously structured as separate nations and unified groups. Struggles to achieve stability continue into the twenty-first century in many parts of the region. Education and mathematics are highly valued as keys to further progress.

Ancient Mesoamerica

Without the benefit of influence from other cultures, the ancient Mesoamericans built large city-states sometimes supporting several hundred thousand people, and extensive empires, with no domesticated large mammals and with no use of the wheel, other than in children's toys. They mastered basic arithmetic, with a concept of zero evident a millennium before European civilizations. They shared a counting system based on 20 rather than 10. Numeral representations included dots for units, bars for five, and a circle or seashell for zero. Ancient ruins show evidence of meticulous accounting of trade and personal lives. The construction of imposing pyramids and other structures aligned to astronomical features and adorned with harmonic geometric design reveal an advanced level of engineering, architecture, and astronomy to rival that found in Europe at the same time. From as early as 2000 B.C.E., the people of the area had sophisticated calendars, which were used in tandem to mark time reflecting both human and solar cycles. Ethnomathematicians continue to study ancient and modern Central America, and many teachers use Mesoamerican mathematics concepts as the basis of lesson plans and assignments.

Modern Central America

Central America is defined by the United Nations to include the modern countries of Belize, Costa Rica, El Salvador, Guatemala, Honduras, Mexico, Nicaragua, and Panama. The countries share ethnic, economic, and geological features. The peoples are primarily Spanish, Amerindian, or Mestizo (a mixture of the two). The climate ranges from mountainous to tropical coastline. While significant portions of the population are centered in large urban areas, much of the population of the region is located in small villages, sometimes isolated by rugged terrain.

Education

Central Americans are continuously improving their education systems, and efforts at reform often include careful inclusion of children from both urban and rural areas with the use of radio, television, and computer technologies. Teacher salaries and the contrasts of management of schools by local or federal administrators are recent areas of research. United Nations data report high participation in formal schooling. Private schools usually are more prestigious than public schools in most areas.

As calendars held power in ancient Mesoamerica, knowledge of mathematics is held to be essential for the people in modern Central America. High school graduates receive extensive content instruction in mathematics and science but historically with little emphasis on mathematics applications. Teachers are encouraged to teach mathematics in context rather than as an isolated, esoteric discipline both for the better understanding and for the application of learning to solve problems and promote progress. Recent research in mathematics from the region includes a diverse range of areas like topology, noncommutative geometry, and applied mathematics.

Mathematics researchers gather for conferences, research seminars, educational forums, and social events. For example, the Sociedad Matemática Mexicana (Mexican Mathematical Society) was founded in 1943. The society's goals include encouraging mathematical research, including cooperation with related scientific disciplines; improving mathematics education at primary, secondary, and college levels; and providing various forums for discussion and dissemination, including journals and conferences.

Further Reading

Evans, Susan. *Ancient Mexico & Central America: Archaeology and Culture History.* London: Thames & Hudson, 2008.

Jimenez, Emanuel, and Sawada, Yasuyuki. "Do Community-Managed Schools Work? An Evaluation of El Salvador's EDUCO Program." *World Bank Review* 13, no. 3 (1999).

Sociedad Matemática Mexicana. http://smm.org.mx/smm.

Valero, Paolo. "Deliberative Mathematics Education for Social Democratization in Latin America." *Mathematics Teaching and Democratic Education, Part 2. ZDM The International Journal of Mathematics Education* 31, no. 1 (1999).

Vegas, Emiliana, and Llana Umansky. "Improving Teaching and Learning Through Effective Incentives: Lessons From Educational Reform in Latin America," *The World Bank*, February, 2005. http://info.worldbank.org/etools/docs/library/242822/day5Improving%20teaching%20and%20learning_Final.pdf.

Judith E. Beauford

America, North

Category: Mathematics Around the World.
Fields of Study: All.
Summary: Mathematics has a long history in North America, including a twentieth and twenty-first-century focus on improving mathematics education.

North America, as defined by the United Nations, includes the United States, Canada, the Danish autonomous country of Greenland, the British overseas territory of Bermuda, and the French overseas territory of Saint Pierre and Miquelon. The United States and Canada have been especially active in the field of mathematics. By the mid-twentieth century, people from around the world were increasingly coming to North America to study and to work in mathematical disciplines. At the beginning of the twenty-first century, mathematicians and mathematics educators continue to explore ways to improve and advance research and teaching. Research and other work done by mathematics organizations in Canada and the United States show that mathematics education is a concern in North America, in part because of international comparisons of student performance. These efforts are also driven in part by the increasingly technical demands of society and the resulting economic and social needs.

Brief Early History

Mathematics played a role in the societies of the earliest native peoples as well as those of settlers from around the world. The prehistoric serpent burial mounds in what is now the state of Ohio have mathematical elements and interpretations.

In the seventeenth century, the first North American colleges began to teach a variety of subjects, including mathematics. North American mathematicians made advances in mathematical theory and contributed to a wide range of inventions.

Canada

One way to explore mathematical efforts and priorities in the twenty-first century is to examine the activities of professional associations like the Canadian Mathematical Society (CMS). The purpose of the CMS is to promote and advance the discovery, learning, and application of mathematics in Canada. According to the CMS Web site, the CMS is currently seeking to "more aggressively reach out to and form new partnerships with the users of mathematics in business, governments, and universities, educators in the school and college systems as well as other mathematical associations; and in doing so, share experiences, work on collaborative projects and generally enhance the perception and strengthen the profile of mathematics in Canada."

The mathematical skills of Canadian students have been a primary concern for Canadian educators and business owners alike. The CMS is particularly interested in reaching out to students who are interested in mathematics and in working with the educational system to improve mathematics education. To that end, the CMS sponsors a variety of educational activities, including national and regional mathematics camps, the Sun Life Financial Canadian Open Mathematics Challenge, and the Canadian Mathematical Olympiad. Additionally, the CMS publishes a journal dedicated to unique and challenging mathematics problems that can be used in secondary and collegiate mathematics classes. The CMS also provides funding for a Public Lecture Series with the goal of promoting public awareness of mathematics. The CMS strongly promotes collaboration between mathematics education and business in an effort to align the education of students with the needs of the business community, and it has developed workshops and publications to broaden participation in mathematics,

United States

World wars, especially World War II, had a notable influence on the evolution of twenty-first-century mathematics, especially in the United States. Many European mathematicians fled their native countries because of violence or oppression and settled in the United States. Military and industrial needs spurred a great deal of mathematics research and applications, which further escalated during the Cold War, spurred by advances like the Soviet Union's *Sputnik* satellite. The growth of universities in the wake of this boom, along with the relative isolation of the Soviet Union, were contributing factors to the rising numbers of students from other countries studying mathematics in the United States. By the beginning of the twenty-first century, the influx of foreign nationals into the United States educational system and workforce had slowed, in part because of change in political policies, including caps on visas; the rising prominence of universities in many other parts of the world; and the efforts of many nations to stem the "brain drain" or emigration of educated individuals.

Within the United States, many mathematical organizations have had a strong impact on the field of mathematics, including the Mathematical Association of America (MAA) and the American Mathematical Society (AMS). Many of the concerns in the United States are similar to those in Canada. There has also been a great deal of concern and discussion regarding the perception that only some students are capable of succeeding at mathematics. Some assert that the No Child Left Behind Act of 2001 was designed to challenge this perception by ensuring that all students could demonstrate grade-level mathematics proficiency. However, this measure was negatively received by many, in part because increased demands on teachers and schools were not always fully funded and criteria used to measure success and improvement were not universally agreed upon as appropriate. A primary focus is on improving the mathematics achievement of public school students in an effort to ensure that more students are "college-ready."

In an effort to address this need, the National Council of Teachers of Mathematics (NCTM) released a series of publications that focus on the idea that mathematics education at every grade level needs to center on in-depth development of a few key mathematical concepts. The MAA and AMS both have made resources available to teachers to aid in this endeavor. Like Canada, the United States also works to recruit a wider demographic of students into mathematical fields.

Further Reading

American Mathematical Society. http://www.ams.org.
Canadian Mathematical Society. http://www.math.ca.
Duren, Peter, Richard Askey, and Uta Merzbach. *A Century of Mathematics in America*. Vols. 1–3. Providence, RI: American Mathematical Society, 1991.
Hankes, Judith, and Gerald Fast. *Changing the Faces of Mathematics: Perspectives on Indigenous People of North America*. Reston, VA: National Council of Teachers of Mathematics, 2002.
Jones, Philip. *A History of Mathematics Education in the United States and Canada (32 Yearbook)*. Reston, VA: National Council of Teachers of Mathematics, 2002.
Mathematical Association of America. http://www.maa.org.

Calli A. Holaway

America, South

Category: Mathematics Around the World.
Fields of Study: All.
Summary: Long before European settlement, mathematics flourished in South America.

South America includes Argentina, Bolivia, Brazil, Chile, Colombia, Ecuador, French Guiana, Guyana, Paraguay, Peru, Suriname, Uruguay, and Venezuela. The history of South American mathematics begins with pre-Columbian developments like the Nazca lines and *quipus* ("KEE-poos") and continues through the astronomy boom of the colonial period to work by modern mathematicians and ethnomathematics studies in Brazil.

Quipus

The Incan empire, with its capital in Cuzco, Peru, dominated pre-Columbian South America. The Incan civilization emerged from the highlands in the early thirteenth century and extended over an area from what is now the northern border of Ecuador, Peru, western and south central Bolivia, northwest Argentina, northern and central Chile, and southern Colombia. The Incas reached a high level of sophistication with remarkable systems of agriculture, textile design, pottery, and administration. Since the Incas had no written records, the *quipu* (or *khipu*) played a pivotal role in keeping numerical information about the population, lands, produce, animals, and weapons.

Quipus were knotted tally cords that consisted of a main cord from which hung a variable number of pendant cords containing clusters of knots. These knots and their clusters conveyed numerical information in base-10 representation. For example, if the number 365 was to be recorded on the string, then five touching knots were placed near the free end of the string followed by a space, then six touching knots for the 10s, another space, and finally three touching knots for the 100s. Specific information was conveyed via the number and type of knots, cluster spacing, color of cord, and pendant array. Inca administrators and accountants employed this complex system for numerical storage and communication. Quipus were mathematically efficient and portable. Unfortunately, the Spanish destroyed many quipus, potentially hiding clues to understanding Incan architectural processes, irrigation, and road systems.

Nazca Lines

The Nazca lines are a set of figures that appear engraved in the surface of the Nazca desert in southern Peru. The lines include hundreds of geometric shapes and renderings of animals and plants, including birds, a spider, a monkey, flowers, geometric figures, and lines—some of them miles long. The Nazca lines, best appreciated from an airplane, are one of the world's enduring mysteries. It is hard to explain how the ancient people of Nazca (900 B.C.E.–600 C.E.) achieved such geometrical precision in an area over 300 square miles. German-born mathematician and archaeologist Maria Reiche spent five decades studying and preserving these lines. She, like many other scientists, believed that the Nazca lines represented an astronomical calendar and observatory, while other theories suggest that they map areas of fertile land.

Mathematics in the Colonial Era

The accidental arrival of navigator Christopher Columbus in the Americas in 1492 marked the beginning of a 300-year period of Spanish and Portuguese colonial rule in South America that ended in the early nineteenth century. Under the Treaty of Tordesillas (1494), Portugal claimed what is now Brazil, and Spanish claims were established throughout the rest of the continent with the exception of Guyana, Suriname, and French Guiana. Roman Catholicism and an Iberian culture were imposed throughout the region, and mathematical systems and practices of ancient cultures were replaced by the Hindu-Arabic decimal system used by the Spanish.

Mathematical activity in Spain between the sixteenth and nineteenth century decisively influenced mathematical thinking and practices in South America. In sixteenth-century Spain, two lines of mathematical thought existed: the arithmeticians (calculators, interested in the uses of mathematics) and the algebraists (abstract or pure mathematicians). Because the European countries used the colonies to enhance their trade and economic resources, the emphasis in South America was on applied mathematics.

Later, the Spanish and the Portuguese established schools—mostly run by Catholic religious orders—which concentrated mathematics teaching on economic applications related to trade. There was also an interest on mathematics related to astronomical observations. The first nonreligious book published in the Americas was an arithmetic book related to gold and silver mining printed in 1556.

Astronomy was a major area of interest in South America in the seventeenth century. In Brazil, research on comets was of major importance, as exemplified by the work of Valentin Stancel (1621–1705), a Jesuit mathematician from Prague who lived in Brazil from 1663 until his death (his astronomical measurements are mentioned in Newton's *Principia*). As in many cultures, most astronomical interpretations attempted to explain divine messages to humankind. Other developments in Brazil included the first aircraft known to fly: the Passarola, invented by Bartolomeu de Gusmão, a Brazilian priest and scientist from Sao Paulo. De Gusmão, also known as the "Flying Priest," studied mathematics and physics at the Universidade de Coimbra in Portugal. The Passarola was an aerostat heated with hot air and flew in Lisbon, Portugal, in 1709.

Mathematics in the Era of Independence

In the first quarter of the nineteenth century, many successful revolutions resulted in the creation of independent countries in South America. Mathematical activity increased throughout Latin America in the twentieth century. For instance, Argentinian mathematician Alberto P. Calderon (1920–1998) developed new theories and techniques in classical and functional analysis. Professor Calderon worked at the University of Chicago for many years. He was awarded the National Medal of Science in the United States.

Research by Professor Ubiritan D'Ambrosio and his students in the slums and indigenous communities in Brazil focused on ethnomathematics—a sub-field of mathematics history and mathematics education. The goal of ethnomathematics is to understand connections between culture and the development of mathematical processes and ideas. Other researchers have explored specific mathematical habits and methods in South American cultures. In the 1980s, Terezinha Nunes and her collaborators studied differences between street mathematics and school mathematics in Brazil by comparing how street vendors (including children) and farmers solve problems compared to those who encounter similar problems in formal school situations.

For example, in their study of young street vendors in Recife, the interviewers acted as customers and asked questions that required the use of arithmetic skills (such as making change). The children did much better in this "real" situation than on a formal test given a week later that used similar numbers and operations. One possible explanation is that the children were better able to keep the meaning of the problem in mind in the "real" situation. Many others, such as Geoffrey Saxe, have found similar results. An implication of these studies is that the essence of school mathematics, which the Recife children were not as successful at, is highly symbolic and possibly devoid of meaning. These studies have been important in advancing the goal of mathematics education that students must initially construct appropriate meanings for the various concepts and methods they encounter.

Further Reading

Ascher, Marcia. "Before the Conquest." *Mathematics Magazine*. 65, no. 4 (1992).

D'Ambrosio, Ubiritan. "Ethnomathematics and Its Place in the History and Pedagogy of Mathematics." *For the Learning of Mathematics* 5, no. 1 (1985).

Nunes, Terezinha, et al. *Street Mathematics and School Mathematics*. New York: Cambridge University Press, 1993.

Ortiz-Franco, Luis, Norma Hernandez, and Yolanda De La Cruz, eds. *Changing the Faces of Mathematics: Perspectives on Latinos*. Vol. 4. Reston, VA: National Council of Teachers of Mathematics, 1999.

Gisela Ernst-Slavit
David Slavit

Arabic/Islamic Mathematics

Category: Government, Politics, and History.
Fields of Study: Algebra; Connections; Geometry; Measurement; Number and Operations; Representations.
Summary: Arabic and Islamic mathematicians popularized the decimal system and Arabic numerals and also developed algebra.

Mathematicians living in Islamic lands and writing in Arabic have played a central role in the development of mathematics, particularly during the 700-year period from around the year 750 c.e to around 1450 c.e. These scholars lived in an area that not only includes the present-day Middle East but stretches into the western parts of India, the major cities of central Asia, all of northern Africa, and most of the Iberian Peninsula. Most of the influential mathematicians of this seventh-century era were Muslim, and most wrote in Arabic. However, the lands ruled by Muslim rulers included many ethnicities, cultures, languages, and religions. Muslims, Christians, Jews, Zoroastrians, Manichaeans, Sabians, Buddhists, Hindus, Persians, Turks, Sogdians, Mongols, Arabs, Berbers, Egyptians, and many others contributed to a remarkable multiethnic, multicultural civilization. Mathematics was not an exception. The full story of mathematics in this era has yet to be told. Hundreds of manuscripts await examination, translation, and a critical edition. Undoubtedly, in the years to come, our understanding of the extent, the import, and the influence of the mathematics of this period will change dramatically.

While their knowledge of what came before them was incomplete and uneven, the mathematicians of the Islamic era were aware of—and in some ways heirs to—ideas, methods, and points of view that originated in India, Persia, and—especially—Greek Alexandria. A remarkable translation movement coupled with a scholarly tradition of writing commentaries on previous works meant that mathematicians of this era were comfortable with the contents and the methodology of the works of, among others, Euclid, Archimedes, Apollonius, Ptolemy, and Diophantus as well as the basics of Indian decimal arithmetic and trigonometry. They also had access to Persian astronomical tables. They accomplished a great deal with this heritage. What the mathematicians of the Islamic era bequeathed to those who came later was very different in content, style, and approach than what had come before them. (A note on names: names of mathematicians and places can be transliterated to English based on their Arabic, Persian, or Turkish versions. For the most part, we have chosen what is currently most common in English. The one exception is that we have often omitted the Arabic definite article "al" that precedes titles and nicknames.)

The Decimal System and the Concept of Number

For Euclid—the preeminent mathematician of Greek Alexandria—"number" meant a rational number. In his work, irrational numbers were called magnitudes and were treated quite differently from numbers. In fact, Euclid's very influential book *Elements* contains few numbers and hardly any calculations. Starting with Khwarizmi of Khwarizm (c. 780–850 c.e.), the principles of the positional decimal system that had originally come from India were organized and widely disseminated. Hence, with the use of 10 symbols it was possible to carry out all arithmetic operations. Over the following centuries, the methods for these arithmetical operations were improved and included working with decimal fractions and with large numbers. In fact, in the process, the Euclidean concept of number was gradually enlarged to include irrational numbers and their representation as decimal fractions. The mathematician Kashani (c. 1380–1429), also known as al-Kashi, worked comfortably with irrational numbers and, for example, was able to produce an approximation that was correct to 16 decimal places. The Arabic texts on the decimal number system were translated to Latin and were the basis for what are now called the Hindu-Arabic numerals.

Algebra

While it is possible to recognize algebraic problems in ancient mathematics, algebra as a discipline distinct from geometry and concerned with solving of equations was developed during the Islamic period. The first book devoted to the subject was Khwarizmi's *Al-kitab al-muhtasar fi hisab al-jabr wa-l-muqabala* (*Compendium on Calculation by Completion and Reduction*).

In this title, "al-jabr"—the origin of the word "algebra"—means "restoration" or "completion" and refers to moving a negative quantity to the other side of an equation where it becomes positive. *Al-muqabala* means "comparison" or "reduction" and refers to the possibility of subtracting like terms from two sides of an equation. While all algebra problems were stated and solved using words and sentences—symbolic algebra did not arise until much later in the fifteenth century in Italy—an algebra of polynomials was developed by Abu Kamil (c. 850–930), Karaji (c. 953–1029), and Samu'il Maghribi (c. 1130–1180+, also known as al-Samaw'al). Powers, even negative powers, of unknowns were considered and many algebraic equations were classified and solved. Khwarizmi gave a full account of second-degree equations, and Khayyam (1048–1131) gave a geometric solution to equations of degree three using conic sections. Here, we give a problem—translated to modern notation—solved by Abu Kamil. Some 300 years later, this exact same problem appeared in Chapter 15 of the 1202 text *Liber Abaci* by Leonardo Fibonacci. Abu Kamil gave a solution to the following system of three equations and three unknowns:

$$x + y + z = 10$$
$$x^2 + y^2 = z^2$$
$$xz = y^2.$$

Abu Kamil first started with the choice of $x = 1$ and solved the latter two equations for y and z. Since, for the latter two equations, any scalar multiple of the solutions continues to be a solution, he then scaled the solutions so that the first equation was also satisfied. He simplified the answer to get:

$$x = 5 - \sqrt{\sqrt{3125} - 50}.$$

Geometry

Geometrical methods and problems were ubiquitous in the Islamic era. While algebraic problems were solved using the newly developed algebraic algorithms (the word "algorithm" itself is derived from *algorismi*, the Latin version of the name of the mathematician al-Khwarizmi), the justification for the algebraic methods was usually given using geometrical arguments and often relying on a distinctively Euclidean style. Guided by problems in astronomy and geography (for example, finding, from any place on Earth, the direction of Mecca for the purpose of the Islamic daily prayers), spherical geometry was developed.

But new work in plane geometry was also carried out. Khayyam and Nasir al-din Tusi (1201–1274), for example, studied the fifth postulate of Euclid and came close to ideas that much later on led to the development of non-Euclidean geometries in Europe. However, as is the case with much of the mathematics of this era, applications play an important role in the choice of questions and problems.

For example, Abu'l Wafa Buzjani (940–997) reports on meetings that included mathematicians and artisans. A problem of interest to tile makers is how to create a single square tile from three tiles. A traditional mathematician, Abu'l Wafa explains, translates this problem into a ruler and compass construction and gives a method for constructing a square of side $\sqrt{3}$.

While logically correct, this construction is of little use to the tile maker, who is confronted with three actual tiles and wants to cut and rearrange them to create a new tile. Abu'l Wafa also gives the customary practical method that is actually used by tile makers to solve this problem, and proves that their method, while practical, is not precise, and the final object is not exactly a square. While stressing the importance of being both practical and precise, and the virtues of Euclidean proofs, he presents his own practical and correct methods for solving this and related problems.

Trigonometry

The origins of trigonometry begin with the Greek study of chords as well as the Indian development of what is now called the "sine function." Claudius Ptolemy's table of chords and Indian tables of sine values were powerful tools in astronomy. However, a systematic study and use of all the trigonometric functions motivated by applications to astronomy, spherical geometry, and geography begins in the Islamic era. Abu'l Wafa had a proof of the addition theorem for sines and used all six trigonometric functions; Abu Rayhan Biruni (973–1048) used trigonometry to measure the circumference of Earth; and Nasir al-din Tusi gave a systematic treatment in his *Treatise on the Quadrilateral* that helped establish trigonometry as a distinct discipline.

Combinatorics

One of the earlier known descriptions and uses of the table of binomial coefficients (also known as the Pascal triangle) is that of Karaji. While his work on the subject is not extant, his clear description of the triangle survives in the writings of Samu'il Maghribi. Binomial coefficients were used extensively, among other applications, for extracting roots. Kashani, for example, used binomial coefficients to give an algorithm for extracting fifth roots. He demonstrated it by finding the fifth root of 44,240,899,506,197. Other combinatorial questions were treated as well. Ibn al-Haytham (c. 965–1039, also known as Alhazen) gave a construction of magic squares of odd order, and Ibn Mun'im (died c. 1228) devotes a whole chapter of his book *Fiqh al-Hisab* to combinatorial counting problems.

Numerical Mathematics

The prominence of applied problems, the development of Hindu-Arabic numerals and calculation schemes, and the development of algebra and trigonometry led to a blossoming of numerical mathematics. One prime example is Kashani's *Miftah al-Hisab* or *Calculators' Key*. In addition to his approximation of 2π and his extraction of fifth roots, he also gave an iterative method for finding the root of a third-degree polynomial in order to approximate the sine of one degree to as close as an approximation as one wishes.

Further Reading

Berggren, J. L. *Episodes in the Mathematics of Medieval Islam*. New York: Springer Verlag, 1986.

Katz, Victor J. *A History of Mathematics. An Introduction*. 2nd ed. London: Addison Wesley, 1998.

Katz, Victor, ed. *The Mathematics of Egypt, Mesopotamia, China, India, and Islam. A Sourcebook*. Princeton, NJ: Princeton University Press, 2007.

Van Brummelen, Glen. *The Mathematics of the Heavens and the Earth: The Early History of Trigonometry*. Princeton, NJ: Princeton University Press, 2009.

Shahriar Shahriari

Asia, Central and Northern (Russia)

Category: Mathematics Around the World.
Fields of Study: All.
Summary: The contributions of central Asia have included algebra and its great houses of wisdom.

Throughout history, countries in Asia have had shifting sociopolitical boundaries. The names of some countries have changed, influenced by the Arab and Islamic empires as well as European colonialism in the eighteenth and nineteenth centuries. Though not widely used, Northern Asia sometimes refers to the part of the Asia occupied by the transcontinental country of Russia, which is commonly included in eastern Europe. Central Asia includes the former Soviet satellites of Kazakhstan, Kyrgyzstan, Tajikistan, Turkmenistan, and Uzbekistan. Mongolia, typically considered part of central Asia by historians, is in the modern world classified as part of Eastern Asia by the United Nations. "Northern Asia" is a term that is not commonly used, thus the transcontinental country of Russia is usually thought of as part of Eastern Europe. Knowledge of the contributions of mathematicians around the world is constantly changing as historians discover and translate written materials in many languages. Further, the breakup of the Soviet Union and shifting alliances have given researchers access to documents from decades in which many Eastern Bloc nations kept themselves in isolation, as well as even older works contained in the libraries and educational institutions of these nations. For example, medieval Islamic texts in Uzbekistan have helped shed light on the rich mathematics culture of central Asia. However, the mathematics contributions and achievements of some people from central Asia may be included in the histories of other areas, countries, or cultures.

In the seventh century, the great Library of Alexandria in Egypt was captured by a Muslim army, and there are some historians who believe some contents of the library were taken into Muslim lands. Many cities in central Asia became famous in the medieval period for their own libraries, which contained original works and translations of texts from Greek and Sanskrit, some of which became the only surviving copies of these earlier works. Houses of wisdom provided places for scholars to gather, as well as scientific centers such as the fifteenth-century Samarkand Observatory in what is now Uzbekistan, which was founded by astronomer Muhammad Taragay Ulughbek. This observatory reputedly served as a model for later observatories in India. Astronomer and mathematician Ala al-Din Ali ibn Muhammed, also known as Ali Kushji, later preserved and disseminated some of the knowledge gathered by the observatory when it was destroyed. This catalogue of stars, containing the most accurate mathematical measurements of location known prior to the invention of the telescope, is still studied.

Significant Central Asian Mathematicians

In the same way that mathematicians in central Asia studied and developed many concepts that were first introduced by other cultures, other concepts and techniques in twenty-first-century mathematics were first brought to Europe by mathematicians who worked in or came from central Asia. The word "algorithm" derives from a Latin transliteration of the name of eighth- and ninth-century mathematician Abu Abdallah Muhammad ibn Musa al-Khwarizmi (sometimes written as Al-Khoresmi). The Khwarizm (or Koresm) region included portions of what are now Turkmenistan and Uzbekistan. The word "algebra" comes from the term *al-jabr*, which was found in al-Khwarizmi's treatise on that subject. Another of his mathematical writings, the *Book of Addition and Subtraction by the Indian Method*, helped promote the Hindu base-10 decimal system within the Arabic world. This system spread to Europe and revolutionized mathematics around the world in subsequent centuries.

Historical evidence suggests that tenth-century astronomer and mathematician Abu Mahmud Hamid ibn al-Khidr Al-Khujandi was born in the city of Khudzhand, in what is now Tajikistan. His mural sextant produced some of the most accurate astronomical observations of the day, and he may have contributed to trigonometry. The tenth- and eleventh-century mathematicians Abu Rayhan al-Biruni and Abu Nasr Mansur are also cited as being natives of Khwarizm. Al-Biruni studied a diversity of topics in mathematics and science, including cartography and map projections, trigonometry, combinatorial analysis, ratio theory, algebraic problem solving, geometry, Archimedes of Syracuse's theorems, conic sections, and spherical triangles. Along with his own prolific body of writings, he was also a translator of Sanskrit texts. Abu Nasr Mansur taught and collaborated with al-Biruni—the two frequently cited one another's contributions to their own work.

Many consider Mansur's primary mathematical contributions to be his commentary on Menelaus of Alexandria's *Sphaerica*, his development of trigonometry, and his tables for numerical solutions to problems in spherical astronomy. In the same time period, Abu Ali al-Husain ibn Abdallah ibn Sina, also known by the Latin name Avicenna, wrote on many topics, including medicine and mathematics. Some of his investigations included ruler and compass constructions, areas of circles, and geometric algebra. He also considered music to be a subdiscipline of mathematics, and some believe that his studies led to musical tuning by the method of just intonation, where the note frequencies are related by ratios of small whole numbers, rather than Pythagorean tuning, named for Pythagoras of Samos.

Beginning in about the twelfth century, central Asia underwent a great deal of social and political disruption, and there is often little surviving evidence regarding mathematics and science during those eras. During the Soviet period, mathematicians from Kazakhstan, Kyrgyzstan, Tajikistan, Turkmenistan, and Uzbekistan may have been drawn to some of the central academic centers in Russia and other parts of the Soviet Union. Since the fall of the Soviet Union, these countries are reestablishing themselves as independent nations, and achievement in mathematics continues. For example, students from central Asia have participated in and won numerous medals in the International Mathematical Olympiad, an annual competition for high school students in which individual medals are awarded based on each student's success in solving a set of mathematics problems. Countries send six-member teams.

Kazakhstan, Kyrgyzstan, and Turkmenistan first participated in 1993, Uzbekistan in 1997, and Tajikistan in 2005. In 2010, Kazakhstan hosted the 51st Olympiad in its capital of Astana. Students from 98 countries around the world participated. Professor Askar Dzhumadildayev noted, "Mathematics is one of the most important indexes of the education level in the country. Gathering the best young mathematicians in Astana is a great honor for us." A news report regarding the Olympiad acknowledged the rich history of central Asia: "... we should not forget that our country is an heiress of the mathematical school founded by great scientists of the Middle Ages.... who greatly contributed to development of mathematics long before the modern countries of the West appeared."

Further Reading

Bobojan, Gafurov. "Al-Biruni: A Universal Genius Who Lived in Central Asia 1000 Years Ago." *UNESCO Courier* (June 1974). http://unesdoc.unesco.org/images/0007/000748/074875eo.pdf.

"The 51st International Mathematical Olympiad." http://www.imo2010org.kz.

Matvievskaya, G. P. "History of Medieval Islamic Mathematics: Research in Uzbekistan." *Historia Mathematica* 20 (1993).

Sarah Boslaugh

Asia, Eastern

Category: Mathematics Around the World.
Fields of Study: All.
Summary: Across eastern Asia, mathematics education is given a high priority, with the goal of continuing the region's tradition of excellence.

Eastern Asia is one of the most populated regions of the world, lagging behind only southern Asia, and includes the Chinese cultural sphere once called the "Far eastern civilizations": China, Hong Kong, Macau, Taiwan, Japan, Mongolia, North Korea, and South Korea. The region is by no means homogeneous but has certainly been influenced to varying degrees by China in its writing systems, its cuisine, its architecture, and its religion. These influences are principally historical, cultural exchange being less centralized now, and influences like the Western world and the Soviet Union (in the case of Mongolia and North Korea) having been significant in the recent past. The technology sector is important in much of this region and mathematics education is a priority. Mathematics education in most of east Asia follows the Confucian model.

Number System

The number system in all Chinese-based east Asian languages centers on the same decimal system as the West but with stricter adherence to simple place-value patterns. For example, employing literal translations, the word for the number 12 is *ten-two*, 20 is *two-ten*, 37 is *three-ten, seven*, and 533 is *five-hundred, three-ten, three*. This system, along with the use of an abacus, facilitates the understanding of place value among east Asian elementary students. east Asian countries also follow the Chinese myriad-grouping system, which groups large numbers by ten thousands, rather than thousands. In other words, these languages have single words for the numbers "ten thousand" and "one-hundred million," but not for "million" or "billion."

Educational Philosophy

Historically, public east Asian mathematics classrooms could be generalized as teachers delivering lectures to large classes of students who are expected to master calculations and grasp theory through repetition and memorization. Inherent in this Confucian approach is the assumption among students, parents, and teachers that mathematical success results more from diligent studying than natural talent. Student-centric and practical applications of mathematics are not a primary focus in east Asia, as they sometimes are in the West. This educational philosophy is true not only of the textbooks, which in east Asia are succinct and cover the minimal core set forth by each of the national governments but also of the classrooms, which must closely follow the textbooks. However, since the international test results illuminated relative weaknesses in problem solving, creativity, and practical applications, the east Asian governments have been working to adapt curricula in various ways.

China

Chinese children's task of memorizing thousands of Chinese characters naturally seems to transfer to the subject of mathematics where memorization of formulas and processes is assumed to lead to understanding and discovery.

While mainland China did not participate in some international comparisons, the Chinese team has performed exceptionally well in the annual International Mathematical Olympiad (IMO), a competition among high school students, where it placed first almost every year between 1990 and 2010. But these achievements in mathematics are not limited to Chinese students; two Chinese mathematicians have received the distinguished Wolf Prize in Mathematics: Shiing-Shen Chern in 1983–1984 and Shing-Tung Yau in 2010.

Hong Kong

The mathematics education system in Hong Kong employs elements both from mainland China and Great Britain. Despite the fact that international test scores ranked students from Hong Kong as years ahead of many Western countries, there is widespread concern about students viewing mathematics as irrelevant beyond testing. This concern has been leading to a curriculum that emulates the Western approach to teaching more mathematics related to problem solving and practical abilities.

Japan

While Japan distinguished itself in mathematics from the other east Asian countries during the Edo period (1603–1868), modern Japanese mathematics carries few remnants of this period. One such remnant is the *soroban*, a Japanese modification of an abacus. Japanese schoolchildren continue to use this beaded calculation device as a means of mastering the decimal system. Like in all east Asian countries, private schools (called *juku*) are attended widely by Japanese students. Japan has produced some of Asia's best mathematicians of the past century, including three winners of the Wolf Prize (Kunihiko Kodaira in 1984–1985, Kiyoshi Ito in 1987, and Mikio Sato in 2002–2003), and three winners of mathematics' most revered award, the Fields Medal (Kunihiko Kodaira in 1954, Heisuke Hironaka in 1970, and Shigefumi Mori in 1990).

Mongolia

Geographically, Mongolia lies between China and Russia. Until the early twentieth century, it was largely under the control of China and was later strongly influenced by Russia and the Soviet Union, adopting a Soviet-style government until 1990. Mongolian teams began participating in the International Mathematical Olympiad in 1964. Ming Antu was a Mongolian mathematician and astronomer, though he has been referred to as Chinese in the past. He worked on infinite series in the eighteenth century, among other accomplishments.

North Korea

While North Korea has the same Confucian background as the other east Asian countries, the former Soviet Union played a significant role in sculpting the modern approach to mathematics education. As do most countries around the world, the North Korean education system upholds mathematics as a central focus for both primary and secondary students, although North Korean story problems tend to be phrased in a nationalistic context. Students who excel in mathematics during their secondary school education may be admitted into the esteemed Kim Il-Sung University. In terms of global rankings, North Korea has sporadically entered a team into the International Mathematics Olympiad, some of which placed in the top 10.

South Korea

From childhood, South Koreans grow up using two separate number systems in their daily lives. The first one, a purely Korean system, is used mainly for counting objects, animals, and people and is no longer used for numbers larger than 99. It is worth noting that the numerals in this

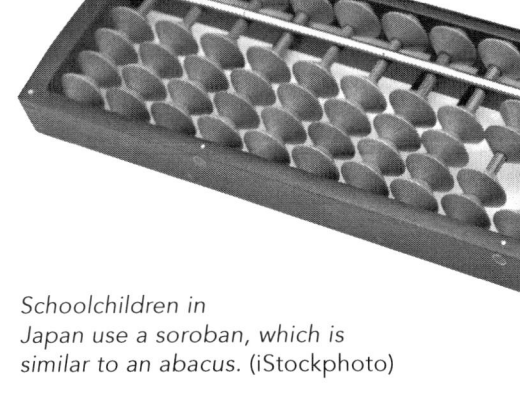

Schoolchildren in Japan use a soroban, which is similar to an abacus. (iStockphoto)

Korean system do not follow the same simple place-holding constructions as the number systems rooted in the Chinese language. The Sino-Korean number system, on the other hand, does follow these rules, and is most commonly used with money and large numbers. In school, many South Korean students receive just as much, if not more, of their mathematics instruction from private tutors or *hagwons* (academies) as from the public school environment. This system stems from the inextricable link between a student's mathematics performance on entrance exams and his or her eventual place in society. Some people cite this pressure as an explanation for why South Korean and Japanese students, despite performing exceptionally well on international tests, also rank the highest in their professed dislike for mathematics.

Taiwan

Private mathematics academies in Taiwan are referred to as *buxiban* (cram schools), suggesting their primary, but not exclusive, role of preparing Taiwanese students for entrance examinations. With electronics as a major industry, there has been a recent overhaul of the Taiwanese education system to focus on practical applications of mathematics instead of only theoretical mathematics.

Further Reading

Kennedy, Peter. "Learning Cultures and Learning Styles: Myth-Understandings About Adult (Hong Kong) Chinese Learners." *International Journal of Lifelong Education* 21 (2002).

Lankov, Andrei. *North of the DMZ: Essays on Daily Life in North Korea*. Jefferson, NC: MacFarland & Company, 2007.

Lee, Jihyun. "Universals and Specifics of Math Self-Concept, Math Self-Efficacy, and Math Anxiety Across 41 PISA 2003 Participating Countries." *Learning and Individual Differences* 19 (2009).

Leung, Frederick K. S., Klaus-D. Graf, and Francis J. Lopez-Real, eds. *Mathematics Education in Different Cultural Traditions—A Comparative Study of East Asia and the West*. New York: Springer, 2006.

Rong, Xin. "The General Solution of Ming Antu's Problem." *Acta Mathematica Sinica, English Series* 20, no. 1 (2004).

Usiskin, Zalman, and Edwin Willmore, eds. *Mathematics Curriculum in Pacific Rim Countries—China, Japan, Korea, and Singapore*. Chicago: Informations Age Publishing, 2008.

Yau, Shing-Tung. *S.S. Chern: A Great Geometer of the Twentieth Century*. Boston: International Press of Boston, 1998.

Daniel Showalter

Asia, Southeastern

Category: Mathematics Around the World.
Fields of Study: All.
Summary: Mathematics in the region has long been intertwined with religion and astrology and in recent generations has been impacted by colonialism.

The United Nations classification of southeastern Asia includes Brunei Darussalam, Cambodia, Indonesia, Lao People's Democratic Republic, Malaysia, Myanmar, Philippines, Singapore, Thailand, Timor-Leste, and Vietnam. Throughout history, the countries of Asia have had shifting political and social boundaries, and the names of many countries have changed over time, especially from the European colonial eras of the eighteenth and nineteenth centuries—when Western historians often began to study and document these countries—into the twenty-first century. For example, Burma became known as Myanmar; Siam became Thailand; Malay or Malaya became Malaysia; the Dutch East Indies or Netherlands East Indies and Java became Indonesia; and French Indochina included Laos, Cambodia, and Vietnam. Singapore was also part of Malaysia for a brief time in the 1960s, and the two regions share many historical developments. China and India, which have long histories of mathematics work and achievement, also had an influence in this region of the world. Therefore, mathematics contributions of some people from southeastern Asia, may be included within the histories of other regions or countries.

Early History

The great architectural feats found in places such as Borobudur, built in the ninth century on the island of Java, now part of Indonesia, and Angkor Wat, constructed three centuries later in Cambodia, suggest to scholars and historians that the architects and the builders must have had considerable mathematics knowledge. Some mathematics was probably brought to the region from India and China, as also happened in Europe and other areas, but there were almost certainly local mathematicians as well. The geometry involved in the design of both Borobudur and Angkor Wat has amazed generations of scholars who have discovered many complex ratios and formulas in the designs. Historians have also discussed the interconnection between religion, astronomy, mathematics, and astrology in southeastern Asia. Often there was little distinction made between mundane and divine matters, and some sequences of numbers (for example, 4, 8, 16, and 32) had religions connotations.

These numbers were used in both government and spiritual matters, such as the number of chiefs and territories in some Malay courts. Numerical systems emerged for the Burmese, Siamese, Cambodian, Laotian, Vietnamese, and Javanese languages. When Europeans began to explore and colonize southeastern Asia, they brought with them their own formal methods of school structure and mathematics teaching, which were documented by historians. Colonial influence saw the Vietnamese language develop a Romanized script, along with Western systems of counting, but the other scripts kept their systems of numerals. The introduction to southeastern Asia of a European-style school education, which replaced previous systems of instruction at pagodas or mosques, was a contributing factor in mathematics education. Much of this education came from the commercial needs of colonial powers to educate boys for work as bookkeepers and businessmen, so Western accounting systems were introduced to these populations—though many merchants continued to use Chinese systems, including the abacus, up through the twenty-first century.

Singapore and Malaysia

Singapore and Malaysia have active mathematics programs. The Raffles Institution in Singapore has a mathematics club whose members compete in events like the Singapore Mathematical Olympiad. The school was established in 1823 and named for (Thomas) Stamford Raffles, who is known as the founder of the British colony in Singapore. The Singapore Mathematical Society was founded in 1952. In the twenty-first century, it organizes participation in events like the national and international mathematics olympiads and the Singapore Mathematics Project Festival, among other educational and professional activities. Singapore first participated in the International Mathematical Olympiad (IMO) in 1988. While many twentieth-century textbooks on mathematics were imported into Singapore, the "Singapore Math Method," first developed in the 1980s and used in the national curriculum in Singapore, is now used in several places in the United States and elsewhere.

One of Singapore's well-known mathematicians is Tony Tan, who completed his doctorate, with a dissertation on "Mathematical models for commuter traffic in cities," at the University of Adelaide, South Australia. He taught mathematics before going into banking, and then into politics, ultimately becoming his country's deputy prime minister. Raffles College in Singapore taught mathematics from the time it started operations in 1928. Relations between Singapore and Malaysia in the twentieth century led to its transformation into the University of Malaya, then the University of Singapore, and the National University of Singapore. Sir Alexander Oppenheim, the vice-chancellor of the University of Malaya 1957–1965, was a prominent mathematician who had taught at Raffles College.

The Malaysian Mathematical Sciences Society, founded in 1970, was formerly known as the Malaysian Mathematical Society. It hosts events like the National Mathematical Olympiad in Malaysia; Malaysia first participated in the IMO in 1995. The Penang Free School, established in Malaysia in 1816, has taught mathematics from its inception. The Institute of Mathematical Sciences at the University of Malaya, founded in 1959 as the Department of Mathematics, continues to provide education for many Malaysian and overseas students and is an important mathematical institute in that country.

Thailand

Historically, Thailand was the only country in southeastern Asia never to be colonized by a foreign nation. Rulers such as the nineteenth-century King Mongkut, the inspiration for the 1946 movie *Anna and the King of Siam* and often called "the father of science and technology," embraced Western innovations. Assumption College, Bangkok, founded in 1885, had an extensive program of mathematics. The Mathematical Association of Thailand publishes the *Thai Journal of Mathematics* and hosts conferences and contests. Thailand has been participating in the IMO since 1989. The Center for Promotion of Mathematical Research of Thailand was established in 1978. Mathematician Yupaporn Kemprasit is an acknowledged world expert on algebraic semigroup theory, ring theory, and algebraic hyperstructure theory.

Cambodia, Laos, and Vietnam

In French Indochina, mathematics was encouraged for commerce. The Quoc Hoc or National Academy, established in 1896, included mathematics in its curriculum, with French as the language of instruction. Until the 1950s, most secondary schools in this region used French and French-language mathematics books—this was done in Cambodia until the early 1970s. Growth in the education system in the late twentieth century produced new native mathematics teachers, including Cambodian Communists Saloth Sar (Pol Pot), Khieu Samphan, and Gaing Kek Ieu (called "Comrade Deuch").

The Vietnamese Mathematical Society was founded in 1965, roughly the same time as one major build-up of American troops during the Vietnam conflict. Many educational institutions were closed for many years because of the war, but the society continued to support regional mathematical research. Vietnam first participated in the IMO in 1974 and hosted the competition in 2007. Mathematics researchers and students from Lao People's Democratic Republic (Laos) also participate in conferences and competitions. For example, in 2010, two high school students won a mathematics prize in a competition that included students from Brunei, Cambodia, Indonesia, Malaysia, the Philippines, Singapore, Vietnam, Thailand, and Laos.

Indonesia

The Dutch in the Netherlands East Indies operated a system of European schools, so-called "native schools," and vocational schools, teaching primarily in Dutch with Dutch-language textbooks. Many of the applied mathematics courses were directed toward engineering. After independence, with the expansion of the education system in Indonesia, there are mathematics departments in all schools and most universities in the country. Indonesia first participated in the IMO in 1988.

Brunei, Myanmar, and the Philippines

Elsewhere in the region there is also mathematical activity. The study of mathematics in the Philippines has been influenced by its close connections with the United States. The Mathematics Society of the Philippines was established in 1973, and the Philippines began participating in the IMO in 1988. Brunei participated in the IMO in 2000. The country of Myanmar has been isolated for much of the period since its independence in 1948. At the start of the twenty-first century, it initiated a 30-year plan for educational reform to address the challenges of the information age. Traditionally, state schools focused on writing, reading, and speaking in Myanmar and English, as well as mathematics, science, and Myanmar geography and history. Newer programs offer increased access to computer skills, as well as courses on information technology, medicine, and engineering, which require more advanced mathematics skills.

Further Reading

Hong, Kho Tek, Yeo Shu Mei, and James Lim. *The Singapore Model Method for Learning Mathematics*. Singapore: EPB Panpac Education, 2009.

Southeast Asian Mathematical Society. http://www.seams-math.org.

Southeast Asian Ministers of Education Organization-Regional Centre for Education in Science and Mathematics. http://www.recsam.edu.my/html/history.html.

Justin Corfield

Asia, Southern

Category: Mathematics Around the World.
Fields of Study: All.
Summary: Southern Asia's history of mathematics reaches back thousands of years and mathematics continues to be a priority.

Southern Asia has a rich tradition in mathematics. Persian, Hindu, and Vedic scholars, among others in this area, contributed to the body of mathematics knowledge. Some of the achievements that have been historically credited to Arabic or Islamic mathematicians may have been influenced by pre-Islamic Persia. From ancient times, the rise and fall of various empires, wars, migration, and colonial influences have resulted in shifting cultural and geographical boundaries. As a result, many countries and regions in southern Asia have changed over time. The United Nations statistical classification for southern Asia contains Afghanistan, Bangladesh, Bhutan, India, Iran (Islamic Republic of), the Maldives, Nepal, Pakistan, and Sri Lanka. In the twenty-first century, these Asian nations continue to make advances in mathematics and mathematics education.

History

Construction of many ancient temples or monuments in southern Asia clearly involved mathematical knowledge, and mathematicians from this time period made various contributions to mathematics. One example is Indian scholar Baudhayana, who lived around 800 B.C.E. and is credited by some with developing the Pythagorean theorem, although others feel he was reflecting Babylonian work. The Vedic priest Katyayana, who lived approximately six centuries later, appears to have been interested in mathematics for religious purposes. Panini (520–460 B.C.E.), born in Shalatula, now part of Pakistan, wrote a scientific theory of Sanskrit. Some historians have theorized that development of algebraic structures and number systems in this region may be tied to the linguistic structure of Sanskrit. Panini's work also influenced computer languages. Aryabhata (476–550) wrote a mathematical text known as the *Aryabhatiya*. It is composed of 123 metrical stanzas, whose organization has been studied by mathematicians because it differs from later mathematical works in many traditions. Some historians believe that it was influenced by Mesopotamia, while others suggest that it might be an anthology of works by earlier mathematicians. Another text, the *Bakhshali* manuscript, discovered in 1881 near Peshawar in present-day Pakistan, is believed to date from the seventh century, although some experts have dated it to up to eight centuries earlier or five centuries later.

By medieval times, Indian mathematicians had developed the notion of zero as a number, the use of negative numbers, and the definitions of sine and cosine. Some early Indian poetry also shows evidence of the binary number system and the use of decimal numbers. Mathematician Abd Al-Hamid ibn Wasi ibn Turk Al-Jili (c. ninth century) is believed to have been born in Iran, Afghanistan, or Syria. He wrote an algebra book. Persian mathematician, poet, and astronomer Omar Khayyam (1048–1141) wrote books on arithmetic and algebra by the age of 25 and contributed to many mathematical areas. Mathematician Nasir al-Din al-Tusi (1201–1274) was born in the city of Tus, now in Iran. He wrote Arabic translations of several Greek mathematical texts and is also credited with developing planar and spherical trigonometry from what many considered an astronomical tool into a separate mathematical discipline. Ghiyath al-Din Jamshid Mas'ud al-Kashi (1380–1429) was born and worked in Kashan, now in Iran. His *Treatise on the Circumference* included a calculation of π, which exceeded any known precision at the time. He also authored a teaching text called *The Key to Arithmetic*.

Education

Mathematics education has long been a focus in southern Asia. Mathematics was a part of *garakula* residential schools in ancient Nepal and India. From the fourteenth century, what became known as the "Kerala school of astronomy and mathematics" emerged in southern India. There was a flourishing of new discoveries, including the use of calculus long before it was developed by Isaac Newton and Gottfried Leibniz. These developments continued under mathematicians such as Citrabhanu (c. 1530) and Jyesthadeva (c. 1500–1575). English scholar Charles Whish (1794–1833) publicized many of the Kerala achievements to the rest of the world. Even then, the work of Whish—primarily a collector of Sanskrit manuscripts—was largely unknown beyond the scholarly community until the Indian mathematicians K. M. Marar and C. T. Rajagopal were able to demonstrate the advances made in Kerala just prior to the establishment of the European colonial empires in India.

British colonialism brought some European teaching styles into areas of southern Asia, and many universities were founded in the nineteenth century. Also in the nineteenth century, some Nepalese students traveled to India to study, where they were exposed to texts like Bhaskaracharya II's (1114–1185) *Siddhanta Siromani*. French mathematics traditions were introduced to southern Asia by Father Racine (1897–1976), a Jesuit missionary who had previously earned a doctorate in mathematics. With Indian colleagues such as

Ramaswamy Vaidyanathaswamy (1894–1960), he promoted "modern" or contemporary mathematics teaching versus solely classical mathematics in the twentieth century. Indo-French collaborations continue to flourish into the twenty-first century and have been cited as contributing to development of areas like algebraic geometry and theoretical partial differential equations in southern Asia. There were other well-known collaborations, such as that between Indian mathematician Srinivasa Ramanujan and British mathematician Godfrey (G. H.) Hardy. In the 1980s, the Maldives introduced a new school curriculum that increasingly emphasized the importance of a variety of subjects, including mathematics.

Researchers in southern Asia have investigated a wide variety of different curricular issues such as gender differences in mathematics in Pakistan. King of Bhutan Jigme Khesar Namgyel Wangchuck noted in 2009:

> In all the countries where progress has been strong in the areas we strive to develop, the strength of the education system has been in Math and Science. In fact in India, the favourite subject for most students is Mathematics. In Bhutan, Mathematics is one of our main weaknesses.

Students from Bangladesh, India, Iran, Pakistan, and Sri Lanka have competed in the International Mathematics Olympiad: Iran since 1985, India since 1989, Sri Lanka since 1995, and Bangladesh and Pakistan since 2005. Mumbai, India, hosted the Olympiad in 1996.

Further Reading

Dauben, Joseph W., and Rohit Parikh. "Beginnings of Modern Mathematics in India." *Current Science* 99, no. 3 (August 10, 2010). http://www.ias.ac.in/currsci/10aug2010/suppl/15.pdf.

Jha, K., P. R. Adhikary, and S.R. Pant. "A History of Mathematical Sciences in Nepal." *Kathmandu University Journal of Science, Engineering and Technology* II, no. 1 (2006). http://www.ku.edu.np/kuset/second_issue/e2/KANAIYA%20JHa-pdf.pdf.

Joseph, George. *The Crest of the Peacock: Non-European Roots of Mathematics*. 3rd ed. Princeton, NJ: Princeton University Press, 2011.

Katz, Victor. *The Mathematics of Egypt, Mesopotamia, China, India, and Islam*. Princeton, NJ: Princeton University Press, 2007.

Waldschmidt, Michel. "Indo–French Cooperation in Mathematics." *Mathematics Newsletter of the Ramanujan Mathematical Society* 19, Special Issue 1 (2010) http://www.math.jussieu.fr/~miw/articles/pdf/IndoFrenchCooperationMaths.pdf.

Justin Corfield

Asia, Western

Category: Mathematics Around the World.
Fields of Study: All.
Summary: The people of western Asia have long studied and influenced mathematics.

Ancient western Asia, including Anatolia, Syria, Mesopotamia, and the Iranian plateau, along with Egypt, is regarded by many as the cradle of civilization. Activities that shaped numerous civilizations are traced historically to this region, including the invention of the wheel, practice of agriculture, first writing systems, and first administrative structures. Many intellectual and scientific disciplines flourished. The development of mathematics followed and was affected by the rise and decline of the civilizations of western Asia. Throughout history, the territory has been settled or invaded by many ethnic groups, including the Babylonian, Persian, Hellenistic, Roman, and Islamic cultures. Some countries were also part of the Soviet Union. It is not always possible to determine the exact origin of historical figures, and, as such, people may be included in the histories of many regions or identified by cultural heritage and the location where they did their work. Further, many of their accomplishments are named for later mathematicians. The twenty-first-century United Nations grouping for western Asia is listed as Armenia, Azerbaijan, Bahrain, Cyprus, Georgia, Iraq, Israel, Jordan, Kuwait, Lebanon, Occupied Palestinian Territory, Oman, Qatar, Saudi Arabia, Syrian Arab Republic, Turkey, the United Arab Emirates, and Yemen.

Babylon

Historical knowledge of Babylonian mathematics is largely limited to translations of the surviving clay tablets that have been unearthed by archaeologists, but even this evidence suggests a rich depth and breadth of mathematics scholarship, largely focused on practical problems. Subsequent cultures that came to the region also left parts of their mathematical legacies. With the emergence of Islam at the end of the sixth century, many of the nomadic tribes living in the Arabian Peninsula joined together to form a significant power.

By the early eighth century, a sociopolitical entity often called the Islamic Empire, which was ruled mostly by a series of government entities known as *caliphates*, spanned from Spain and north Africa to southeastern Anatolia, Persia, and the western portion of central Asia. On the east, the region shared a long border with India, and hence many Muslim intellectuals were also cognizant of Indian culture and mathematical accomplishments. Many local rulers encouraged scholarship, building on the legacy left by the Hellenic and Roman periods.

The House of Wisdom in Baghdad, in what is now Iraq, became the main hub of research and intellectual activity, rivaling Alexandria at its zenith. Works of Hellenistic mathematicians were translated into Arabic—the only surviving copies of certain works. Mathematicians also extended and introduced new ideas and fields. Social factors were another motivating influence in mathematics scholarship in Muslim lands, such as the calculation of the local daily prayer times, the direction of the prayer (toward Mecca), and the determination of the local first day and the end of the holy month of Ramadan. Since the commonly used lunar calendar was 11 days shorter than the solar year, this problem added complexity for numerous peoples and religions in the area. Observing the heavens and predicting the astronomical events was a major field of research for mathematicians and astronomers.

Ottoman Empire and Turkey

Wars brought turmoil to the area, and scholarly activities suffered. Following the conquest of Istanbul in 1453, Ottoman Sultan Mehmed-II built madrasas (buildings used for teaching Islamic theology and religious law, often including a mosque) and encouraged scholars to congregate. However, later events negatively impacted mathematics in western Asia; for

Mathematics was used in Muslim lands to calculate the direction of prayer toward Mecca. (Photos.com)

example, the destruction of centers of learning such as the Istanbul observatory and the spread of religious scholasticism (a philosophy of teaching that follows a relatively a narrow set of traditional methods heavily influenced by religious teachings), which also occurred in medieval Europe. Some scholars indicate the passage of mathematical leadership over to Europe after about the fifteenth century.

Ottoman Empire efforts of the early nineteenth century reenergized mathematics efforts. Vidinli Hüseyin Tevfik Pasa (1832–1901) contributed to linear algebra and Mehmet Nadir (1856–1927) worked on the theory of Diophantine equations, named for Diophantus of Alexandria. The Ottoman Empire faded after World War I, but the Turkish Republic continued its efforts. A well-known Turkish mathematician is Cahit Arf (1910–1997), known for the Arf invariant in algebraic topology, Arf semigroups, and Arf rings, among others. The *Turkish Journal of Mathematics* is one of the many scientific journals published by the Scientific and Technological Research Council of Turkey. The Turkish Mathematical Society was founded in 1948, and the country is a member of the International Mathematical Union (IMU), a worldwide association that promotes mathematics research and activity. In 1978, Turkey began participating in the International Mathematical Olympiad (IMO), a competition for high school students. Turkey hosted the IMO in 1993.

Israel

Mathematical activity in Israel dates back to antiquity, and it is one of the countries in western Asia with a thriving mathematics community. This fact is due in part to researchers like algebraist Shimshon Avraham Amitsur, who was one of the 1963 founders of the *Israel Journal of Mathematics*. Some other notable Israeli-born mathematicians include Oded Schramm, Saharon Shelah, and 2010 Fields Medal winner Elon Lindenstrauss. The Einstein Institute of Mathematics, named for Albert Einstein, was founded in the 1920s. Israel is a member of the IMU, and Israeli high school students first participated in the IMO in 1979. The Israel Mathematical Union is an organization that offers opportunities for students, teachers, and researchers. In the twenty-first century, there were some calls to boycott Israeli scholars over disputed territories. In response, numerous mathematical organizations worldwide, including the IMU, passed resolutions that stressed the importance of open international scientific exchange.

Other Countries

A revitalization of mathematical activity took place in many other western Asian countries in the twentieth century often connected with professional organizations or national institutes of science. For example, the development of contemporary mathematics in Armenia is tied to the 1944 beginnings of the Institute of Mathematics of the National Academy of Sciences of the Republic of Armenia. The country began participating in the IMO in 1993, the same year as Azerbaijan.

The first issue of the *Azerbaijan Journal of Mathematics* was published in January 2011. The Kuwait Foundation for the Advancement of Sciences supports the Kuwait Mathematics Program at the University of Cambridge, which underscores the relationships between western Asia and universities in other areas of the world. Kuwait began participating in the IMO in 1982.

In 2010, the editor of the *Arab Journal of Mathematics and Mathematical Sciences* was from Jordan. The Cyprus Mathematical Society was founded in 1983 and hosts activities like the Cyprus Mathematical Olympiad. Cyprus began participating in the IMO in 1984, Bahrain in 1990, the United Arab Emirates in 2008, and Syria in 2009. Saudi Arabia first participated in the IMO in 2004. It is also a member of the IMU, and mathematicians gather through the Saudi Association for Mathematical Sciences. Oman is an associate member of the IMU. Countries such as Qatar have developed mathematics standards for grades 1–9. Some countries in western Asia continue to be affected by the area's ongoing sociopolitical volatility. Georgia declared its independence from the Soviet Union in 1991 and is redeveloping many aspects of its national identity. It began participating in the IMO in 1993 and is a member of the IMU through the Georgian National Mathematical Committee. Iraq is also rebuilding itself after the turmoil of the late twentieth century and early twenty-first-century wars.

Some countries in the region participated in the Trends in International Mathematics and Science Study (TIMSS). In 2003, the study included fourth graders from the Republic of Yemen; eighth graders from Bahrain, Israel, Jordan, Lebanon, the Palestinian National Authority, the Syrian Arab Republic, and Saudi Arabia; and both fourth and eighth graders from Armenia and Cyprus. In 2007, even more countries from this region participated, including Armenia, Bahrain, Cyprus, Georgia, Israel, Jordan, Kuwait, Lebanon, Oman, the Palestinian National Authority, Qatar, Saudi Arabia, the Syrian Arab Republic, Turkey, and Yemen. In 2011, Armenia, Azerbaijan, Bahrain, Georgia, Israel, Jordan, Kuwait, Lebanon, Oman, the Palestinian National Authority, Qatar, Saudi Arabia, the Syrian Arab Republic, Turkey, the United Arab Emirates, and Yemen are included with benchmarking participants from this region listed as including Abu Dhabi, UAE, and Dubai, UAE.

Further Reading

Carr, Karen. "West Asian Mathematics." History for Kids. http://www.historyforkids.org/learn/westasia/science/math.htm.

Inonu, Erdal. "Mehmet Nadir: An Amateur Mathematician in Ottoman Turkey." *Historia Mathematica* 33, no. 2 (2006).

Irzik, Gurol, and Güven Güzeldere, eds. *Turkish Studies in the History and Philosophy of Science (Boston Studies in the Philosophy of Science)*. New York: Springer, 2005.

Mathematics in Israel: Historical and Current Affairs. http://imu.org.il/#mathinisrael.

Supreme Education Council. "Qatar Mathematics Standards: Grade 9." http://www.education.gov.qa/CS/en_math/9.pdf.

Trends in International Mathematics and Science Study. http://timss.bc.edu/timss2003.html.

Dogan Comez
Sarah J. Greenwald
Jill E. Thomley

Atomic Bomb (Manhattan Project)

Category: Government, Politics, and History.
Fields of Study: Algebra; Data Analysis and Probability; Number and Operations; Representations.
Summary: The atomic bomb was made possible by Einstein's discovery of energy-mass equivalence.

Influenced by a letter from the famous German-American theoretical physicist Albert Einstein and other prominent scientists, U.S. President Franklin D. Roosevelt authorized the establishment of the Manhattan Project (the code name given to an elaborate effort to design, construct, and detonate an atomic bomb) in mid-1942. The project was directed by physicist J. Robert Oppenheimer, and his group of scientists, mathematicians, and engineers conducted secretive, pioneering research that led to the development of the first nuclear weapons.

Among the scientists who worked on the Manhattan Project were Italian physicist Enrico Fermi, American theoretical physicist Richard Feynman, Hungarian-American mathematician John von Neumann, Hungarian-American theoretical physicist Edward Teller, and Polish-American mathematician Stanislaw Ulam (Einstein also worked as a consultant throughout the project). Notably, several of these scientists, including Einstein and Ulam, were of Jewish decent and eventually resided in America because of Nazi persecution.

Through much trial and tribulation, the first nuclear bomb detonation test titled "Trinity" was successfully conducted on July 16, 1945, in Alamogordo, New Mexico. The Manhattan Project ultimately produced two types of atomic bombs; the plutonium implosion device (the plutonium or implosion bomb), and the uranium bomb (the uranium "gun" bomb). The plutonium bomb was the more difficult of the two to construct and required testing, whereas the uranium bomb was comparatively more simplistic and remained untested until the war.

Following Trinity, the U.S. government attempted to end World War II by detonating its uranium bomb nicknamed "Little Boy" over Hiroshima, Japan, on August 6, 1945. The blast destroyed approximately one-third of the city and caused about 140,000 causalities. Japan's reluctance to surrender prompted the United States to drop its plutonium bomb nicknamed "Fat Man" over Nagasaki, Japan, three days later. This blast killed about 70,000 people, destroyed about one-third of the city, and subsequently ended the war.

The revolutionary science of the Manhattan Project—namely the process of creating atomic explosions—was seemingly insurmountable, and paved the way for significant advancements in physics, chemistry, and mathematics. However, the historical impact of the atomic bombs dropped on Japan, as well as the philosophical and ethical ramifications, is an issue still debated today. In this regard Oppenheimer said, "It is a profound and necessary truth that the deep things in science are not found because they are useful; they are found because it was possible to find them."

The First Nuclear Reactions

Nuclear fission is the splitting of the nucleus of a heavy atom into smaller pieces, which releases a gigantic amount of energy. When this type of reaction is self-sustaining (it stimulates further reactions), it is called a "chain reaction." A critical mass is the minimum mass of fissionable material needed to ensure that a nuclear reaction sustains a chain reaction. Achieving a critical mass and, ultimately, a chain reaction was the essential challenge in developing both Little Boy and Fat Man.

The Little Boy design utilized the "gun method," which was detonated by firing a mass of uranium-235 down a cylinder into another mass of uranium-235 to produce a chain reaction. Fat Man was an implosion-type device that used plutonium-239. In this design, plutonium was placed in the center of a hollowed-out sphere of high explosives, and a number of detonators located on the high explosive's surface were simultaneously fired pressurizing the core and increasing its density—creating an implosion that resulted in a chain reaction. The Trinity test bomb was similar and was nicknamed "The Gadget." Little Boy produced a blast of approximately 12,500 tons of trinitrotoluene (TNT). Fat Man had the explosive power of about 22,000 tons of TNT and The Gadget had a blast yield of around 15–20 tons of TNT.

A tremendous amount of engineering, chemistry, physics, and mathematics was involved in the development and deployment of the atomic bombs. Among these fields was a branch of theoretical physics called "quantum mechanics" (the set of scientific principles that describe the behavior of matter and energy predominating at both the atomic and subatomic levels), which at the time was in its infancy. Quantum mechanics was developed under the assumption that energy is not infinitely divisible but rather composed of quanta (small increments).

Unlike classical or Newtonian mechanics, which describes the motion of objects we encounter every day at the macrocosmic level, quantum mechanics deals with uncertainty in many of its results and is statistical and probabilistic in nature. Although initially this branch of physics was not readily accepted, it nonetheless proved an essential tool in the development of the atomic bomb as it provided many of the insights necessary for its construction. In regard to the science and mathematics utilized in the development of the bomb Stanislaw Ulam said, "It is still an unending source of surprise for me to see how a few scribbles on a blackboard or on a sheet of paper could change the course of human affairs."

The Energy-Mass Equivalence

One of the most imperative concepts in the development of the atomic bomb was the mathematical formulation of the energy-mass equivalence, which was derived by Einstein. He established that mass and energy are, in fact, both different manifestations of the same thing. This idea was a counterintuitive and revolutionary result that spawned from his 1905 special theory of relativity. Einstein's formulation implied that very minute amounts of mass can be converted into excessively large amounts of energy. For example, this very encyclopedia is, in actuality, a form of energy in storage, which could equivalently be called rest energy or mass. If this encyclopedia could be completely converted into energy, it would yield a gigantic amount of energy indeed. This energy-mass equivalence concept is depicted symbolically through one of the world's most famous equations:

$$E = mc^2.$$

This equation is interpreted as the rest energy E of an object being equivalent to the mass m of the object multiplied by the square of the speed of light c in a vacuum. Alternatively, $E = mc^2$ can be construed as the equation that allows one to determine the amount of mass needed to produce a certain amount of energy—assuming all of the mass can be converted completely into pure energy.

To better understand how this famous simple equation was crucial in the development of the atomic bomb, one needs to understand its nature. First, $E=mc^2$ is a "direct proportion" (E is directly proportional to m), and is symbolically expressed as $E \propto m$. In general, a direct proportion has the form of

$$x \propto y \text{ or equivalently } x = ay$$

where a is the proportionality constant. As a simple example:

$$4 \propto 2 \text{ or equivalently } 4 = 2a.$$

In this case, the proportionality constant is $a=2$, whereas in the energy-mass equivalence, the proportionality constant is c^2. According to the International Bureau of Weights and Measures, the value of c is 299,792,458 meters per second (m/s), or about 186,282.4 miles per second (mi/s). For computational simplicity, c is often rounded to 300,000,000 m/s (186,000 mi/s), except when performing experiments that require exact values for light speed. Now, taking $c = 300,000,000$ m/s $= 3 \times 10^8$ m/s one can compute that 1 kilogram of plutonium could theoretically turn into

$$E = 90,000,000,000,000,000 \text{ kg m}^2/\text{s}^2 = 9 \times 10^{16} \text{ kg m}^2/\text{s}^2$$

Therefore, one can intuitively understand why a small amount of uranium or plutonium can generate explosions as massive as the ones produced by Little Boy and Fat Man.

It is interesting to note that for the Trinity test, the mushroom cloud expanded to nearly 300 meters (about 984 feet) in .053 seconds.

Further Reading

Bird, Kai, et al. *American Prometheus: The Triumph and Tragedy of J. Robert Oppenheimer.* New York: Random House, 2005.

Coster-Mullen, John. *Atom Bombs: The Top Secret Inside Story of Little Boy and Fat Man.* Self-Published, 2003.

Groves, Leslie M. *Now It Can Be Told: The Story of the Manhattan Project.* New York: Harper, 1962.

Kelly, Cynthia C., et al. *The Manhattan Project: The Birth of the Atomic Bomb in the Words of Its Creators, Eyewitnesses, and Historians.* New York: Black Dog & Leventhal Publishers, 2007.

Rhodes, Richard. *The Making of the Atomic Bomb.* New York: Touchstone, 1986.

Serber, Robert. *The Los Alamos Primer: The First Lectures on How to Build an Atomic Bomb.* Berkeley: University of California Press, 1992.

Sullivan, E. T. *The Ultimate Weapon: The Race to Develop the Atomic Bomb.* New York: Holiday House, 2007.

Daniel J. Galiffa

Babylonian Mathematics

Category: Government, Politics, and History.
Fields of Study: Algebra; Connections; Geometry; Measurement; Representations.
Summary: Babylon had an advanced utilitarian mathematics from which we inherited sexagesimal timekeeping.

Our knowledge of Babylonian mathematics (2100–200 b.c.e.) is based on extensive mathematical calculations found on clay tablets in the area of Mesopotamia (now Iraq), surrounding the ancient city of Babylon between the Tigris and Euphrates rivers. Because only a fraction of the tablets have survived—and only a small fraction of those have been translated—our knowledge of the depth and breadth of Babylonian mathematics is limited. Mathematics historian Otto Neugebauer likens the situation to tearing a few random pages out of a few textbooks and then trying to reconstruct a representation of modern mathematics. Nonetheless, Babylonian mathematics did involve complicated mathematics, and was used primarily to solve practical problems. These mathematical problems ranged from arithmetic calculations, to algebraic rules, to geometrical formulas, to numerical ideas.

Babylonian Number System

The Babylonian number system was sexagesimal, using both a place value notation based on powers of 60 and a base-10 grouping system for numbers between 1 and 59 within each place value.

Traces of their sexagesimal notation remain today in the recording of time (hours, minutes, seconds) and the measurement of angles (degrees, minutes, seconds). Their numbers were written in cuneiform, or the use of a triangular stylus to make wedges on a clay tablet. A vertical line represented unity and a horizontal wedge mark represented a 10.

For example, within each place value, the number 57 would be represented by 5 horizontal wedges and 7 vertical lines. Expanding the example, a cuneiform number represented in modern form as "3, 4, 57" was equivalent to

$$3(60^2) + 4(60^1) + 57(60^0)$$
$$= 3(3600) + 4(60) + 57(1) = 11097.$$

The Babylonians had neither a symbol for zero as a placeholder nor a symbol to designate the "decimal" point in their sexagesimal fractions. Writing and reading numbers required the Babylonian mathematician to understand the problem's context and the use of a space to represent either an "empty" place value or shift to fractional place values. Thus, the previous number, "3,4,57," possibly was equivalent to:

$$3(60^4) + 0(60^3) + 4(60^1) + 57(60^0)$$
$$\text{or } 3(60^4) + 4(60^3) + 57(60^2)$$
$$\text{or } 3(60^1) + 4(60^0) + 57(60^{-1})$$

or even $3(60^{-1}) + 4(60^{-2}) + 57(60^{-3})$.

To avoid ambiguity, modern translations of these numbers would be first "3,0,4,57" or "3,4,57,0,0" or "3,4;57" or "0;3,4,57" respectively, where the semicolon separates whole numbers from fractional numbers. Tablets from the Seleucid period (300 B.C.E.) did include a special symbol that played the double role of a placeholder (zero) and the separator between two sentences.

Babylonian Arithmetic

Using the sexagesimal system, the Babylonians were able to add, subtract, multiply, and divide numbers. Their computations were complemented by the use of extensive tables. Their multiplication tables had products ranging from 1 × 1 through 59 × 59, and seeming somewhat unusual, they had access to multiplications tables for "1,20" (or 80), "1,30" (or 90), "1,40" (or 100), "3,20" (or 200), "3,45" (225), and even "44,26,40" (or 160,000). Some of this can be explained by looking at their tables of reciprocals for working with fractions. For example, one table includes the deceptive notation 1 ÷ 1,21 = 44,26,40, with the latter value actually being "0;0,44,26,40."

The Babylonians produced extensive tables of squares and cubes, tables of square sides and cube sides (square and cube roots), and sums of squares and cubes. When a table side-value was not available, the Babylonians approximated roots using an interpolation process based on averaging and division; this process was quite fast, producing 26-decimal accuracy in five iterations.

Babylonian Algebra

Though without an algebraic notation, the Babylonians solved numerous types of algebraic equations. Each solution involved the replication of a formulaic prescription represented by a step-by-step list of rules. In effect, their prescriptions invoked algorithms, which were usually specific to a stated problem and not generalized to a class of problems.

For example, consider this Babylonian problem: the area and two-thirds of the side of my square have I added and it is 0;35. In modern notation, their step-by-step solution was: 1, the unit, you take; two-thirds of 1, the unit, is 0;40: Its half is 0;20 and 0;20 you multiply 0;6,40, you add 0;35 to it and 0;41,40 has 0;50 for its square root. 0;20 that you multiplied with itself, from 0;50 you subtract and 0;30 is the side of the square.

In modern mathematics, this same problem would involve solving the quadratic equation:

$$x^2 + \frac{2}{3}x = \frac{35}{60}.$$

The steps in this problem also can be interpreted using geometrical algebra, where the square is "completed" in a manner similar to the derivation of our general quadratic formula.

In their solution of special types of algebraic equations, the Babylonians made extensive use of their tables of the sums of squares and cubes, especially if the equation was of the third or fourth degree. Some of their solutions to algebraic problems were quite sophisticated. For example, one problem involved a system of equations of the form

$$xy = n \quad \text{and} \quad \frac{mx^2}{y} + \frac{py^2}{x} + q = 0.$$

Its solution using substitution would normally lead to a single-variable equation involving x^6, but the Babylonians solved it by viewing it as a quadratic equation in x^3.

Babylonian Geometry

Dominated by their work with algebraic ideas, the Babylonians' geometry focused on practical measurements such as the calculation of lengths, areas, and volumes. Again, the Babylonians used prescriptive formulas. For example, to calculate a circle's circumference, they multiplied the diameter by 3, implying their value of π was 3. For the circle's area, they squared the circumference and divided by 12, which is equivalent to our modern formula $A = \pi r^2$ if the correct value of π had been used.

Mathematics historians credit the Babylonians with the division of a circle into 360 degrees. Neugebauer suggests it is related to their Babylonian mile, a measure of long distance equal to about 7 miles. This measure evolved into a time unit, being the time it took to travel this distance. After noting that 12 of these time units equaled a full day or one revolution of the sky, the Babylonians subdivided their mile into 30 equal parts for simplicity, leading to 12 × 30 = 360 units in a full circle.

The Babylonians computed areas of right triangles, isosceles triangles, and isosceles trapezoids, as well as the volumes of both rectangular parallelepipeds and some prisms. They had difficulties with certain three-dimensional shapes, being unable to compute correctly the length of the frustum of a pyramid (they claimed it was the product of the altitude by the average of the bases).

The Babylonians did know some general geometric relationships. For example, they knew that perpendiculars dropped from the vertex of an isosceles triangle bisected the base, that corresponding sides of similar triangles were proportional, and that angles inscribed in a semicircle are right angles. The Babylonians used this knowledge to solve difficult geometrical problems, such as their determination of the radius of a circle circumscribing an isosceles triangle.

Evidence suggests that they knew a precursor of the Pythagorean formula. One cuneiform tablet (c. 1700 B.C.E.) includes sexagesimal numbers written along a square's side (30) and diagonal ("42,25,35" and "1; 24, 51, 10").

The latter number is both the product of the other two numbers and a good approximation of the square root of 2 (1.414214). Also, in the Plimpton 322 collection, some of the tablets contain tables of Pythagorean triples ($a^2 + b^2 = c^2$), arranged with increasing acute angle of the associated right triangle.

Signs of Advanced Mathematical Thinking

For the most part, Babylonian mathematics was utilitarian, being tied to solving practical problems. Nonetheless, interpretations of some of the tables on the clay tablets suggest that the Babylonians occasionally explored theoretical aspects of mathematics. Examples include their tables of Pythagorean triples and tables of exponential functions (which perhaps were used to compute compound interest in business transactions). Also, the Louvre tablet (300 B.C.E.) includes two series problems

$$1 + 2 + 2^2 + 2^3 + \ldots + 2^8 + 2^9 = 2^9 + 2^9 - 1$$

and

$$1^2 + 2^2 + 3^2 + 4^2 + \ldots 9^2 + 10^2 = \left(1\frac{1}{3} + 10\frac{2}{3}\right)(55) = 588$$

but historians do not suggest the Babylonians knew general series formulas such as

$$\sum_{k=0}^{n} r^k = \frac{r^{n+1} - 1}{r - 1}.$$

Specific to number theory, mathematics historians point to the cumbersome nature of the Babylonians' sexagesimal system, making it difficult to explore ideas such as factors, powers, and reciprocals. Some suggest that this is symptomatic of the Babylonian's reasonable choice of 3 for π, rather than the fraction

$$\frac{22}{7}$$

equal to the more complicated repeating expression "3; 8, 34, 17, 8, 34, 17, ..."

Further Reading

Aaboe, Asger. *Episodes From the Early History of Mathematics.* Washington, DC: Mathematical Association of America, 1975.

Friberg, Jöran. *Unexpected Links Between Egyptian and Babylonian Mathematics.* Singapore: World Scientific Publishing, 2005.

Katz, Victor J., ed. *The Mathematics of Egypt, Mesopotamia, China, India, and Islam: A Sourcebook.* Princeton, NJ: Princeton University Press, 2007.

Van der Waerden, Bartel Leendert. *Science Awakening.* Oxford, England: Oxford University Press, 1985.

———. *Geometry and Algebra in Ancient Civilizations.* Berlin, Germany: Springer, 1983.

Jerry Johnson

Bankruptcy, Business

Category: Business, Economics, and Marketing.
Fields of Study: Data Analysis and Probability; Number and Operations; Problem Solving.
Summary: The value of a business entering bankruptcy is determined by the asset, income, or market approach and creditors are repaid according to their risk.

Bankruptcy of a business occurs when the business is legally declared insolvent (its assets are less than its liabilities). If the debtor files a bankruptcy petition, it is called a voluntary bankruptcy. However, if creditors force the debtor into bankruptcy, then it is called an involuntary bankruptcy. Most bankruptcies are voluntary. In either case, the value of the business needs to be determined for legal purposes. The standard of the value used in the valuation is the fair market value (the value of the price of the firm that a rational buyer is willing to pay to a willing seller in a free market). There are three basic approaches for valuating the business: the asset approach, the income approach, and the market approach. The hierarchy of the creditor in a bankruptcy is determined by the amount of risk the creditor bears: the creditor who bears least amount of risk will have priority to receive payment after liquidation.

Asset Approach

The asset approach determines the value of a company by adjusting its book value of assets to the current market value. It is based on the economic principles of substitution: a rational investor will not pay more for a business asset than the price of a different asset that provides similar utility. There are two methods associated with the asset approach: the adjusted book value method and the replacement cost method.

In the adjusted book value method, the assets and liabilities on the balance sheet are examined item by item by professionals to determine the business's current market value. Once the assets and liabilities have been adjusted to the current market value, the value of the company is calculated as the difference.

In the replacement cost method, the value of each asset and liability on the balance sheet is first determined as the cost to replace it. Then, the value the company is determined as the difference of its assets and liabilities.

The asset approach is not reliable for companies with significant intangible assets because the approach involves professional judgment. It is more suitable for companies that have many tangible assets and few intangible assets.

Income Approach

The income approach determines the fair market value of a firm by discounting its expected cash flows at an appropriate discount rate assuming the firm will continue to operate without liquidation. The discount rate is often chosen to be the firm's weighted average cost of capital (WACC). The procedure is completely analogous to that of determining the net present value of a firm in corporate finance theory. Mathematically, the fair market value under the income approach can be written as

$$\text{FMV} = \frac{E(C)}{\text{WACC}}$$

where FMV is the fair market value, $E(C)$ is the expected cash flows under the assumption that the firm will continue to operate, and WACC is the weighted capital of cost.

In corporate finance theory, WACC is often calculated as the weighted average of the cost of debt of the firm and the cost of equity of the firm

$$\text{WACC} = (1 - T_c) r_D \frac{B}{B+S} + r_E \frac{S}{B+S}$$

where T_c is the corporate tax rate, r_D is the cost of debt, r_E is the cost of equity, B is the market value of the firm's bonds, and S is the market value of the firm's stocks.

WACC takes into consideration the facts of leverage and taxes and thus is the appropriate discount rate used for income approach. The income approach assesses the value of the debtor to the creditors. However, it fails to take account of the value inherent in the flexibility of decision making, which is often valued using a mathematical tool called "decision tree."

Market Approach

The market approach assesses a company's value by comparing it with similar companies in the market. The rationale behind this approach is that the price of the subject company should be very close to the values of the similar companies in the market. There are two methods associated with the market approach: the guideline public company method, and the comparable transaction method. In the guideline public company method, a peer group of public companies with similar sizes, natures, operations, and financial characteristics is first selected. Next, the enterprise value of each company in the group is calculated as

$$\text{EV} = P_S \times N_S + D - C_E$$

where EV is the enterprise value, P_S is the stock price per share, N_S is the number of outstanding shares, D is total debt, and C_E is excess cash.

Then market multiples, such as enterprise value/revenue and enterprise value/earning before interest and tax, will be calculated using the enterprise value. Finally, the value of the subject company is determined by applying the calculated market multiples. For example, if the enterprise value/revenue is used, then the value of the subject company can be calculated as

$$V = EV \times R$$

where V is the value of the subject company and R is the revenue of the subject company.

In the comparable method, the value of the subject company is determined in a similar fashion as in the guideline public company method. In other words, market multiples are derived, and then they are applied to the subject company to determine its value. However, in the comparable method, public data of comparable transactions are used to calculate the market multiples.

Thus, the comparable method also consists of three steps: selecting a group of comparable transactions, calculating market multiples, and applying the market multiples.

The biggest drawback to the guideline public company method is that it is not applicable for nonpublic companies. The challenge with the comparable method is finding appropriate and reliable comparable transactions.

Paying Creditors

When a company declares bankruptcy, its creditors must be paid, but the creditors receive only some of the money they are owed. For example, if a bankrupt company is ordered to pay 10 cents on the dollar, this means for every dollar the company owes a creditor, it will pay only 10 cents. This is a proportional solution that is easy to arrive at using simple algebra. However, this is not the only payout strategy. There are several mathematical methods that can be used to determine how much money each creditor should receive. In the total equality method, available capital is simply divided equally among debtors, regardless of how much they are owed. A variation, traced back to medieval philosopher Moses Maimonides, proposes giving every debtor as equal a share as possible but never more than they are owed. In modern terms, this is a constrained optimization problem that can be solved using methods such as linear programming. Other decision methods are logically and analytically more complex, like the Shapely value, which considers paying a sequence of creditors their full amounts owed, to the extent of available funds, for all possible orderings. This game-theory approach is named for American mathematician and economist Lloyd Shapely.

Further Reading

Brealey, Richard A., Stewart C. Myers, and Allen Franklin. *Principles of Corporate Finance*. 9th ed. New York: McGraw-Hill, 2008.

Copeland, Thomas E., Fred J. Weston, and Shastri Kuldeep. *Financial Theory and Corporate Policy*. 4th ed. Upper Saddle River, NJ: Pearson Education, 2005.

Newton, Grant W. *Practice and Procedure*. Vol. 1, Bankruptcy and Insolvency Accounting. Hoboken, NJ: Wiley, 2010.

Ratner, Ian, Grant Stein, and John C. Weitnauer. *Business Valuation and Bankruptcy*. Hoboken, NJ: Wiley, 2009.

Liang Hong

Bankruptcy, Personal

Category: Business, Economics, and Marketing.
Fields of Study: Data Analysis and Probability; Number and Operations; Problem Solving.
Summary: Personal bankruptcy can be caused by exponentially increasing debt, and mathematics is used to calculate payments or to divide assets among creditors.

Personal bankruptcy is a legal proceeding intended to provide relief for the debtor. Personal bankruptcy essentially results from huge debts, which can be caused many factors including unexpected medical bills, huge credit card debts, poorly managed loans, unemployment, and divorce. The fundamental formula that lies behind most large debts is exponential growth.

Legal Procedure

Personal bankruptcy in the United States is usually a court-supervised procedure that provides the debtor with the opportunity for a fresh financial start. The earliest personal bankruptcy law in the United States can be traced to 1800. The most recent personal bankruptcy law passed by the U.S. Congress is the Bankruptcy Code of 1978. Under this law, an individual may file a voluntary petition under either Chapter 7 (liquidation) or Chapter 13 (Reorganization).

If a personal bankruptcy case is filed under Chapter 7, a court-supervised procedure begins. The debtor's assets will be classified as either exempt or nonexempt according to the state law. A trustee will then collect the nonexempt assets of the debtor. The debtor is allowed to keep all the exempt assets provided such an asset is not secured by any property. For example, a mortgage is secured by the house. Thus, debtors can still lose their houses if mortgaged payments fall behind. The debts of the debtor will be wiped out except certain non-dischargeable debts including alimony, child support, student loans, taxes, and any fines resulting from criminal conviction. The record of personal bankruptcy could stay on the debtor's credit history for up to 10 years. In summary, under Chapter 7, the debtor is discharged most of the debts and surrenders all possessions except those necessary for living. However, not everyone is qualified for Chapter 7 bankruptcy. To qualify, the debtor must complete Official Form 22A (Chapter 7) to pass the means test. The personal bankruptcy involves balancing the conflicting interests of the creditors and the debtor. While a qualified debtor can wipe out most debts under Chapter 7, some creditors will not receive any payment. The 2005 Bankruptcy Abuse Prevention and Consumer Protection Act was enacted to prevent the abuse of Chapter 7 and makes it more difficult for a debtor to file under Chapter 7.

If a personal bankruptcy petition is filed under Chapter 13, then the debtor is required to propose a repayment plan that will pay the debts during a specified period of time (typically three to five years). The plan must be reasonable and meet certain requirements. It must be approved by the court. Although the debts of the debtor cannot be written off immediately under Chapter 13, the debtor is protected from debt-collecting actions from the creditors while the repayment plan is in effect. Thus, the Chapter 13 bankruptcy is often chosen by those who have a stable income.

Exponential Growth

Personal bankruptcy results from unmanageably large debts that can be caused by many factors such as medical costs because of under-insurance and uninsured status, compulsive buying habits, loss of job, or irresponsible loans. The fundamental formula that leads to a large debt is the law of exponential growth, which occurs when the growth rate of a quantity is proportional to the current value. In mathematics, exponential functions generally involve the constant e. Mathematically equivalent forms with different bases may be used in order to more intuitively correspond to the parameters of a real-life problem, such as interest calculations. The traditional way of calculating interest on a loan is called compound interest, under which the interest earned during each interest measurement period (month, quarter, or year) will automatically be added to the principal to earn additional interest during the next interest measurement period. Mathematically, this can be expressed as

$$B_t = B_0 (1+r)^t$$

where B_0 is the principal amount, B_t is the balance of the loan at the end of t-th interest measurement period, and r is the interest effective per interest measurement period.

The loan balance under the compounding interest grows rapidly over a relatively long period, even if the interest rate is not high. For example, consider a person who takes a loan of $10,000 from a bank at a monthly interest rate of 1.5%. The loan balance after one, five, and 10 years will be $11,956, $24,432, and $59,693, respectively.

Mathematical Division of Assets

The ideas of dividing and choosing have existed as long as mankind. The mathematical theory of fair division dates back to World War II, to Polish mathematicians Hugo Steinhaus, Bronisław Knaster, and Stefan Banach. The classic bankruptcy problem in game theory addresses fairness in one way. It involves allocating some amount of resources among two or more individuals who have a claim on them, assuming that any division of the assets is allowable and that there are not enough resources to satisfy all claims. Real-life examples include someone who has declared Chapter 13 bankruptcy and therefore must repay some creditors, or dividing a deceased person's estate among several heirs—especially when the estate cannot satisfy all the deceased's commitments.

Assets may be divided equally (with or without ensuring no claimant receives more than his or her claim), proportionally according to the relative size of the claim, or by other more complex strategies. The cake-cutting problem also tackles the issue of fair allocation but includes more subjective measures of valuation that must be modeled mathematically, and sometimes an asset pool with constraints on the ways in which it may be divided. Cake-cutting problems typically require iterative algorithms to solve.

Further Reading

Anosike, Benji O. *How to Declare Your Personal Bankruptcy Without a Lawyer.* 3rd ed. Newark, NJ: Do-It-Yourself Legal Publishers, 2004.

Haman, Edward A. *The Complete Chapter 7 Personal Bankruptcy Guide.* Naperville, IL: Sphinx, 2007.

Kellison, Stephen G. *Theory of Interest.* 3rd ed. New York: McGraw-Hill, 2009.

Vaaler, Leslie Jane, and James W. Daniel. *Mathematical Interest Theory.* 2nd ed. Washington, DC: MAA, 2009.

Liang Hong

Bar Codes

Category: Business, Economics, and Marketing.
Fields of Study: Algebra; Number and Operations; Representations.
Summary: Bar codes encode numerical data visually for product identification and other purposes.

A bar code is a visual representation of information intended to be decoded by an optical scanner called a bar code reader. The reader illuminates the bar code, thus allowing its light sensor to detect the patterns of dark and light bars. The sequence and width of dark and light bars represents a unique sequence of numbers and letters.

Origins

It took 26 years for the idea of bar codes to be successfully implemented in the retail industry. In 1948, two graduate students at Drexel University, Norman J. Woodland and Bernard Silver, overheard a conversation in which the president of a local supermarket chain in Philadelphia wished to automate the checkout process. At that time, a cashier in a supermarket would have to type into a cash register the price of all items in a purchase—a time-consuming and error-prone task. Woodland and Silver filed a patent application in 1949, obtaining the patent in 1952, for an optical device that would read information automatically. The first prototype was produced by IBM but was impractical because of both its size and the heat generated by the 500-watt light bulb used by the bar code scanner. The patent was sold in 1952 to the Philadelphia Storage Battery Company (Philco), which was also unable to produce a viable prototype, and sold the patent the same year to the Radio Corporation of America (RCA). Bernard Silver died in a 1963 car accident, before the bar code system was implemented in practical settings. The invention of lasers and integrated circuits in the 1960s allowed the manufacture of small, low-energy bar code readers. RCA developed a modern version of bar codes in 1972 in a Kroger store in Cincinnati, but the code was printed in small stripes that were easily erased or blurred by employees who had to attach them manually to each item. Norman J. Woodland was an employee at IBM at the time and led a team that produced bar codes according to a standard known as Universal Product Code (UPC) still in use today. Bar codes are used in nearly all retail products worldwide. The applications of bar codes have also reached far beyond the retail industry; they are now used in such disparate applications as patient identification, airline luggage management, and document management, as well as purchase receipts.

Figure 1. Zones of a bar code.

The Mathematics of Bar Codes

The most ubiquitous form of bar codes consists of a visual pattern of long lines (hence the "bar" in "bar code"), which has four well-defined zones (see Figure 1): (1) quiet zone, or empty zone, located in the left and right zones of the code; (2) initial character (right) and final character (left) are standard bars that appear on all bar codes, and indicate where the information begins and ends; (3) variable-length character chain, which contains as many characters as needed to encode the message; and (4) checksum, which is a number that is computed algebraically from the other characters using modular arithmetic, and is used to verify that the characters have been correctly transmitted and interpreted. The digits are either simply added or are weighted. For example, the 10-digit International Standard Book Number (ISBN-10) uses weights based on digit position and modulus 11 arithmetic.

Each digit is encoded by two white and two black bars. The bars have widths of 1 to 4 units, and the total width for each digit is always seven units. Bar code readers are designed to read bar codes irrespective of their size; a magnified bar code encodes the same information as a reduced-size bar code. This property is mathematically known as scale invariance.

Further Reading

Adams, Russ, and Joyce Lane. *The Black & White Solution: Bar Code & the IBM PC*. Dublin, NH: Helmers Publishing, 1987.

Palmer, Roger C. *The Bar Code Book: A Comprehensive Guide to Reading, Printing, Specifying, Evaluating, and Using Bar Code and Other Machine-Readable Symbols*. 5th ed. Victoria, BC: Trafford Publishing, 2007.

Wittman, Todd. "Lost in the Supermarket: Decoding Blurry Bar codes." *SIAM News* 37, no. 7 (September 2004).

Juan B. Gutierrez

Budgeting

Category: Business, Economics, and Marketing.
Fields of Study: Measurement; Number and Operations.
Summary: Creating a viable budget requires mathematical analysis.

The word "budget" originally meant a small pouch. By the end of the sixteenth century, people used the word to refer to the contents as well as the bag. The connection with finance dates back to at least 1733. In general, a budget is a balanced plan for spending and saving that includes expected incomes and expenditures. Individuals or families use budgets to manage earnings; pay bills; save for events like retirement, college, or vacations; and to plan for large purchases like a car or a home.

Businesses manage revenues and expenses for materials, taxes, advertising, and payroll using budgets. They may also have smaller budgets for individual projects. City, state, and national governments use budgets to distribute incomes from taxes and other sources among expenditures like infrastructure, social programs, national defense, and debt. Mathematicians play a large role in developing mathematically sound budgets at all levels, especially accountants and actuaries. In the past, budgeting in classroom settings was confined largely to home economics classes, but now budgeting activities are often used to teach various mathematical principles in context.

Some budgets are created using known amounts. Other times, the values are forecasts of income or expenditures based on data or mathematical models. Budgets themselves can also be used for modeling and production. For example, a static budget is a fixed budget created before any input and output values are known, while a flexible budget can be adjusted based on information about actual activity. A metric called "flexible budget variance" compares flexible budgets to actual results to determine the effects of economic variables on business operations. Sales volume variance compares flexible budgets to static budgets to determine the effect a company's activity had on its operations. Budget accuracy ratios also quantify differences between various budgets or actual production. These can be used to create more accurate future budgets and to plan operations. Budgeting concepts can also apply to resources other than money. Lisa Sullivan, a senior budget analyst working for the U.S. Department of the Navy, regularly uses algebra, statistics, mathematical modeling, and operations research to explore resource allocation problems that affect budgeting. She often works on unique mathematical problems that do not occur in private industry, such as determining the optimal number of Navy surgeons needed in wartime.

Budgeting Basics

Creating a spending plan can be complicated; however, the easier the plan, the more likely it is to be followed. One of the simplest budgets used is the 10-10-80 principle. John D. Rockefeller, the first person in the world to amass a fortune of $1 billion and the wealthiest American ever when adjusted for inflation, is reported to have used the 10-10-80 principle. The crux of the principle is simple: give 10% of your earnings to charity, save 10% of your earnings, and live on the remaining 80%.

Anytime you receive income (for example, paycheck, gift, or prize), first multiply that amount by 10%. Finding 10% of an amount is a relatively easy process: move the decimal point to the left one place value. For example, if you received earnings of $342.57, multiplying by 10% would yield $34.257 (rounded to $34.26). Based on the 10-10-80 principle, you should first give $34.26 away to charity. Many people donate this money to religious institutions or charities such as the Red Cross or the United Way. One argument for giving, besides being altruistic, is to show ourselves that we have control of our money. By freely and willingly giving some of it to others instead of tightly holding onto it, we gain confidence that we have enough and can therefore live on what we are given.

The next 10%, or $34.26 in this example, is given to yourself into some sort of savings vehicle like a savings account or a money market fund. Ideally, this money is never needed as it becomes part of your long-term savings. This money may go toward retirement or an emergency fund in case of job loss or major disaster. Many people are tempted to use this long-term savings for expenses like taking vacations, buying a car, or replacing an appliance. However, these foreseeable expenses should be budgeted as part of the remaining 80%.

Once you have given 20% of your income away (10% to others and 10% to yourself), the remaining 80% can be used for living expenses (including short-term savings). How that 80% is spent can vary depending on many factors including how many people are being supported (for example, you do not need to buy as much food for a single adult as you do for a family of five). Usually the biggest expenditures are for housing and transportation.

Combined, these two categories should not account for more than 50%, or half, of your income. Of course, the less you spend on these the more you have to spend on other areas. Housing, by itself, should account for less than 35% of your income. In the previous example, 35% of $342.57 is $119.91. Set aside this $119.91 to cover any housing expenses you have.

The 10-10-80 principle is to give 10% to charity, save 10%, and live on the remaining 80% of your earnings. (Photos.com)

Housing expenses include not only the obvious rent or mortgage but also utilities (heat, electric, plumbing, sanitation, telephone, and Internet), insurance, property taxes, and property maintenance (property maintenance is usually about 5% of the property value each year).

If housing and transportation together should be 50% (or less) of your income, then 15% should be used on transportation. In the example, 15% of $342.57 is $51.39. This amount becomes earmarked to cover all transportation expenses. These expenses include car payments, insurance, license, gasoline, parking, and maintenance (car maintenance is usually about 10% of your transportation costs).

If you spend 50% of your income on housing and transportation, this leaves a mere 30% for everything else. If you have been spending more than you earn, you probably have credit card debt or other personal debt. Ideally, your debts (not including housing or transportation debts) do not account for more than 5% to 10% of your income. What remains should be used to pay for food, life insurance, medical insurance, medical and dental expenses, clothing, entertainment, short-term savings (for vacations and replacement costs), and other miscellaneous spending.

Further Reading

Johnson, Kay. *The Mathematics of Budgeting: Mathematics for Everyday Living*. Erie, PA: Meridian Creative Group, 1999.

Joshi, Mark. *The Concepts and Practice of Mathematical Finance*. 2nd ed. Cambridge, England: Cambridge University Press, 2008.

Shim, Jae. *The Art of Mathematics in Business: Analyzing Facts and Figures for Smart Business Decisions*. Sterling, VA: Global Professional, 2009.

CHAD T. LOWER

Census

Category: Government, Politics, and History.
Fields of Study: Data Analysis and Probability; Measurement; Representations.
Summary: Conducting a valid and reliable census depends on mathematical and statistical methods.

The term "census" comes from the Latin word *censere*, meaning "to assess." A census is a systematic collection of data about an entire population of interest. Usually the population is people but historically it has also been done for land, livestock, and trade goods. Sometimes a census is a one-time event, or it may be repeated periodically, like the decennial census in the United States.

There are also two primary philosophies of data collection that can affect the outcome of a census: *de jure* and *de facto*. *De jure* counts people at their usual place of residence, while *de facto* counts people where they are on the day of the census.

For example, one biblical account of the birth of Jesus involves a census in which individuals were required to return to their town of origin rather than being counted where they lived, as opposed to the U.S. census, which is centered about people's permanent residences. Archaeological records indicate that many ancient civilizations conducted censuses, the purpose of which was often taxation or military recruitment. The constitutionally stated purpose of the U.S. census is to determine each state's congressional representation, though it has grown to include additional descriptive and predictive activities. The U.S. Census Bureau is one of the largest employers of mathematicians and statisticians, who not only collect and analyze data but also lead the way in developing new data collection and analysis methods.

Statisticians work internationally as well. For example, in 1949, British statistician Frank Yates was appointed to the United Nations Commission on Statistical Sampling and published *Sampling Methods for Censuses and Surveys*, which is widely acknowledged to have been influential in establishing sound principles and technical terminology. Overall, mathematical and statistical procedures improve the quality, reliability, and representation of census data, and the methods used by census-takers are constantly evolving.

The History of the Census

The practice of completing a census for an entire population occurred in many ancient civilizations. Records suggest that the Babylonians conducted a census in about 3800 B.C.E., and that Egyptians did so in the second millennium B.C.E. Male Roman citizens had to register for a census every five years and declare both family and property. Elected censors oversaw and coordinated the census process. The censors would then summon every tribe in the country to appear before them so they could record the relevant details. In ancient Rome, the census recorded the names of the family members, along with details of any property or land they owned. This provided the leaders of the country at the time the ability to tax their citizens according to the amount they owned. William the Conqueror carried out a census in Britain in 1086 C.E. for taxation purposes. This census took years to complete and attempted to compile a comprehensive list of all land and property in Norman Britain. Such a comprehensive exercise was previously unheard of in Europe, and it preceded an early example of a modern census by nearly 600 years.

Instructed by King Louis XIV in 1666, Jean Talon, a French colonial administrator, conducted a census in order to expand the colony in New France, North America. Talon used the *de jure* method and visited many of the colonial settlers personally, compiling data on settlers' names, age, sex, and occupation. The aim of this census was to help the colony settle by using the statistics to decide how best to develop agriculture, trade, and manufacturing industry. In all, Talon managed to compile details of 3215 inhabitants and paved the way for the development of a number of further censuses in the New World. In Britain, a 1798 paper written by demographer Thomas Malthus discussed the possibility that not knowing the population size and growth rate (demographics), of a country could lead to food shortages and overuse of other resources, resulting in famine and disease as the population is unable to sustain itself. These revolutionary modeling ideas led the British government to pass through parliament the Census Act of 1800. The first modern British census took place in 1801; the process has been repeated decennially since then, except for in 1941 during World War II.

The Modern-Day Census—Data Collection

The U.S. census is required by constitutional law to take place every 10 years and involves sending forms to every residence in the United States and Puerto Rico. The data are then analyzed in order to determine how each state is represented in the U.S. House of Representatives and to provide the correct resource allocation for the current population, that is, how much of the federal fund is given to hospitals, schools, and other public services. Individual responses to the U.S. census are kept confidential for 72 years.

A similar process is used in the United Kingdom, although the census details are kept confidential for 100 years. A key difference in the census forms in the United Kingdom (UK) and the United States is that the U.S. form has just 10 questions and is two pages long. The UK census form for 2011 contained 43 questions in a 32-page booklet.

In Canada, a national census is taken every five years. Each household receives a census questionnaire, to either be filled out online or returned in the post.

Census Controversies

While the aim of collecting census data is to provide complete record of data on a population, there can be many difficulties in obtaining such comprehensive data. Past problems have included members of the population objecting to the potentially intrusive nature of such a full-scale inquiry, which has the potential to be misused, and difficulties reaching the entire population.

This second problem was especially problematic in the 1990 decennial census and spurred a great deal of developmental activity with regard to statistical survey methods. Even further in the past, there was a heated debate among the U.S. Founding Fathers about how to account for slaves in the U.S. census, as these counts had the potential to dramatically shift the balance of representative power between the northern and southern colonies. Even now, evolving social constructs and definitions of significant demographic variables, like race, can be a controversial topic.

Practical Problems With Census Taking

A number of problems can arise when attempting to take a census of an entire population. For the data to be useful, the characteristics of the whole population need to be reflected. This requirement means that any nonresponse could jeopardize the quality of the data. Nonresponses can happen, for example, when an address list is not comprehensive, or people fail to fill in their forms fully and return them. There are a number of measures used to prevent this, including following up with nonrespondents in a face-to-face interview, and setting fines for nonrespondents.

A statistical technique called "imputation" was used by the U.S. Census Bureau in its 2000 census to create data using the nearest neighbor "hot deck" method. Where a gap in the counting (for example, an entire household's data was missing) was identified, data from the geographically closest neighbor were used instead. Where a household had not completed every question fully, the missing data were imputed from a nearest neighbor record where the households are of the same size. Where a respondent within a household gave incomplete data, the characteristics were imputed from the characteristics of other household members. This technique enabled the U.S. Census Bureau to produce a more complete set of data on the U.S. population.

In order to overcome the obstacle of an incomplete address list, a number of different address lists can be combined to get a more complete list—thus ensuring a wider population is reached, and improving the reliability of the data. Alternatively, another statistical method, called "sampling," can be used to estimate features of the population. A forward-thinking statistical sampling plan was proposed by many mathematicians and statisticians after the 1990 census turned out to be particularly problematic in terms of issues such as undercoverage of certain subpopulations. The U.S. Supreme Court refused to permit sampling to completely substitute for counting. These mathematical methods are used, however, for other types of estimation and to gauge how much undercoverage or other biases might exist.

Analysis of Census Data

A number of mathematical and statistical techniques can be used to draw the most descriptive and predictive information possible from raw census data. For example, to identify resource need, data can be ranked in such a way to identify areas where there are more children, thus enabling the government to plan where to locate schools. Alternatively, areas with a high percentage of elderly people could be identified and provided with more social care. Since the 1990s, census data have become a major resource for both amateur and professional genealogists now that older records are being digitized. Census data are also used to find ways to make future collections efforts better.

Edna Lee Paisano grew up on a Nez Perce Indian Reservation in Sweetwater, Idaho. Talented in both mathematics and science, she attended the University of Washington and earned a graduate degree in social work, studying statistics in the process. In 1976, she was hired by the U.S. Census Bureau to work on issues regarding Americans and Alaskan Natives, and was the Bureau's first full-time Native American employee.

Using data from both the 1980 census and a survey she developed, Paisano discovered that Native Americans in some locations were undercounted. This was a serious issue, as allocation of federal funds to tribal units is based on census figures. She used statistical methods to improve the accuracy of the census and encouraged others in the Native American community to become educated in mathematics-related fields such as computer science, demography, and statistics. The 1990 census showed a 38% increase in U.S. residents counted as American Indians.

Further Reading

Aly, Gotz, and Karl Heinz Roth. *Nazi Census: Identification and Control in the Third Reich*. Philadelphia: Temple University Press, 2004.

Anderson, Margo. *The American Census: A Social History*. New Haven, CT: Yale University Press, 1990.

Kertzner, D. *Census and Identity: The Politics of Race, Ethnicity, and Language in National Censuses*. New York: Cambridge University Press, 2002.

Wright, Tommy, and Joyce Farmer. *A Bibliography of Selected Statistical Methods and Development Related to Census 2000*. Washington, DC: U.S. Bureau of the Census, 2000.

Amy Everton

Chinese Mathematics

Category: Government, Politics, and History.
Fields of Study: Algebra; Connections; Geometry; Representations.
Summary: Chinese mathematicians have a long history of investigation and discovery, sometimes predating similar findings in other cultures.

Chinese mathematics has a very long history, and its development is quite independent of other civilizations before the thirteenth century. Roughly speaking, it has four periods of developments before the middle of the Qing dynasty, namely

- *The early development period*: from ancient times to the Qin dynasty (2700–200 B.C.E.)
- *The foundation period*: from the Han dynasty to the Tang dynasty (200 b.c.e–1000 C.E.)
- *The golden period*: from the Sung dynasty to the Yuan dynasty (1000–1367)

- *The east meets west period:* from the Ming dynasty to the middle of the Qing dynasty (1367–1840)

Early Chinese mathematics is problem based and is motivated by various practical problems, including astronomy, trade, land measurement, architecture, and taxation.

The Early Development Period

It is written in *Yi Jing* (*I-ching* or *Book of Changes*) that, "In early antiquity, knotted cords were used to govern with. Later, our saints replaced them with written characters and tallies." In other words, the ancient Chinese used knotted cords to record numbers. Later, written symbols and tallies were used instead. In the Shang dynasty (1600–1050 B.C.E.), numerals were invented and inscribed on oracle-bones or tortoiseshells for recording numbers. It was a decimal system and was widely used at the time.

In the Zhou dynasty (1050–256 B.C.E.), mathematics was one of the *Six Arts* (*Liu Yi*), which were taught by teachers at schools. The other five arts were rites, music, archery, charioteering, and calligraphy. From this dynasty onward, the ideas of *Taichi, Ying Yang, Trigrams,* and *Hexagrams* largely influenced the developments of sciences, mathematics, philosophy, arts, architecture, and many other areas in Chinese culture. For example, *Luoshu* (3×3 magic square) is closely related to the eight *Trigrams*. It has both ceremonial and metaphysical importance, which plays a significant role in Chinese philosophy for several thousands of years.

From the Kingdoms of Spring and Autumn (720–480 B.C.E.) to the period of Warring States (480–221 B.C.E.), the Chinese used counting rods to do calculations. Numbers were expressed by nine symbols, and blanks were used to denote zeros. The numeration system was already a decimal place-valued system.

The first definitive work on geometry in ancient China was the *Mo Jing*, which was compiled after the death of Mozi (470–390 B.C.E.). Many basic concepts of geometry can be found in this book. For example, the *Mo Jing* defines a point to be the smallest unit that cannot be divided, and points on a circle to be equidistant from the center. The book also mentions the definitions of endpoints, straight lines, parallel lines, diameter, and radius.

In the Qin dynasty, the famous Great Wall and many huge statues, tombs, temples, and shrines were built, which required sophisticated skills and mathematical knowledge for calculating proportions, areas, and volumes. Unfortunately, not much is known about the actual mathematical development in the Qin dynasty now, because of the burning of books and burying of scholars ordered by Emperor Qin Shi Huang.

The Foundation Period

In 1984, a Chinese mathematics text called *Suan Shu Shu*, completed at about 200 B.C.E., was discovered in a tomb at Zhangjiashan of the Hubei Province. It is about 7000 characters in length, and is written on 190 bamboo strips. Its content is mainly concerned with basic arithmetic, proportions, and formulas of areas and volumes. The next complete surviving text is the *Zhou Bi Suan Jing*, written between 100 B.C.E. and 100 C.E. Although it is a book on astronomy, it contains a clear description of the *Gougu Theorem* (the Chinese version of the Pythagorean theorem), which is very useful in solving problems in surveying and astronomy. This work is perhaps the earliest recorded proof of the Pythagorean theorem.

After the book burning in 212 B.C.E., the Han dynasty (202 B.C.E.–220 C.E.) began to edit and compile the mathematical works lost in the Qin dynasty. The most important one is the *Nine Chapters on the Mathematical Art* (*Jiuzhang Suan Shu*), completed at around 179 C.E. Although the editor is unknown now, this book had a great impact on the mathematical developments in China and its neighboring countries, such as Japan and Korea. It contains a collection of 246 mathematical problems on agriculture, engineering, surveying, partnerships, ratio and proportion, excess and deficit (the method of double false positions), simultaneous linear equations, and right-angled triangles.

The general method of solutions is provided, but no proof is given in the Greek sense. Most of the methods are of computational nature, and they can be applied to solve problems algorithmically. For instance, square roots, or cubic roots, can be found in a finite number of steps by using a procedure called *Kai Fang Shu*. For skillful users of this method, the answers can be computed efficiently by manipulating the counting rods. For circular measurements, the approximated value of π is taken as 3. Some problems are expressed in terms of a system of linear equations and then solved by algebraic techniques. For instance, a problem in Chapter Eight leads to the system

$$3x + 2y + z = 39$$

$$2x + 3y + z = 34$$

$$x + 2y + 3z = 26$$

which can be solved by a method like the matrix approach described in modern textbooks.

In the third century, Liu Hui wrote his famous commentary on the *Nine Chapters on the Mathematical Art*. He also wrote a book called *The Sea Island Mathematical Manual* (*Haidao Suan Jing*), to demonstrate how to apply the Gougu Theorem. He was the first Chinese mathematician to deduce that the value of π lies between 3.1410 and 3.1427, by repeatedly doubling the number of sides of a regular polygon inscribed in a circle. It is called the method of dissection of a circle. Liu Hui also discovered the Cavalieri's Principle and used it to find the volume of a cylinder. About two centuries later, Zu Chongzhi (430–501) and his son, Zu Geng, found that the value of π lies between 3.1415926 and 3.1415927, based on the pioneer works of Liu Hui. He also obtained the remarkable rational approximation 355/113 for π, which is correct to six decimal places. Working with Zu Geng, he successfully applied the Cavalieri's Principle to deduce the correct formula for the volume of the sphere by computing the volume of a special solid called *Mouhe Fanggai* (the double vault) as proposed earlier by Liu Hui.

Unfortunately, his own work called *Zhui Shu* was discarded from the syllabus of mathematics in the Song dynasty and was finally lost in the literature. Many believed that *Zhui Shu* probably describes the method of interpolation and the major mathematical contributions by Zu Chongzhi and Zu Geng.

At the beginning of the Tang dynasty, Wang Xiaotong (580–640) wrote the *Jigu Suanjing* (*Continuation of Ancient Mathematics*), a text with only 20 problems that illustrate how to solve cubic equations. His method was a first step toward the *Tian Yuan Shu* (the method of coefficient array), which was then further developed by other mathematicians in the Sung and Yuan dynasties.

In the sixth century, mathematics was a subject being included in the civil service examinations. Li Chunfeng (602–670) was appointed by the Chinese emperor as the chief editor for a collection of mathematical treatises for both teachers and students. The collection is called the *Ten Classics* or the *Ten Computational Canons*, which include the *Zhou Bi Suan Jing*, the *Jiuzhang Suan Shu*, the *Haidao Suan Jing*, the *Sunzi Suan Jing*, the *Wucao Suan Jing*, the *Wujing Suan Shu*, the *Shushu Jiyi*, the *Xiahou Yang Suan Jing*, the *Zhang Qiujian Suan Jing*, and the *Jigu Suan Jing*. The book *Zhui Shu* by Zu Chongzhi had been included in the *Ten Classics* at the beginning, but it was later replaced by the *Shushu Jiyi* because of it being lost in the Sung dynasty.

The Golden Period

No significant advances in mathematics were made between the tenth century and the eleventh century. However, Jia Xian (1023–1050) improved the methods for finding square roots and cube roots, and also extended them to compute the numerical solutions of polynomial equations by means of the *Jia Xian Triangle* (the Chinese version of the Pascal Triangle).

The golden period of mathematical development in China occurs in the twelfth and the thirteenth centuries, which is called the "Renaissance of Chinese mathematics" by some authors. Four outstanding mathematicians appeared in the Sung dynasty and the Yuan dynasty, namely Yang Hui (1238–1298), Qin Jiushao (1202–1261), Li Zhi (also called Li Yeh, 1192–1279), and Zhu Shijie (1260–1320). Yang Hui, Qin Jiushao, and Zhu Shijie all used the Horner–Ruffini method to solve quadratic, cubic, and quartic equations. Li Zhi, on the other hand, revolutionized the method for solving problems on inscribing a circle inside a triangle, which could be formulated as algebraic equations, and solved by using the Pythagorean theorem. Another mathematician, Guo Shoujing (1231–1316), worked on spherical trigonometry for astronomical calculations. Therefore, much of the modern mathematics in the West had already been studied by Chinese mathematicians in this period.

Qin Jiushao (1202–1261) invented the symbol for "zero" in Chinese mathematics. Before this invention, blank spaces were used to denote zeros. Qin Jiushao also studied indeterminate problems and generalized the method of Sunzi to become the now-called "Chinese Remainder Theorem." He wrote the *Shushu Jiuzhang* (*Mathematical Treatise in Nine Sections*), which marks the highest point in indeterminate analysis in ancient China.

Yang Hui was an expert in designing magic squares. He discovered elegant methods for constructing magic squares with an order greater than three. Some of the orders are as high as 10. He was also the first in China to give the earliest clear presentation of the *Jia Xian Triangle* in his book *Xiangjie Jiuzhang Suanfa*.

The famous work of Li Zhi is the *Sea Mirror of the Circle Measurements* (*Ce Yuan Hai Jing*). It is a collection of some 170 problems. He used *Tian Yuan Shu* or the Method of Coefficient Array to solve polynomial equations of degree as high as six. He also wrote the book *Yi Gu Yan Duan* (New Steps in Computation) in 1259, which is an elementary book related to solution of geometric problems by using algebra.

The most important text in the thirteenth century is the *Precious Mirror of the Four Elements* (*Si Yuan Yujian*), written by Zhu Shijie in 1303. This book marks the peak of the development of algebra in China. The unknowns that appeared in equations are called the four elements, namely heaven, Earth, man, and matter. This book describes how to solve algebraic equations of degrees as high as 14. The method is the same as the Horner–Ruffini method. Zhu Shijie also used the matrix methods to solve systems of equations. He was also an expert in summation of series. Many formulas on summation of series can be found in the *Precious Mirror of the Four Elements*. He also wrote an elementary mathematics text called the *Introduction to Computational Studies* (*Suanxue Qimeng*) in 1299, which had a significant impact on the development of Japanese mathematics later.

The East Meets West Period

In the Ming dynasty, not much original mathematics work emerged in China. Even the famous work *Suanfa Tongzong* (*General Source of Computational Methods*) by Cheng Dawei (1533–1606) was an arithmetic book for the abacus only. Its style and content were still influenced very much by the *Nine Chapters on the Mathematical Art*. It was only when the Italian Jesuit Matteo Ricci (1552–1610) came to China in 1581 that the development of mathematics in China was influenced by the West from this time onwards. For instance, Xu Guangqi (1562–1633) and Matteo Ricci translated a number of Western books on sciences and mathematics into Chinese, including the famous *Euclid's Elements*, the influence of the Western culture on China became more apparent.

However, the Chinese mathematicians also did an excellent job in editing and recording their traditional mathematics and science works in the early Qing dynasty, so that much of them can come down to us now. For example, Mei Juecheng (1681–1763) edited the famous mathematical encyclopedia *Shuli Jingyun* in 1723, and Ruan Yuan (1764–1849) edited the *Chouren Zhuan* (*Biographies of Astronomers and Mathematicians*) in 1799. Both of these works are very valuable and useful references for historians to study the mathematical developments in China before the middle of the Qing dynasty.

Achievements in Chinese Mathematics

After the decline of Greek mathematics in the sixth century, Western Europe was undergoing the period of Dark Ages. On the other hand, many of the achievements of Chinese mathematics predated the same achievements before and shortly after the Renaissance. For instance, before the fifteenth century, China was able to (1) adopt a decimal placed-value numeral system, (2) acknowledge and use negative numbers, (3) obtain precise approximations for π, (4) discover and use the Horner–Ruffini method to solve algebraic equations, (5) discover the *Jia Xian Triangle*, (6) adopt a matrix approach to solve systems of linear equations, (7) discover the Chinese Remainder Theorem, (8) discover the method of double false position, and (9) handle summation of series with higher order. It was only after the fourteenth century that the development of Chinese mathematics began to decline and lag behind the Western mathematics in the Ming and Qing eras. However, it is worthy to note that the traditional Chinese mathematics still can find its contribution in mechanized geometry theorem proving in the twentieth century, because of its algorithmic characteristics.

Further Reading

Cullen, Christopher. *Astronomy and Mathematics in Ancient China: The Zhou Bi Suan Jing*. Cambridge, England: Cambridge University Press, 1996.

Martzloff, J. C. *A History of Chinese Mathematics*. Corrected ed. New York: Springer-Verlag, 2006.

Shen, Kangshen, John N. Crossley, and W. C. Lun Anthony. *The Nine Chapters on the Mathematical Art: Companion and Commentary*. Oxford, England: Oxford University Press, 1999.

Swetz, Frank. *Legacy of the Luoshu: The 4000 Year Search for the Meaning of the Magic Square of Order Three.* Wellesly, MA: A K Peters, 2008.

Yiu-Kwong Man

Civil War, U.S.

Category: Government, Politics, and History.
Fields of Study: Algebra; Geometry; Measurement; Number and Operations; Problem Solving.
Summary: The U.S. Civil War saw numerous advances in firearms, cryptography, and strategy.

The U.S. Civil War, also sometimes known as the War Between the States, was a conflict fought from 1861 to 1865 between 11 southern U.S. states that seceded from the Union to form the Confederate States of America and the remaining United States. Precipitating causes of the war centered on economic issues and states' rights versus federal power, often symbolized by the central dividing issue of slavery. More than 600,000 men on both sides died, which is greater than the combined U.S. losses in all subsequent wars and military conflicts through the beginning of the twenty-first century, though World War II exceeds this count if the metric is combat deaths versus deaths from all causes. Some also refer to the Civil War as the first "industrialized war" because of the extensive use of the telegraph, railroads, and mass-manufactured goods and weaponry.

Mathematics was instrumental in this war in many ways. Introduction of the Spencer repeating rifle has been cited by many as the turning point toward the eventual Union victory. Ciphers and code-breaking efforts were important in communicating military strategies and plans. The U.S. Army Signal Corps, founded in 1860, used both telegraphy and line-of-sight methods, such as the wig-wag signaling system in which the left, right, or upward positions of a single flag represented the numbers 1–3 and specific number combinations corresponded to letters. The ranks of leaders on both sides were filled with mathematically educated graduates of schools like the Unites States Military Academy at West Point. Mathematics education and research were also impacted by the war.

Weaponry

Changes in small arms and artillery that were occurring in Europe and the United States at this time had a tremendous impact on the war. Many different types of smoothbore or rifled artillery were used during the Civil War. One way in which they were distinguished was by their bore size, which was the diameter of their barrels, usually expressed in inches. Another differentiating feature was the weight of the projectiles they fired, in pounds. Different classes of weapons also had different trajectories. Cannons known as "guns" had relatively flat trajectories. Mortar rounds followed a steeply arcing path. Howitzers fell between the other two because the possible angles of inclination and powder charges could be varied more than the other two types.

The most common artillery piece was the Napoleon, a howitzer named after Napoleon Bonaparte, an avid student of mathematics who had revolutionized infantry and artillery warfare. Artillery ammunition included solid shot or balls, grape, canister, shell, and chain shot. Canister and grape shot could be particularly devastating to humans, since they disintegrated into smaller, scattering projectiles along a number of trajectories when fired. At the start of the war, both sides relied primarily on muzzle-loading rifled muskets such as the 58 Springfield, though some smoothbore muskets were also in use. Rifled weapons had greater range—nearly half a mile—versus the 100-yard range for smoothbores, which affected infantry tactics. The breech-loading Spencer repeating rifle was a major innovation, and was considered the most advanced weapon of its time. It used metal cartridge ammunition and it could hold seven cartridges at a time, which greatly increased rate of fire and accuracy, though the Union was initially concerned about the corresponding increase in the demand for ammunition.

Revolvers also replaced muzzle-loading pistols, with similar effects. Minié ball ammunition, named for French military officer Claude Minié, was used extensively in the Civil War. Previously, rifles had been difficult to load because the bullets fit tightly in the bore of the weapon, which was necessary for them to be propelled effectively by explosive powder charges. Despite being called a "ball," the Minié projectile was conical. It was smaller in diameter than older ammunition, and also had grooves that allowed it to fall smoothly and quickly into the barrel of the rifle. A hollow indenta-

A 12-pdr. howitzer gun captured by Butterfield's Brigade of the 12th Maine Infantry in May 1862. (Library of Congress)

tion extending from the base caused it to expand to the size of the gun's bore when fired, optimizing the combination of loading time and accuracy.

Cryptography

One problem addressed by mathematical problem-solving approaches was the terrible problem of "hacking." Increasingly, communications were being relayed by the newest technology—the telegraph and Morse code. It was relatively easy for someone to climb the telegraph pole, connect a telegraph key to the wires, and intercept messages that were being relayed back and forth from the front lines to the base camp. Messages needed to be coded so that interceptors could not interpret them, which was not a new problem. The problem of encoding military messages can be dated to at least Julius Caesar and earlier to the Spartan military.

The governors of Ohio, Indiana, and Illinois were close enough to the Confederate border that they felt the need to have their messages encoded. Governor William Dennison of Ohio asked Anton Stager to prepare a cipher (a code) that could be used by these three governors. Stager adapted a transposition coding system that had been used in Great Britain years earlier.

Words were rearranged into a grid. The first word of the message was the "key" to indicate how many columns were to be formed and in what order they were to be read. Instructions for these codes were printed on cards about the size of a standard index card, which were the precursors of codebooks. Included on the cards were the route, the keys, the code words, and words used to check the cipher.

This system underwent a number of modifications, and Stager's route cipher was eventually adopted as the Union's official cipher. Increasingly sophisticated ciphers were created during the war and, as a result, the instructions could no longer fit on cards. Some of the resulting codebooks were 48 pages long. The messages were intercepted by the Confederacy and sent to Richmond, Virginia, the capital of the south. By twenty-first-century coding standards, they appear to have been relatively easy codes to break, but evidence suggests that the south did not have either sufficient manpower or mathematical knowledge to decode them.

By contrast, the Union forces had a team to work in breaking the southern codes, including many times President Abraham Lincoln. The codes used by the south at the beginning of the war were not standardized, resulting in many messages that were unreadable. The Confederacy eventually settled on a code from 1587, the Vigenère cipher, named for Blaise de Vigenère—although others before him, such as Giovan Bellaso, are also noted as having invented it. This code consisted of a tableau of staggered alphabets. The ease of this code was that the code did not have to change if a coded message was captured, only the code phrase. The problem with the Vigenère code came in errors in transmission over the telegraph. Even though the code was harder to break than the Union cipher, it was more difficult to implement because a missed letter would result in an incoherent message. For instance, General Edmund Smith reportedly spent 12 hours trying to decode a message from General Joseph Johnston during the Vicksburg campaign. The message requested reinforcements, but Smith was unable to read it. He eventually sent a courier, but it was too late for reinforcements; Johnston's army was already cut off. Revisions to the code to avoid this problem in the future made deciphering easier as well.

Mathematically Educated Leaders

Many of the military leaders for both the north and the south were graduates of the U.S. Military Academy at West Point, which was the United States' only engineering school for an extended period of time. The Civil War was fought with 359 generals who graduated from West Point. They served on both sides, 217 for the Union and 142 for the Confederacy. This list of elite officers and leaders includes many well-known officers. Ulysses S. Grant had plans to return to West Point to teach mathematics, and these plans were changed by the outbreak of the Mexican War. Robert E. Lee graduated second in his class in 1829 and served as an assistant professor for mathematics for his first two years at West Point. Edmund Kirby Smith (1845) taught mathematics at the University of the South after the war, where he joined another West Point graduate, Josiah Gorgas. Other well-known graduates who served both sides during the Civil War included Confederate President Jefferson Davis, Braxton Bragg (1837), John Bell Hood (1853), Thomas Jonathan "Stonewall" Jackson (1846), Albert Sidney Johnston (1826), James Longstreet (1842), George E. Pickett (1846), J. E. B. Stuart (1854), William Tecumseh Sherman (1840), George G. Meade (1835), George McClellan (1846), Joseph Hooker (1837), Abner Doubleday (1842), George Armstrong Custer (1861), and Don Carlos Buell (1841).

All these men were mathematically educated, which was unique for that point in U.S. history and likely played a role in many aspects of the war. For example, the maps and messages of the military in the Civil War show the influence of what is now referred to as "descriptive geometry," which was created by Gaspard Monge and was incorporated into the curriculum after engineer Claudius Crozet brought it to the Military Academy. Mathematics textbooks used by most of the leaders on both sides of the Civil War include those written by mathematicians Charles Davies and Albert Church, some of which were adaptations of earlier French works. This education of the leaders of both sides of the conflict may have had a great deal to do with the long length of the conflict and allows historians the opportunity to study other differences in the two sides.

It was not just the military leaders during the Civil War who made use of mathematics. Lincoln purportedly had a great reverence for Euclid of Alexandria and geometry. Some historians assert that he kept a copy of Euclid's *Elements* in his saddlebag and studied it by lamplight to develop his logic and reasoning skills. Phrasing in Lincoln's well-known 1863 address at the Gettysburg battlefield has sometimes been compared to Euclid.

Education

While mathematics undoubtedly influenced the course and outcome of the war, the Civil War also affected mathematics education and research. Antebellum college curriculum in schools such as The Citadel or Harvard consisted of classes in mathematics that were filled with "practical applications," such as mercantile transactions, navigation, surveying, civil engineering, mechanics, architecture, fortifications, gunnery, optics, astronomy, geography, history, and "the concerns of Government." These topics were all expanded in one of the common textbooks of the day, *An Introduction to Algebra*, by Jeremiah Day. Geometry and trigonometry were also commonly taught, and analytic geometry, conic sections, and calculus were often optional classes.

The problems discussed and worked in these classes, both in surveying and in navigation, were carefully chosen and adapted to make them easily done but not extremely realistic. Thus, the navigators and the surveyors being prepared for the Army were ultimately ill prepared to handle the realistic situations of making measurements under fire or in harsh seas. According to the work of Andrew Fiss, the Union army regulations required that surveyors plot the best course for the army to take. The topographical engineers worked so slowly that many of the generals took to asking local citizens for the best directions. By two years into the war the topographical engineers were incorporated into the Army Corps of Engineers. Likewise, the Navy found that the U.S. Naval Academy, founded near the middle of the nineteenth century, could prepare navigators better than a mathematics department. However, the academy was negatively impacted by the temporarily relocation from Maryland to Rhode Island during the war.

After the war, many universities started offering higher level mathematics courses, and some increasingly focused on research. Harvard and John Hopkins University graduated doctorates in mathematics within the next decade.

Further Reading

Antonucci, Michael. "Code Crackers: Cryptanalysis in the Civil War." *Civil War Times Illustrated* 34, no. 3 (1995). http://www.eiaonline.com/history/codecrackers.htm.

Benac, T. J. "The Department of Mathematics." United States Naval Academy. http://www.usna.edu/MathDept/website/mathdept_history.pdf.

Fiss, Andrew. "The Effects of the Civil War on College-Level Math Education." Talk given at the meeting of the History and Pedagogy of Mathematics—Americas Section. March 13, 2010, Washington, DC. http://www.hpm-americas.org/wp-content/uploads/2010/04/Fiss_Civil_War.pdf.

Parshall, Karen Hunger, and David E Rowe. *The Emergence of the American Mathematical Research Community, 1876–1900: J. J. Sylvester, Felix Klein, and E. H. Moore*. Providence, RI: American Mathematical Society, 1994.

Plum, William Rattle. *The Military Telegraph During the Civil War in the United States*. Reprinted in 2 Volumes. Charleston, SC: Nabu Press, 2010 (1882).

Rickey, V. Frederick, and Amy E. Shell. "The Mathematics Curriculum at West Point in the Nineteenth Century." http://www.math.usma.edu/people/Rickey/papers/WP19thCentury/WP19thCentCurr.htm.

Rickey, V. Frederick. "201 Years of Mathematics at West Point." In *West Point: Two Centuries and Beyond*, Lance A. Betros, ed. Abilene, TX: McWhiney Foundation Press, 2004 (1974).

Sauerberg, James. "Route Ciphers in the Civil War." http://www.mathaware.org/mam/06/Sauerberg_route-essay.html.

David C. Royster

Cold War

Category: Government, Politics, and History.
Fields of Study: All.
Summary: The Cold War had a broad influence on mathematics, including education, coding theory, game theory, and many applied fields.

The Cold War was a 45-year-long period of bitter competition between two large groups of nations. It lasted from the end of World War II in 1945 to the collapse of the Soviet Union in 1991. The two groups never came to direct combat—hence the term "cold"—but it was a war in every other way, fought with deadly ferocity in the political, economic, ideological, and technological arenas. It was also a time of unprecedented investment in new mathematical ideas, driven in part by the desire of each side to dominate the other through nuclear intimidation, economic strength, espionage, and political control. The Cold War had a great impact on mathematics education, on the study of codes and code-breaking algorithms, and the development of new fields such as game theory. More generally, the term "cold war" can be applied to any fight-to-the-death competition between nations in which the two sides avoid direct military combat.

In the original Cold War (1945–1991), the two groups of nations divided along ideological lines. One side, the Soviet bloc, adhered to the communist political and economic philosophy of Karl Marx and Vladimir Lenin. The other side, the Western bloc, adhered to the older free-market capitalist philosophy originated by Adam Smith.

The two sides of the Cold War were essentially forced to avoid military conflict by the recent invention of the atomic bomb, because neither side wanted to risk combat that might give rise to an unstoppable military escalation. The inevitable result of such an escalation would have been worldwide nuclear war, with most large cities destroyed in an instant by nuclear warheads, followed by massive clouds of radioactive ash circling the globe and causing the death of hundreds of millions of innocent people.

Mutually Assured Destruction (MAD)

Prior to the development of the atomic bomb, there were no weapons capable of destroying the population of an entire city in a single blow. Wars were fought as purely military conflicts, without risking the life of civilization itself. This nature of conflict changed forever with the advent of nuclear weapons.

Prior to the Cold War, the dominant mathematical model of warfare was a simple predator–prey model invented by Frederick Lanchester during World War I. Perhaps not surprisingly, given the slow, grinding progress of World War I, the Lanchester model places primary emphasis on the rates of attrition of the military forces. The side that survives this deadly attrition process wins the battle. In a nuclear-armed battle, however, it is survival rather than attrition that is the vital concern, and the Lanchester equations are irrelevant.

In the earliest years of the Cold War, only the United States possessed the theory and technology to construct an atomic bomb. The presence of the bomb in the arsenal of one side but not the other made possible a strategy known as "nuclear blackmail." The owners of the atomic bomb could threaten to use the bomb if their adversaries did not comply with their demands. For example, newly declassified documents have revealed that in 1961, Great Britain threatened China with nuclear retaliation if China were to attempt a military invasion of the British Crown Colony of Hong Kong. The United States backed up this threat, and China refrained from invading Hong Kong.

Clearly, the ownership of the secret of the atomic bomb by just one nation in 1945 had destabilized the military balance of power among the victors of World War II. Great Britain and France allied themselves with the United States and were given access to atomic secrets. The Soviet Union chose to develop its own versions of the atomic bomb, or to steal the secrets through espionage. Thus arose the great division of the Cold War, between the respective allies of the Soviet Union—known as the Warsaw Pact—on one side, and the United States—through the North Atlantic Treaty Organization (NATO)—on the other.

The Soviet Union tested its first nuclear weapon in 1949, ending its four-year period of vulnerability to nuclear blackmail. The military doctrine that took its place was known as mutually assured destruction (MAD). As long as each side in the Cold War could assure the other that it would be utterly destroyed in a nuclear exchange, then—so it was hoped—military conflict could be prevented. MAD did indeed prevent the two nuclear powers from directly attacking each other, but it had two unfortunate consequences: the people of both sides lived in terror of nuclear annihilation, and both superpowers engaged in so-called proxy wars, using much smaller nations as their proxies in localized military conflicts.

Albert Wohlstetter was an influential and controversial strategist who was a major force behind efforts to deter nuclear war and avoid nuclear proliferation. He worked as a consultant to the RAND Corporation's mathematics division starting in 1951. Initially, he collaborated on problems related to modeling logistics, but then he was asked to turn his skills to a problem posed by the U.S. Air Force regarding the assignment and location of bases for Strategic Air Command (SAC). On the surface, it was a common logistics problem, but ultimately SAC's method of basing its medium-range manned bombers, which were one of the country's major deterrents against a Soviet invasion of western Europe, had far greater implications. This work drew him into global strategy. He and his wife, Roberta Wohlstetter, a historian and intelligence expert, received the Presidential Medal of Freedom in 1985. Wohlstetter was also reputedly one of the inspirations for the film *Dr. Strangelove*.

The Arms Race as a Nash Equilibrium

From its very outset, neither of the Cold War's two superpowers—the United States and the Soviet Union—believed that they could stop developing new and ever more powerful nuclear weapons. The MAD doctrine applied only as long as the forces of each side could pose a credible nuclear threat to the other. Therefore, each side worked to create new weapons as fast as possible. Throughout the 1950s and well into the

Cold War 57

Old Soviet anti-aircraft missile rockets, first deployed in 1957. Generations of children grew up with the ever-present threat of war during the Cold War (1945–1991) and were taught bomb threat procedures. (iStockphoto)

1960s, both nations tested ever more powerful nuclear weapons. This became known as the "arms race."

In the mathematical theory of games, the military arms race brought about by the MAD doctrine is an example of a Nash equilibrium, named for mathematician John Nash. In the decades-long arms race between the two Cold War competitors, each side could be seen as playing a simple noncooperative game. Each player in this game has a choice: to construct new and more terrible nuclear weapons, or not. If either player chose not to develop further weapons, while the other did, then the first player would face the very real risk of eventually facing nuclear blackmail. Each player understood the other's dilemma all too well, and so both continued to develop new weapons as fast as possible.

The persistence of this behavior comes from the fact that neither player can benefit by changing strategy unilaterally. When this occurs in a game, then it is in a form of equilibrium whose existence was first proved in the general case by Nash in 1950.

Game theory itself was a child of the Cold War, having been created in 1944 by John von Neumann, a mathematician who also played a key role in the development of the first atomic bomb, and Oskar Morgenstern, an economist. Throughout the Cold War, the theory of games was studied and elaborated, both by the military and by economists, as a means for better understanding the fundamental nature of competition, cooperation, negotiation, and war.

The fundamental irrationality of the nuclear arms race, in which each side became able to kill every single person on the planet many times over, was apparent to almost everyone. This realization did little to stop the arms race, because of the power of the Nash equilibrium to trap the players of the game into modes of behavior that, individually, they deplored.

In some critical respects, an arms race resembles a famous game known as the Chicken: two cars race toward each other down a narrow road, with the driver who first swerves away to avoid a crash being the loser.

The key to winning a game of chicken is to act in such a way that your opponent comes to believe that you are so irrational as to be willing to die before swerving. In other words, the "rational" solution to the game is to be utterly and convincingly irrational. The same principle holds in a nuclear arms race.

The game of Chicken and other apparent paradoxes of rationality within the theory of games led to the development in the 1970s of meta-game analysis. This and other mathematical forms of strategic analysis played an important role in the eventual winding down of the arms race with a series of strategic arms agreements between the major powers of the Cold War.

Political Competition in the Cold War

The bitter competition of the Cold War was at least as much political and economic as military, and new mathematical ideas contributed mightily to this competition. In the economic arena, the Cold War was fought between the proponents of multiparty, free-market economies on the Western side, and the proponents of single-party, command economies on the Soviet side.

Both sides claimed to be democratic in the Cold War, but they used different meanings for the word. In the West, the word "democracy" retained its historic meaning, a political system in which leaders are chosen in free elections. In the Soviet system, "democracy" meant a "dictatorship of the proletariat" in which all political power rested in a hierarchy of labor councils, known as *soviets*, and the supreme soviet could dictate any aspect of public affairs. Soon after the Russian Revolution, however, the Communist Party seized control of the soviets, and after that, no election in the Soviet system was free.

The intense political competition between these two systems of government led to great interest in the West in how to conduct elections in the fairest possible way. A large body of mathematical theory of elections emerged, much of it devoted to the study of election systems that come the closest to meeting a measure of fairness known as the Condorcet criterion. In an election, the Condorcet winner is the candidate who can beat any of the other candidates in a two-person run-off election. Many forms of preference balloting, in which voters rank the candidates, come quite close to the Condorcet criterion, but none is without problems. Arrow's Paradox, discovered and proved by Kenneth Arrow in 1950, states that when voters have three or more choices, then no voting system can convert the ranked preferences of the voters into a community-wide ranking that meets a particular beneficial set of criteria. This Cold War mathematical discovery is the starting point of the modern theory of social choice, the foundation of the mathematical theory of political science.

Economic Competition in the Cold War

There are many forms of socialism known in economic theory, but the form practiced by the Soviet bloc of nations was particularly severe. In its purest form, Soviet socialism entailed state ownership of all means of economic production: all industrial plants, all commercial businesses, all farms, and all financial institutions. Soviet socialism was a command economy, meaning that the state had to tell every plant, business, and institution how much to produce, and at what price they should sell their goods and services.

In order to come up with the enormous number of production and price commands that had to be sent out every month and year, the Soviet system of government employed a vast bureaucracy. The system used by these bureaucrats was developed in the 1920s, during the early years of the Soviet Union, without the benefit of mathematics. Known as the "method of balances," this system attempted to function so that the total output of each kind of goods would match the quantity that its users were supposed to receive.

In practice, the Soviet "method of balances" functioned very much like the U.S. War Production Board during World War II, and by its counterparts in the war economies of Great Britain and Germany. The first production decisions were made with respect to the highest priority items (ships, tanks, airplanes) and those were balanced with the available amounts of strategic resources (iron, coal, electricity), and so on down to the lowest-priority items. The command system was thought to be crude and error-prone, and its mistakes and imbalances were widely noticed.

In the West, the response of mathematicians to these failures of the wartime command economy was the development of the field of engineering known as operations research. The mathematical technique known as "linear programming"—originally a little-known Russian discovery—was successfully developed by George Dantzig and John von Neumann in 1947 to optimize production quantities under linear constraints on supplies. The Soviet economy was very slow to adopt these ideas, preferring for ideological reasons to stick with the inefficient and error-prone method of balances until very late in the Cold War.

After World War II, the nations of the West ended their wartime command economic systems and reverted back to using the free market to make price and production decisions. The Soviet Union and its allies, however, continued to rely on a large army of bureaucrats to make all economic decisions without the aid of good operational theory.

Wassily Leontief, a Russian economist working in the United States, solved one of the fundamental problems of a command economy in 1949 with his method of input–output analysis. This method required the creation of a very large matrix showing the contribution of each component or sector of the economy to every other component. When properly constructed, the required inputs of raw materials to the economy can be calculated from the desired outputs by matrix inversion. The Soviet Union failed to quickly see the significance of Leontief's achievement, and did not incorporate his ideas into its planning system for many decades.

It is one of the great ironies of the Cold War that the mathematical theories that were required to make a command economy function properly were perfected in the West, where they are now universally employed within industrial corporations—some now larger than the entire economy of the old Soviet Union—to run their operations in the most efficient way possible. In the end, the economy of the Soviet Union and its satellites was not able to keep pace with the West, and in 1991, it suffered catastrophic political and economic collapse.

Further Reading

Arrow, Kenneth. *Social Choice and Individual Values.* 2nd ed. New Haven, CT: Yale University Press, 1963.

Erickson, Paul. "The Politics of Game Theory: Mathematics and Cold War Culture." Ph.D. dissertation, University of Wisconsin-Madison, 2006.

Howard, Nigel. *Paradoxes of Rationality.* Cambridge, MA: MIT Press, 1971.

Johnson, Thomas. "American Cryptography During the Cold War." National Security Agency. http://www.gwu.edu/~nsarchiv/NSAEBB/NSAEBB260.

Karp, Alexander. "The Cold War in the Soviet School: A Case Study of Mathematics Education." *European Education* 38, no. 4 (2006).

Kort, Michael. *The Columbia Guide to the Cold War.* New York: Columbia University Press, 2001.

Leontief, Wassily. "The Decline and Rise of Soviet Economic Science," *Foreign Affairs* 38 (1960).

Loren Cobb

Comparison Shopping

Category: Business, Economics, and Marketing.
Fields of Study: Geometry; Measurement; Number and Operations.
Summary: Both simple and complex algorithms are used to compare consumer prices and contextualize mathematics instruction.

The globalization of the marketplace has resulted in a plethora of choices for any given item at both the local store and via the Internet. People comparison shop for both very expensive items like a car or a plane ticket and fairly inexpensive purchases like a box of cereal. Comparison shopping is perhaps one of the most widely used applied mathematics lessons, both in K–12 and lower-level college courses.

Mathematics is at the forefront of comparison shopping through unit pricing, which makes use of division and fractions. Geometric methods can be used to compare volume or weight. Notions from pre-algebra and algebra model financial decisions such as purchasing a cell phone plan or taking out a car loan. Students explore parameters in order to make balanced and informed choices. Mathematics educators not only use these examples in classrooms, but they also study their effectiveness. Researchers and online shopping agents take advantage of mathematical methods to extract, compare, and mine huge amounts of data. Comparison techniques also include data envelopment analysis and multiple regression.

Unit Pricing

One method of comparing differently priced items in different sized containers is through unit pricing. Dividing the price by the quantity or amount of items, such as how many ounces, will yield a cost per unit term that can be used for comparison purposes. For example, an 11.5 oz box of cereal might cost $4.49, while a 24 oz box of cereal costs $4.99. The unit price of the first box is $4.49/11.5 ≈ $0.39 per ounce, while the second box is $4.99/24 ≈ $0.21 per ounce. Some items are already priced by their weight, like meats, fruits, vegetables, or coffee, and others are priced according to their volume, that is, by the container size. For those items that are not priced by weight or volume, unit pricing is listed on the shelf tag in many stores. However, the unit price is not the only important feature in comparison shopping. Personal preferences and other important factors must also be taken into consideration, like whether one will be able to use up a larger quantity before the expiration date. Unit pricing examples proliferate in lessons on fractions and in classes like pre-algebra and developmental mathematics. Students also compare scenarios in which sales occur or other discounts are applied.

Debt and Interest

Another common classroom scenario is found in comparing house and car purchases in financial mathematics segments. For instance, students can use the loan payment formula to calculate the monthly payment R in terms of the monthly interest rate r, the loan amount P, and the number of months, n

$$R = \frac{rP}{1 - (1+r)^{-n}}.$$

Then they can calculate the total interest by multiplying the monthly payment and the number of months and subtracting the loan amount. One comparison scenario is determining how the monthly payment and total interest change as the price of the car or house changes or the interest rate fluctuates. Another is determining whether one should take out a smaller loan versus paying loan points to buy down the interest rate. Students also compare car prices to income level using the debt-to-income ratio. The debt-to-income ratio is the debt divided by the income, which is the percentage of debt. Banks use the debt-to-income ratio in making decisions about mortgage or car loans. From the Great Depression in the 1930s until the deregulation of banking restrictions in the 1970s, an upper limit of 25% was typical. However, that level rose after deregulation and with the increase in consumer credit card debt. In the twenty-first century, it is common for an upper limit to range between 33% and 36%. Given a monthly car payment, house payment, and other monthly debts, students can add up the total debt and solve for the necessary income level in order to stay below 36%. They can also compare the way that debt and the needed income change as the interest rates vary.

Contextualizing Instruction

Mathematics educators use purchasing scenarios in the classroom and study and debate their effectiveness. Some studies have found that the contextualization of mathematics using examples from shopping helps students. Terezinha Nunes, Analucia Schliemann, and David Carraher compared the mathematical abilities of children who were selling items in Brazilian street markets to questions in school. They found that the closer to the real-life situation, the more successful the student. Other studies have also found that there can be a disconnect between performance in the supermarket and performance in school. Some researchers assert that the contextualization may disguise the mathematics and be problematic in elucidating the underlying mathematical processes.

Marketplace globalization has resulted in increased choices at both local stores and via the Internet. (Photos.com)

Mathematical Models for Comparison Shopping

Businesses and researchers employ a variety of mathematical techniques in order to compare large shopping data sets. Online shopping agents use mathematical methods in situations such as a Web search for airplane ticket prices or hotel rooms. Historically, dating back to at least the nineteenth century, travel agents sold vacations to consumers on behalf of suppliers. Travel agencies grew in popularity with the increase in commercial aviation after World War I. At the end of the twentieth century, the Internet vastly changed the way in which consumers compared and purchased vacation travel. Airlines, hotels, and other vacation companies offered online services directly to consumers, bypassing travel agents. In response, some travel agencies created travel Web sites that would compare options. Their computer programs extracted comparative price data from Web sites in order to build comparison shopping engines. Researchers continue to develop advanced comparison shopping techniques including methods in data mining, data envelopment analysis, and multiple regression. They create sophisticated algorithms to analyze data and find patterns. In data envelopment analysis, networks can be viewed as decision-making units, and efficient configurations are selected. In multiple regression, several variables are combined in an attempt to create a meaningful predictor or measure. Mathematical methods are also important in predicting shopping preferences and consumer behavior.

Further Reading

Berry, Michael, and Linoff, Gordon. *Data Mining Techniques for Marketing, Sales and Customer Support.* Hoboken, NJ: Wiley, 1997.

Boaler, Jo. "The Role of Contexts in the Mathematics Classroom: Do They Make Mathematics More 'Real'?" *For the Learning of Mathematics* 13, no. 2 (1993).

Devlin, Keith. *The Math Instinct: Why You're a Mathematical Genius (Along With Lobsters, Birds, Cats, and Dogs).* New York: Basic Books, 2005.

Herzog, David. *Math You Can Really Use—Every Day.* Hoboken, NJ: Wiley, 2007.

McKay, Lucia, and Maggie Guscott. *Practical Math in Context: Smart Shopping Math.* Costa Mesa, CA: Saddleback Educational Publishing, 2005.

Nunes, Terezinha, Analucia Schliemann, and David Carraher. *Street Mathematics and School Mathematics.* New York: Cambridge University Press, 1993.

SARAH J. GREENWALD
JILL E. THOMLEY

Coupons and Rebates

Category: Business, Economics, and Marketing.
Fields of Study: Algebra; Measurement.
Summary: Mathematical differences between coupons and rebates provide different rewards to consumers.

Offering price reductions through coupons and rebates is a popular means of increasing the number of sales of a product, attracting customers to retail stores (both physical and online), and promoting public awareness of a brand name or product. One of the first known instances of a coupon was in 1894, when the Coca-Cola Company gave out handwritten tickets for samples of its new soft drink. The next year, Charles Post, of Post Cereal, started issuing coupons to help sell groceries. By the 1930s, these coupons were increasingly popular for saving money during the Great Depression. Some researchers claim that by the mid-1960s, half of American households used coupons.

In the twenty-first century, coupons are available on the Internet, or as permanent discount cards, in addition to their traditional paper form. In the age of online shopping, coupons for free shipping are cited by some as one of the most important factors in determining where to shop. For the customer, coupons and rebates bring savings on regularly purchased items and provide an incentive to try new products or services. In general, coupons are small discounts (a few dollars or cents) redeemed at the time of payment. The term "rebate" generally refers to larger reimbursements or discounts where the price reduction is either applied at the time of sale (an instant rebate) or reimbursed after required documents are mailed in by the customer. The specific type of coupon or rebate affects the calculation of both the discount and any applicable sales tax.

Coupons

Coupons may be issued by a product manufacturer or a store, and the redemption is somewhat different for the two types. When a customer presents a manufacturer coupon to a retailer, the customer pays any applicable tax on the full price of the item before the coupon is applied. For example, if a retailer charges $50 for a product with an 8% sales tax and a customer presents a manufacturer coupon for $10, then the cost to the customer at the time of purchase would be $44; that is, the original $50, plus sales tax of $4 ($50 × 0.08), minus the $10 coupon. Typically, the manufacturer reimburses the retailer for the amount of the coupon plus handling.

Sometimes a retailer like a grocery store, pizza restaurant, or automobile detailer will offer its own store coupons or rebates on its products or services. When a customer uses a store coupon, tax is computed on the balance after the coupon is deducted. If the $10 manufacturer's coupon is replaced by a $10 store coupon, the cost to the customer would be less: $43.20 versus $44.00.

Many retailers issue plastic cards that customers present to take advantage of weekly card specials or to receive a certain percentage discount on purchases made with the card. These cards not only allow shoppers to save money, but also the data collected when these cards and the associated purchases are scanned allow stores to better track their sales and inventory, and sometimes offer additional discounts tailored to a specific buyer's purchasing patterns. Sometimes these cards are free, but other times they require an initial or annual fee.

A retailer may offer a card at a cost of $10 that can be used for a 10% discount on all purchases at that store for one year. If a first-time customer checks out with a balance of $110 before tax, the customer can determine whether to purchase the card and take the 10% reduction. Although the card costs $10, the customer would save $11 on the initial balance (10% of the $110 total), resulting in a final cost of $109. The tax would be marginally less as well, since the total was reduced. Thus, the card would pay for itself at the first purchase, even before any other savings occur.

Another form of coupon is a card that is stamped each time the customer purchases a specified type of product, until a certain number of stamps are accrued. The customer then receives the next purchase of the specified product type free of charge, except for—possibly—sales tax. This form of coupon may be offered by certain restaurants, food markets, coffee shops, or bookstores.

Rebates

For a manufacturer rebate, an electronics retailer may sell a computer for $1,500, together with a free $100 printer after a mail-in rebate. The customer pays the tax on both the computer and the printer, and the manufacturer reimburses the customer $100 after the rebate is processed. With 8% sales tax, the cost to the customer after the rebate would be $1,628 (where the sales tax was 8% of $1,600, or $128).

Historically, economists have viewed consumer spending as a function of income. Politicians often cite this principle when pushing for tax rebates, believing they will increase consumption. However, there is little empirical evidence to support this notion, and in some cases there is contrary evidence. In 2001, the U.S. Congress enacted a tax rebate, giving $300 to anyone who had paid income taxes the previous year ($600 for couples). Economic indicators showed no associated increase in spending but rather a spike in saving. A survey of a sample of households that received a rebate reported that roughly one in five of those asked said they would spend the money. The *Wall Street Journal* ran the headline "Rebates Boost Incomes, But Not Spending." A study of the 2008 rebate found similar results.

Coupon Collector's Problem

There is a classic probability problem known as the Coupon Collector's Problem, which has been explored by a number of mathematicians, including the prolific Paul Erdos. The problem supposes that there is some number of different coupons (n) a person needs to collect to win a prize and asks how many coupons will he or she have to acquire, one at a time, to get a complete set. Usually, the coupons are equally likely to be drawn, and getting one of the n coupons does not prevent another of the same type from being drawn. Solutions to the problem can be found in a number of ways, including harmonic numbers, probability generating functions, and simulation. Extensions of the Coupon Collector's Problem are very useful in manufacturing quality control, for situations in which a number of product types must be sampled.

Further Reading

Better Business Bureau. "Mail-In Rebates: Now Available in Paper or Plastic." http://www.bbb.org/us/article/mail-in-rebates-now-available-in-paper-or-plastic-13249.

Spencer, K., and S. Rose. *How to Shop for Free: Shopping Secrets for Smart Women Who Love to Get Something for Nothing.* Philadelphia: Da Capo Press, 2010.

Barbara A. Shipman

Credit Cards

Category: Business, Economics, and Marketing.
Fields of Study: Algebra; Number and Operations; Data Analysis and Probability.
Summary: Credit card issuers use mathematical models to determine credit lines and interest rates, as well as to detect fraud and analyze offers.

Credit card issuers use statistical analysis in a wide variety of ways. Statistical models of risk help the banks decide whom to approve for card membership and what interest rate to charge. Models also help issuers manage the risks of their existing customers and detect fraudulent transactions. Credit card issuers use designed experiments to help decide which offers have the largest potential to be profitable. Typically, the bank tries out the new offer on a sample of people (while leaving others in a control group) before deciding whether the new offer will be successful if given to the entire customer base. Data mining techniques help banks look at customers' past transactions in order to model future uses of the card and to help decide which customers are most likely to want which other products and services that the bank offers.

History

The first credit card was born when businessman Frank McNamara realized that he had forgotten his wallet at a New York City restaurant. After his wife rescued him by bringing cash to the restaurant, he vowed he would never face that embarrassment again. The Diners Club card was born a few months later in 1950 and became the first widespread alternative to cash. The first businesses honoring Diners Club purchases were charged

(Photos.com)

Fraud Detection

Credit card banks use statistical algorithms to detect fraudulent use of credit cards. During the few seconds that it takes to approve or deny a credit card transaction at a merchant's site, information about the card is sent to a processing center. Typically at this point, only cards that are known to be stolen, fraudulent, delinquent, or other states that can be looked up will cause a denial. After the transaction has been approved, algorithms examine transactions to see if the pattern is suspicious. The cardholder may be contacted, usually by telephone, to verify that the transaction was made by the cardholder. The algorithms that identify a suspicious transaction can be quite sophisticated and are based on the past behavior of the cardholder.

7% of each transaction (typical costs are now 2% to 5%), and subscribers were charged $3 per year.

Bank of America pioneered its BankAmericard program in Fresno, California, in 1958, and American Express issued the first plastic card in 1959. Carte Blanche was another early card. The idea of a credit "card" really gained momentum when a group of banks formed a joint venture to create a centralized system of payment. National BankAmericard, Inc. (NBI) took ownership of the credit card network in 1970 and for simplicity and marketability changed its name to Visa in 1976. (One reason for the name "Visa" is that it is pronounced nearly the same way in every language.) That year, Visa processed 679,000 transactions—a volume that is processed on average every four minutes today. The Visa system is currently able to handle a load of about 6800 transactions per second, a capacity nearly exceeded on December 23, 2005, during the height of the Christmas shopping season. Visa is the largest merchant network, although MasterCard, American Express, and others process many transactions as well.

The Fair Isaacs Company (FICO) has grown in parallel with the credit card industry. It was founded in 1956 by mathematician Earl Isaac and engineer Bill Fair with the idea that data, used intelligently, can be used to make better business decisions. The next year, Conrad Hilton hired FICO to design and implement a complete billing system for his Carte Blanche card. FICO next developed the methodology to "score" the credit rating of customers but was unable to sell the idea to credit card banks until the 1970s. By the early 1990s, nearly every credit card bank was using some form of credit card scoring to help decide which customers to approve for credit and at what price. In 1995, both Fannie Mae and Freddie Mac, the two largest mortgage brokers in the United States, recommended using FICO scores for use in evaluating U.S. home mortgages. Today, U.S. citizens can access their various credit scores through online credit bureaus and, in fact, the U.S. government developed a policy allowing consumers to find out their scores once a year for free.

Credit Scoring

Credit bureaus use statistical analysis on past transactions, as well as income and other demographic information, to generate a credit score, usually referred to as a FICO score. This number is on an arbitrary scale that generally runs from 350 to 850 (with slight variations). The three main credit bureaus are Experian, TransUnion, and Equifax. Credit scores on the same individual may differ among the credit bureaus because of slight variations in the statistical model used to generate the number and slightly different data reported to the various bureaus. In all cases, the credit score is a prediction of how likely a borrower is to pay back the loan. For credit card companies, the score is used to decide both whether to issue the card, and what price (annual percentage rate) to charge on a balance that's carried over from month to month.

Data Mining

Credit card transactions, while vital to the running of the credit card bank, also contain information on the cardholder's spending patterns. These databases are very large, containing the records of tens of millions of customers, and dozens to hundreds of transactions per record. Using statistical models (often logistic regression models), banks can use these vast data repositories to identify the customers who are predicted to have the highest probability of enrolling for a new product or service. These offers may be made via a number of different channels. The offer may be given while the cardholder is calling a call center (800 number) with an issue concerning his or her card (in which case, the statistical algorithm will notify the operator that this customer should get the specific offer), by e-mail, by an outbound telemarketing call, by a targeted ad that pops up while the customer is visiting the issuer's Web site, or as direct marketing (so-called junk mail).

Experimental Design

To evaluate whether a new type of offer (the so-called "challenger") will be more effective (as measured by higher enrollment, revenue, profit, or other criteria) than the current offer (the "champion"), banks often use statistically designed experiments. The simplest such experiment is randomized at two levels, also known as a champion/challenger design. In this design, a sample is selected at random from the entire customer database. A proportion of those are chosen as the control group. They receive the current offer (the champion), and the rest are chosen to receive the challenger. The data are then collected, and the differences in response between the two groups are evaluated. The design can be complicated by blocking (stratification) on card type, region, income, or other demographic variables. Designs can be complicated by adding more factors, more levels, and by asymmetries introduced by infeasible treatment combinations. In the credit card industry, analysis is also complicated by the fact that one cardholder may be getting more than one experimental treatment (offer) simultaneously from different groups within the same organization and from different organizations. Capital One Bank claims to run upward of 40,000 such experiments a year on its cardholders.

Further Reading

Box, G. E. P., J. S. Hunter, and W. Hunter. *Statistics for Experimenters*. 2nd ed. Hoboken, NJ: Wiley Interscience, 2005.

McNamee, Mike. "Credit Card Revolutionary." *Stanford Business* 69, no. 3 (2001).

Paterson, Ken. "Credit Card Issuer Fraud Management." Mercator Advisory Group, 2008. http://www.sas.com/new/analyts/mercator_fraud_1208.pdf

Richard De Veaux

Currency Exchange

Category: Business, Economics, and Marketing.
Fields of Study: Algebra; Measurement; Representations.
Summary: Mathematical models seek to price financial products in the foreign exchange market.

The term "currency exchange" refers to the business transaction that trades one currency for another. Such a transaction happens in the foreign exchange (FX) market and is measured by foreign exchange rates, which are often called exchange rates. Exchange rates fluctuate all the time. There are many factors that influence the movements of exchange rates. After all, foreign exchange rates are largely determined by the supply and demand in the FX market. Numerous mathematical models have been proposed by financial mathematicians and financial engineers to price different financial products in the FX market. Some of them have been used successfully by practitioners.

Exchange Rate Definition

There are many different currencies in the world. A measurement of the value of one currency in terms of another is called a (foreign) "exchange rate" or a "currency rate." In simple terms, an exchange rate of K currency X to currency Y means the value of K units of currency X is equivalent to the value of 1 unit of currency Y. It is often quoted as the price of currency X divided by currency Y is K. For example, the price of "euros/U.S. dollars is 1.3578" denotes an exchange rate of 1.3578 U.S. dollars to euros. In other words, it means the value of 1 euro is the same as that of 1.3578 U.S. dollars.

Types of Exchange Rates

A fixed exchange rate (also known as "pegged rate") means one currency is pegged to a major currency such as the U.S. dollar. Usually, the government or the central bank of a country will intervene in the market to peg its currency to a major currency to maintain a fixed exchange rate.

In contrast, a floating exchange rate is determined by the market forces of demand and supply.

Exchange Rate Fluctuation

Fluctuation of exchange rates, like fluctuation of stock prices, interest rates, and many other economic indices, is a ubiquitous phenomenon. Many factors drive the exchange rates up and down. These factors include but are not limited to capital flows, international trades, speculation, political factors, government or central bank intervention, and interest rates. However, the fundamental driving force is the invisible hand—the demand and supply—of the market.

Besides those quantifiable drivers of the FX market, there are other nonquantifiable ones such as the expectation of the investors. Attempts have been made by economists to account for those driving forces as well. Some economists have put the theory of exchange rate into a behavioral finance framework. Others used information theory and game theory.

FX Markets and FX Financial Products

The FX market is where the currency exchange happens, and is one of the largest financial markets in the world. Its major participants include commercial banks, investment banks, companies, investors, hedgers, speculators, traders, governments, and central banks. A variety of financial instruments are traded in the FX market, including currencies, currency forward contracts (also known as "FX forward contracts"), currency futures contracts (also known as "FX futures contracts"), currency options (also known as "FX options") and currency swaps (also known as "FX swaps"). Thus, the FX market has several important submarkets: the FX spot market, the FX forward market, the FX futures market, the FX options market, and the FX swaps market.

Although hundreds of financial products exist in the FX market, the basic ones are currencies, currency forward contracts, currency futures contracts, currency options, and currency swaps. Currencies are priced by the exchange rates. Both currency forward contracts and currency futures contracts are agreements made between two parties to exchange a specified amount of currency for a specified price at a specific future date. The main difference is that a currency forward contract is traded over the counter, whereas a currency futures contract is traded on an exchange. They both are financial derivatives. Their prices can be determined using simple algebra and are expressed in terms of exponential functions. Currency options and currency swaps are also financial derivatives. A currency call/put option gives one party the right—but not the obligation—to buy or sell a specific amount of the currency at a price (called "strike price") at a specific time in the future.

A European option can be exercised only at maturity, whereas an American option can be exercised at any time up to maturity. The cash flows of currency options are more complicated than those of the currency forward and currency futures contracts. The pricing requires sophisticated mathematical tools from stochastic calculus. Fisher Black, Myron Scholes, and Robert C. Merton made fundamental contributions in option pricing by giving the basic pricing formulas of European options. Scholes and Merton were awarded the Nobel Prize in Economics for this accomplishment in 1997 (Black was not awarded the prize because he had passed away).

A currency swap is an agreement between two parties to exchange the principal and interests of one currency at an interest rate for the principal and interests of another currency at another interest rate for a certain period of time. For example, suppose party A enters into a currency swap contract with party B today. For the next five years, party A will pay party B the interest of a principal of $1 million at an annual interest rate of 5%. In return, party B will pay party A the interest of a principal of 95 million Japanese yen at an annual interest rate of 4.5%. The two parties will also exchange the principals at the end of the fifth year. Like currency forward and currency futures contracts, the currency swap can also be priced using simple algebra.

Further Reading

De Grauwe, Paul. *The Exchange Rate in a Behavioral Finance Framework*. Princeton, NJ: Princeton University Press, 2006.

Driver, Rebecca, Peter Sinclair, and Christoph Thoenissen. *Exchange Rates, Capital Flows and Policy*. New York: Routledge, 2005.

Having exchange rates for national currencies allows us to consistently express the value of an item across borders of countries and cultures. (Photos.com)

Hull, John C. *Options, Futures and other Derivatives.* 7th ed. Upper Saddle River, NJ: Pearson Education, 2008.

McDonald, Robert L. *Derivatives Markets.* 2nd ed. Upper Saddle River, NJ: Pearson Education, 2006.

Rosenberg, Michael R. *Exchange Rate Determination: Models and Strategies for Exchange-Rate Forecasting.* New York: McGraw-Hill, 2003.

Weithers, Tim. *Foreign Exchange.* Hoboken, NJ: Wiley, 2006.

Liang Hong

Data Mining

Category: Business, Economics, and Marketing.
Fields of Study: Data Analysis and Probability; Measurement; Number and Operations.
Summary: Data mining is the relatively recent practice of using algorithms to distill patterns, summaries, and other specific forms of information from databases.

Advances in technology in the latter half of the twentieth century led to the accumulation of massive data sets in government, business, industry, and various sciences. Extracting useful information from these large-scale data sets required new mathematical and statistical methods to model data, account for error, and handle issues like missing data values and different variable scales or measures. Data mining uses tools from statistics, machine learning, computer science, and mathematics to extract information from data, especially from large databases. The concepts involved in data mining are drawn from many mathematical fields such as fuzzy sets, developed by mathematician and computer scientist Lotfi Zadeh, and genetic algorithms, based on the work of mathematicians such as Nils Barricelli. Because of the massive amounts of data processed, data mining relies heavily on computers, and mathematicians contribute to the development of new algorithms and hardware systems. For example, the Gfarm Grid File System was developed in the early twenty-first century to facilitate high-performance petascale-level computing and data mining.

History
Data mining has roots in three areas: classical statistics, artificial intelligence, and machine learning. In the late 1980s and early 1990s, companies that owned large databases of customer information, in particular credit card banks, wanted to explore the potential for learning more about their customers through their transactions. The term "data mining" had been used by statisticians since the 1960s as a pejorative term to describe the undisciplined exploration of data. It was also called "data dredging" and "fishing." However, in the 1990s, researchers and practitioners from the field of machine learning began successfully applying their algorithms to these large databases in order to discover patterns that enable businesses to make better decisions and to develop hypotheses for future investigations.

Partly to avoid the negative connotations of the term "data mining," researchers coined the term "knowledge discovery in databases" (KDD) to describe the entire process of finding useful patterns in databases, from the collection and preparation of the data, to the end product of communicating the results of the analyses to others. This term gained popularity in the machine learning and AI fields, but the term "data mining" is still used by statisticians. Those who use the term "KDD" refer to data mining as only the specific part of the KDD process where algorithms are applied to the data. The broader interpretation will be used in this discussion.

Software programs to implement data mining emerged in the 1990s and continue to evolve today. There are open-source programs (such as WEKA, http://www.cs.waikato.ac.nz/ml/weka and packages in R, http://www.r-project.org) and many commercial programs that offer easy-to-use graphical user interfaces (GUIs), which can facilitate the spread of data mining practice throughout an organization.

Types of Problems
The specific types of tasks that data mining addresses are typically broken into four types:

1. Predictive Modeling (classification, regression)
2. Segmentation (data clustering)
3. Summarization
4. Visualization

Predictive modeling is the building of models for a response variable for the main purpose of predicting the value of that response under new—or future—values of the predictor variables. Predictive modeling problems, in turn, are further broken into classification problems or regression problems, depending on the nature of the response variable being predicted. If the response variable is categorical (for example, whether a customer will switch telephone providers at the end of a subscription period or will stay with his or her current company), the problem is called a "classification." If the response is quantitative (for example, the amount a customer will spend with the company in the next year), the problem is a "regression problem." The term "regression" is used for these problems even when techniques other than regression are used to produce the predictions. Because there is a clear response variable, predictive modeling problems are also called "supervised problems" in machine learning. Sometimes there is no response variable to predict, but an analyst may want to divide customers into segments based on a variety of variables. These segments may be meaningful to the analyst, but there is no response variable to predict in order to evaluate the accuracy of the segmentation. Such problems with no specified response variable are known as "unsupervised learning problems."

Summarization describes any numerical summaries of variables that are not necessarily used to model a response. For example, an analyst may want to examine the average age, income, and credit scores of a large batch of potential new customers without wanting to predict other behaviors. Any use of graphical displays for this purpose, especially those involving many variables at the same time, is called "visualization."

Algorithms

Data mining uses a variety of algorithms (computer code) based on mathematical equations to build models that describe the relationship between the response variable and a set of predictor variables. The algorithms are taken from statistics and machine learning literature, including such classical statistical techniques as linear regression and logistic regression and time series analysis, as well as more recently developed techniques like classification and regression trees (ID3 or C4.5 in machine learning), neural networks, naïve Bayes, K-nearest neighbor techniques, and support vector machines.

One of the challenges of data mining is to choose which algorithm to use in a particular application. Unlike the practice in classical statistics, the data miner often builds multiple models on the same data set, using a new set of data (called the "test set") to evaluate which model performs best.

Recent advances in data mining combine models into ensembles in an effort to collect the benefits of the constituent models. The two main ensemble methods are known as "bootstrap aggregation" (bagging) and "boosting." Both methods build many (possibly hundreds or even thousands of) models on resampled versions of the same data set and take a (usually weighted) average (in the case of regression) or a majority vote (in the case of classification) to combine the models. The claim is that ensemble methods produce models with both less variance and less bias than individual models in a wide variety of applications. This is a current area of research in data mining.

Applications

Data mining techniques are being applied everywhere there are large data sets. A number of important application areas include the following:

1. *Customer relationship management* (CRM). Credit card banks formed one of the first groups of companies to use large transactional databases in an attempt to predict and understand patterns of customer behavior. Models help banks understand acquisition, retention, and cross-selling opportunities.
2. *Risk and collection analytics*. Predicting both who is most likely to default on loans and which type of collection strategy is likely to be successful is crucial to banks.
3. *Direct marketing*. Knowing which customers are most likely to respond to direct marketing could save companies billions of dollars a year in junk mail and other related costs.
4. *Fraud detection*. Models to identify fraudulent transactions are used by banks and a variety of government agencies including state comptroller's offices and the Internal Revenue Service (IRS).
5. *Terrorist detection*. Data mining has been used by various government agencies in an attempt to help identify terrorist activity—

although concerns of confidentiality have accompanied these uses.
6. *Genomics and proteomics.* Researchers use data mining techniques in an attempt to associate specific genes and proteins with diseases and other biological activity. This field is also known as "bioinformatics."
7. *Healthcare.* Data mining is increasingly used to study efficiencies in physician decisions, pharmaceutical prescriptions, diagnostic results, and other healthcare outcomes.

Concerns and Controversies
Privacy issues are some of the main concerns of the public with respect to data mining. In fact, some kinds of data mining and discovery are illegal. There are federal and state privacy laws that protect the information of individuals. Nearly every Web site, credit card company, and other information collecting organization has a publicly available privacy policy. Social networking sites, such as Facebook, have been criticized for sharing and selling information about subscribers for data mining purposes. In healthcare, the Health Insurance Portability and Accountability Act of 1996 (HIPAA) was enacted to help protect individuals' health information from being shared without their knowledge.

Further Reading
Berry, M. A. J., and G. Linoff. *Data Mining Techniques For Marketing, Sales and Customer Support*. Hoboken, NJ: Wiley, 1997.

De Veaux, R. D. "Data Mining: A View From Down in the Pit." *Stats* 34 (2002).

———, and H. Edelstein. "Reducing Junk Mail Using Data Mining Techniques." In *Statistics: A Guide to the Unknown*. 4th ed. Belmont, CA: Thomson, Brooks-Cole, 2006.

Piatetsky-Shapiro, Gregory. "Knowledge Discovery in Real Databases: A Workshop Report." *AI Magazine* 11, no. 5 (January 1991).

Richard De Veaux

Egyptian Mathematics

Category: Government, Politics, and History.
Fields of Study: Connections; Geometry; Measurement; Number and Operations; Representations.
Summary: Ancient Egyptians were adept at engineering and geometry and deeply dependent on accurate measurements of the annual Nile flood.

Our knowledge of Egyptian mathematics (3000–1000 B.C.E.) is based on hieroglyphic writings found on stone or as script (hieratic and demotic) in multiple papyri. Preserved in tombs and temples in the Nile valley, a papyrus is a narrow scroll of paper, about 15 feet in length, made by interweaving tiny strips of a water reed called *papu*. The key documents are the Moscow, Rhind, Rollin, and Harris papyri. These works are generally thought to be textbooks used by scribes to learn mathematics and solve problems.

In ancient Egypt, mathematics was used for many purposes necessary to everyday life: measuring time, drawing straight lines, measuring and recording the level of the Nile floodings, calculating land areas, and managing money and taxes. The Egyptians were also one ancient culture that came closest to determining the true length of Earth's year with mathematics. Perhaps most well known to the modern world are the fantastic tombs, pyramids, and other architectural marvels constructed using mathematics. Though their knowledge ranged from arithmetic calculations to algebraic rules to geometrical formulas to numerical ideas, historians consider the Egyptians' mathematical achievements to be somewhat less advanced compared to the Babylonians.

Egyptian Number System
Egyptian numbers are written using a simple grouping system whose symbols denote powers of 10. Their symbols included a vertical staff (10^0), heel bone (10^1), scroll (10^2), lotus flower (10^3), pointing finger (10^4), tadpole (10^5), and astonished man (10^6):

1	10	100	1000	10,000	100,000	1,000,000

Using these symbols, a number was expressed additively. For example, the base-10 number 4501 was represented by a visual collection of 4 lotus flowers, 5 scrolls, and 1 vertical staff. As no place-value system is involved, these symbols can be written in any order or arrangement visually—they equal a numerical value as a group. Though able to represent large values of numbers with these symbols, the Egyptians' lack of place values deterred their ability to calculate proficiently using algorithms.

Again represented by hieroglyphic symbols, Egyptian fractions were restricted to unit fractions (numerator of 1) except for the special fraction 2/3. For example, the unit fraction 1/3 was represented by an ellipse (or dot) placed visually over 3 vertical staffs. The Egyptians had no symbol for zero as a place holder but such was not really needed because of their simple grouping system and use of distinct symbols for each power of 10.

Egyptian Arithmetic

Addition and subtraction are quite easy using the Egyptian numbers, involving only the union or removal of the grouped symbols. In addition, a symbol that appeared 10 times was replaced by the next higher level symbol; for example, 10 vertical staffs could be replaced by 1 heel bone. Similarly, in subtraction, a symbol could be traded in for 10 of the next lesser symbol if such was necessary. For example, to perform 23−8, a heel bone could be traded for 10 vertical staffs so that 8 vertical staffs could be taken from the 13 vertical staffs.

Egyptian multiplication involved repeated addition, using a doubling process along with a counter. For example, to multiply 23 × 13, their process (in modern notation) would look like the following, with the counter on the right:

23	1*
46	2
92	4*
184	8*

Using the starred counters (1 + 4 + 8 = 13), the product is obtained by adding the associated numbers (23 + 92 + 184 = 299). The key to this multiplication is the distributive process, since

$$23 \times 13 = 23 \times (8 + 4 + 1) = (23 \times 8) + (23 \times 4) + (23 \times 1)$$
$$= 184 + 92 + 23 = 299.$$

Thus, base two notation also is the underlying principle, since

$$13 = (1)(2^3) + (1)(2^2) + (0)(2^1) + (1)(2^0).$$

These processes of duplation and mediation (doubling and halving) remained as standard algorithms in Western mathematics until the 1500s.

Division required an inversion of the multiplication process. For example, to divide 299 by 23, the Egyptian scribe determined what number times 23 would produce 299, using a process like the following (in modern notation):

23*	1
46	2
92*	4
184*	8

Using the starred sums, 23 + 92 + 184 = 299, the desired factor (or quotient) is obtained by adding the associated numbers, or 1 + 4 + 8 = 13. The division process becomes complicated when no combination of the starred numbers equals the desired sum (for example, 300 divided by 23), requiring the use of unit fractions:

23*	1
46	2
92*	4
184*	8
1*	1/23

For more difficult divisions (for example, 301 divided by 23), considerable creativity was needed.

To aid in their computations, the Egyptians created tables for doubling and halving numbers, complemented by special 2/*n* tables that would help avoid odd-number situations. For example, the Rhind papyrus had a 2/*n* table for the odd numbers 5–101.

Egyptian Algebra

Though without an algebraic notation, the Egyptians solved numerous types of algebraic equations, known as "aha" calculations. The majority of their problems were linear equations with one unknown (called the "heap"). Their solution process involved the method of false position, where an initial guess is made, examined, and then adjusted to obtain the correct solution. This same process is now fundamental to the area of numerical analysis and is used extensively for scientific computing using computers.

Consider this Egyptian problem, "Heap and a seventh of the heap together give 19." In modern notation, the associated linear equation is $x + x/7 = 19$, while their step-by-step solution was the following:

Make a guess for heap, for example, 7

$$\text{Then } 7 + \frac{7}{7} = 8$$

$$\text{But } \left(2 + \frac{1}{4} + \frac{1}{8}\right)(8) = 19$$

$$\text{Thus, heap } = (7)\left(2 + \frac{1}{4} + \frac{1}{8}\right) = 16 + \frac{1}{2} + \frac{1}{8}.$$

The processes of multiplication and division, as well as the law of associativity, play very important roles:

$$\left(2 + \frac{1}{4} + \frac{1}{8}\right)(8)$$
$$= \left(2 + \frac{1}{4} + \frac{1}{8}\right)\left(7 + \frac{7}{7}\right)$$
$$= \left(2 + \frac{1}{4} + \frac{1}{8}\right)\left[(7)\left(1 + \frac{1}{7}\right)\right]$$
$$= \left[\left(2 + \frac{1}{4} + \frac{1}{8}\right)(7)\right]\left(1 + \frac{1}{7}\right)$$
$$= \left(16 + \frac{1}{2} + \frac{1}{8}\right)\left(1 + \frac{1}{7}\right)$$
$$= 19.$$

The majority of the Egyptians' "aha" problems created practical situations requiring the use of ratios and proportions, such as determining feed mixtures or combinations of grains to make bread. In some instances, the Egyptians did use special hieroglyphic symbols as part of their algebraic work, including "plus" (legs walking left to right), "minus" (legs walking right to left) and other ideograms for "equals" and the "heap."

Egyptian Geometry

The Egyptians' geometry was rooted in an algebraic perspective, devoid of any evidence of generalization or proof. Approximately one-fourth of the problems found in the papyri are geometrical—focusing on practical measurements, such as the calculation of land areas, or volumes of storage containers. Similar to the Babylonians, the Egyptians used prescriptive formulas. For example, they viewed a circle's area as equal to that of a square erected on 8/9 of the diameter. That is,

$$A = \left(\frac{8}{9}(2r)\right)^2 = \frac{256}{81}r^2$$

implying their value of π approximated 3.160493827.

Historians agree that the Egyptians knew key formulas for computing the area of a triangle, the volume of a cylinder, some curvilinear areas, and even the volume of the frustum of a square-based pyramid. These formulas were apparently put to great use by the Egyptians in their accurate construction of the pyramids, feats that required a solid understanding of ratios, proportions, dihedral angles, and even astronomy. No evidence suggests the Egyptians knew of the relationships described by the Pythagorean theorem. Some of their geometrical prescriptions were also incorrect. For example, the area of a general quadrilateral (with ordered side lengths a, b, c, d) was calculated by the formula

$$A = \frac{1}{4}(a+c)(b+d)$$

which is correct only if the quadrilateral is a rectangle or square.

Signs of Advanced Mathematical Thinking

Egyptian mathematics was utilitarian in its direct ties to the solution of practical problems. Also, because their numeration system involved simple grouping with no place values, it is not reasonable to expect that the Egyptians had explored ideas such as factors, powers, and reciprocals. This limitation perhaps explains why no record has been found of tables involving Pythagorean triples. Nonetheless, they did apparently use some number tricks; when multiplying a number by 10, they merely replaced each hieroglyphic symbol by the symbol representing the next higher power of 10 (that is, replacing each vertical staff with a heel bone, each heel bone with a scroll, and so forth).

Problem 79 in the Rhind Papyrus suggests that the Egyptians did some recreational mathematics that had no real-world applications. The problem states, "7 houses, 49 cats, 343 mice, 2401 ears of spelt, 16,807 hekats." Historians assume that the scribe was creating a problem involving seven houses, each with seven cats, each of which eats seven mice, each of which had eaten seven ears of grain, each of which had sprouted seven grains of barley…wanting to know the total number of houses, cats, mice, ears of spelt, and grains. Mathematically, the solution of this problem would require some knowledge of powers of 7 and geometric progressions.

Further Reading

Aaboe, Asger. *Episodes From the Early History of Mathematics*. Washington, DC: Mathematical Association of America, 1975.

Friberg, Jöran. *Unexpected Links Between Egyptian and Babylonian Mathematics*. Singapore: World Scientific Publishing, 2005.

Katz, Victor J., ed. *The Mathematics of Egypt, Mesopotamia, China, India, and Islam: A Sourcebook*. Princeton, NJ: Princeton University Press, 2007.

Van der Waerden, B. L. *Science Awakening*. Oxford, England: Oxford University Press, 1985.

———. *Geometry and Algebra in Ancient Civilizations*. Berlin: Springer, 1983.

Jerry Johnson

Elections

Category: Government, Politics, and History.
Fields of Study: Communication; Data Analysis and Probability; Number and Operations.
Summary: Mathematics can help explain and predict elections.

Long the domain of economists, political scientists, and philosophers, systems of government has emerged as a field ripe for the application and study of mathematics. Elections are typically classified under an emerging branch of mathematics called "social choice theory," though there are historical connections and applications in a number of areas, such as combinatorics and probability theory. Economist Duncan Black's 1958 book *The Theory of Committees and Elections* is credited with helping to revive modern interest in using mathematics to study election questions.

In a democratic society, such as the United States, elections are the primary vehicle for providing citizens a fair and equal voice in the machinations of federal, state, and local governments. As such, it is fundamentally important that elections be conducted in a manner that is perceived to be fair by the citizenry; that is, a governing body derives its legitimacy from the equitable interpretation and application of the voting power of the public.

Beyond the widely known popular elections (electing the candidate with the most first-place votes) there are a number of alternative voting methods; many of these allow voters to express more information about their preferences of various candidates. Since it is possible for different methods to produce different winners given the same voter preferences, a number of voting properties have been postulated. Each property states a desired outcome or effect that a voting system should express. For example, a voting system should be "anonymous" in that individual voters should be able to exchange ballots without affecting the outcome; in other words, one person's ballot should not have special significance. A more challenging property is "independence from irrelevant alternatives," which requires the relative outcome of an election to remain unaffected if candidates are added or removed from consideration (provided this addition or removal does not change the relative way voters feel about the other candidates). Economist Kenneth Arrow demonstrated mathematically in his doctoral dissertation that no voting system can satisfy all the desired properties. Arrow's Impossibility Theorem was later published in his 1951 book *Social Choice and Individual Values*.

A particular type of voting system, weighted voting, arises when voters are assigned different numbers of votes. This system is usually employed to reflect a situation where some voters should have greater say or representation than others. The Banzhaf Power Index, named after John Banzhaf, is a tool that elucidates the voting power enjoyed by the voters in a weighted voting scheme and reveals that voting power is not always

Townspeople lined up to vote in rural areas of Guatemala for the 2007 national elections. In a democratic society, elections provide citizens with a voice in the workings of federal, state, and local governments. (USAID/Maureen Taft-Morales)

commensurate with a voter's number of votes. It is also sometimes called the Penrose–Banzhaf Power Index to include its original inventor, Lionel Penrose.

The U.S. Electoral College, an example of a weighted voting system, is used to elect a winner in U.S. presidential elections. The U.S. Electoral College illustrates a drawback of weighted voting in that a winning presidential candidate may not have received a majority of popular votes. This has sparked much interest in replacing the U.S. Electoral College in favor of the popular vote method but smaller states that enjoy more voting power with the U.S. Electoral College are likely to block attempts at Constitutional reform.

Exit polling, invented by statistician Warren Mitofsky, allows social demographers to understand the dynamics of an election and to predict the winner. Exit polling has become an increasingly important tool for media and news outlets as they scramble to retain and inform viewers on the eve of an important election. A number of studies have investigated the influence of exit polling while an election is taking place; for instance, polls broadcast in real time may influence voters who have yet to vote and hence possibly change the outcome of an election. Exit polling has also garnered interest in recent presidential elections when erroneous predictions caused media sources to prematurely, or incorrectly, identify a winning candidate.

The Ballot Box Problem is an interesting mathematical puzzle, proposed by Joseph Bertrand, which seeks answers about how an election may unfold as ballots are removed from the ballot box and counted. The solution to Bertrand's theorem is a Catalan number, named for Eugène Catalan. An elegant proof was derived by Désiré André.

Types of Elections

Though most people are familiar with the plurality election (also known as "popular vote") in which the candidate with the most votes (most first-place votes) wins, there are a number of alternative election methods. One of the most prominent is the Borda method, named for Jean-Charles de Borda, where voters are required to rank all candidates from their first choice to their last; points are then assigned to each candidate based on the candidate's rank on the each ballot. The sum of a candidate's total points is used to determine the winner. This method allows voters to specify more information about how they view the candidates, other than merely selecting their favorite.

In the Sequential Pairwise method, two of the candidates vie in a head-to-head competition (an imaginary election with only the two candidates) where the losing candidate is eliminated and the winner proceeds forward to battle another candidate. Again, voters rank candidates in preference listings, which are used to determine the winner between a particular pair of candidates. The winner can be inferred from the preference lists by assuming each voter would select the candidate that is higher on his or her list. A drawback of this method is that the order in which the candidates are selected for the individual competitions can change the ultimate outcome of the election.

A Condorcet Winner is a candidate who beats every other candidate in a head-to-head election. When one exists, a Condorcet Winner will obviously win the Sequential Pairwise election but not all sets of voter preference rankings produce a Condorcet Winner. The method is named for Marie Jean Antoine Nicolas de Caritat, Marquis de Condorcet.

In an Instant Run-off election, a plurality vote is taken and the candidate with the least number of first-place votes is eliminated. Then the election is repeated with the remaining candidates until only one winner remains. Again, voter preference rankings can be used to simulate the repeated elections in order to determine the winner without holding a series of actual elections.

Weighted Voting

Much of rationale behind the U.S. system of government is based on the principle of "one person, one vote" (each citizen should have equal say in the system of government). There are times, however, when it is appropriate to give certain individuals (or groups) more voting power than others. This type of voting system, often called "yes–no voting" or "weighted voting," occurs when voters are assigned a different number of votes or "weights" to their votes. Elections are between two alternatives; the winner is selected if the vote total exceeds a predetermined threshold. Each voter must use all available votes toward the same candidate or choice—votes cannot be split between the candidates or choices.

An example of a weighted voting system was the European Economic Community (EEC) established in 1958 as a precedent to the current European Union. The original six members were assigned votes in proportion to their population size:

Country	# Votes
France	4
Germany	4
Italy	4
Belgium	2
Netherlands	2
Luxembourg	1

A threshold is established to determine the number of total votes necessary to win an election. Though this threshold is often simple majority, in the EEC example, a threshold of 12 (of the total 17 votes) was established to pass certain types of legislation.

An interesting question arises as to the dynamics of weighted voting systems and, more specifically, an entity's ability to influence the outcome of an election. Several theorists have shown that voting power is not necessarily proportional to an entity's vote count. For example, it would be misleading to assume that France enjoys 23.5% (4/17) of the voting power in the EEC example.

Banzhaf applied a power index to argue a landmark case in Nassau County, New York, in 1965. His voting power calculations demonstrated the disenfranchisement of certain entities within weighted voting schemes and thus questioned the system's constitutionality.

Banzhaf's computation is based upon the notion of a winning coalition (a collection of voters whose vote total exceeds the threshold). Such a coalition (or "voting block") can win an election by all voting the same way. A voter is critical to a winning coalition if by removing that voter, the coalition no longer exceeds the threshold. A voter's Banzhaf Power Index (BPI) is related to the number of times that voter is a critical member of a winning coalition.

In the EEC example, France, Germany, Italy, and Belgium form a winning coalition since their vote total of 14 exceeds the threshold of 12. France, Germany, and Italy are all critical members because the coalition ceases to win without their votes. However, Belgium is not a critical member since France, Germany, and Italy together still form a winning coalition. The number of times each voter appears as a critical member of some winning coalition is computed as follows:

Country	# Critical	BPI
France	10	10/42 = 23.8%
Germany	10	10/42 = 23.8%
Italy	10	10/42 = 23.8%
Belgium	6	6/42 = 14.3%
Netherlands	6	6/42 = 14.3%
Luxembourg	0	0/42 = 0%

Each country's BPI is the number of times it is critical compared to the total number of critical instances. Here, there are 42 total instances where an entity is critical; Belgium has 6 of them and thus 14.3% (6/42) of the voting power. Thus, Belgium commands 14.3% of the voting power even though it has 11.8% of the votes. In this scheme, Luxembourg has no voting power—it is not able to influence the outcome of any possible election. It is common in weighted voting schemes of smaller size (20 or fewer members) for entities with a greater number of votes to possess greater voting power, while small entities (with a fewer number of votes) possess less voting power. As the number of voters increase, voting power tends to better approximate the proportion of votes. But such weighted voting systems are subject to arbitrary swings of voting power as new voters are added or removed, or as seemingly subtle changes to the weights are made.

An equally popular voting power computation was proposed by Lloyd Shapely and Martin Shubik in 1954. Instead of critical members in winning coalitions, their system identifies pivotal voters as the ones who enter a coalition and cast the deciding vote by doing so. A similar calculation ensues in which voting power is correlated with the percentage instances in which each entity plays the pivotal role.

U.S. Electoral College

The voting system responsible for electing the president of the United States, the U.S. Electoral College, is essentially a weighted voting scheme. A state's electors (or "votes") arise from the sum of their congressional representation: one vote for each of a state's two senators and one vote for each representative to the House of Representatives. The District of Columbia receives three electors to form a total of 538 (100 senators, 435 representatives, and three from Washington, D.C.). A presidential candidate needs a majority of the electoral votes—at least 270—to claim victory.

Under such a system, it is possible that the winning candidate need not garner a majority of first-place votes. In fact, U.S. presidential elections in 1824, 1876, 1888, and 2000 all produced a winner who lost the popular vote total.

Those elections and other issues have created an endless interest in reforming or removing the U.S. Electoral College and replacing it with a popular vote system. As recently as 2004, the Every Vote Counts Amendment proposed to replace the U.S. Electoral College with a popular vote initiative. Such a reform requires a Constitutional change and thus approval of 75% of the states.

It is unlikely such a measure would ever be adopted because small states enjoy significantly more voting power in the U.S. Electoral College than they would in a popular vote system. A state with few votes, such as South Dakota, would likely be ignored by campaigners since the voting population is too small to make a difference under a popular vote election.

The National Popular Vote Compact is an alternative attempt at election reform. In this compact, individual states would cast their electoral votes according to the national popular vote, not simply the tallies within the state. This has the effect of choosing a president elected by popular vote within the Electoral College system and thus bypassing the hurdle of constitutional reform. To date, this compact has been adopted by five states (61 electoral votes) with a number of others considering the compact in state legislature—enough states to compile 270 electoral votes would have to sign on to the compact in order to have the intended effect of electing a president by popular vote.

Exit Polling

An important factor associated with elections is the attempt to predict election outcomes through the surveying of voters as they leave the voting areas, a procedure known as "exit polling." This procedure contrasts with pre-election polls in that actual voters who have (presumably) just cast a vote are being sampled and thus results are typically more accurate than surveying people prior to an election who are "likely" to vote, or who may change their mind between being polled and actually casting a vote.

Although the science of predicting election outcomes has been around as long as elections themselves, it is at the beginning of the twenty-first century—with widespread electronic media coverage and more sophisticated polling techniques—that exit polling has garnered more national attention. A number of papers have been written about the effects of exit polling being broadcast in real time; the researchers hypothesize that exit polling influences voter behavior primarily by making an election seem closer or not closer than was previously perceived. This effect is especially true in the United States where, as a function of different time zones, voters in western states have access to more complete results of a national election unfolding across the country.

Exit polling has garnered an additional spotlight with the controversial presidential elections of 2000 and 2004. In both cases, especially the 2004 election, exit polling differed significantly from the actual vote tally, causing many media outlets to incorrectly, or prematurely, announce a victor.

Ballot Problem

There are several interesting mathematical puzzles based on elections and voting; perhaps the most well known of them is the Ballot Problem, originally presented by Joseph Bertrand in the late nineteenth century. Consider an election between two people, Alice and Bob, where Alice has received A votes and Bob B votes. Let $A>B$ so that Alice wins the election. The puzzle arises from the counting of the votes: what is the probability that as the votes are pulled randomly from the ballot box and tallied one by one, that Alice and Bob are tied in their vote total at some point after the first vote is read?

The puzzle's solution is a creative argument based on combinatorics and probability. Sequences, a listing of votes as they are pulled from the ballot box, can be identified as those with ties and those without. The following is a sequence from an election with nine voters ($A=5$, $B=4$):

$$b\,b\,a\,b\,a\,a\,b\,a\,a.$$

In this sequence, the first tie occurs with the reading of the sixth vote, though there is also a subsequent tie. There is also a "matching" partial sequence in which the a's and b's exchange places up through the point of the first tie:

$$a\,a\,b\,a\,b\,b\,b\,a\,a.$$

Every such sequence of strings that produces a tie somewhere in the intermediate vote tally comes in matching pairs as shown. Out of each pairing, one sequence must start with an a while its match starts with a b. Since Alice wins the election, some of the sequences starting with an a will result in a tie but not all of them. However, every sequence that starts with a b must at some point achieve a tie since ultimately there will be more as than bs. There are three categories of sequences:

- sequences that start with an a but never have a tie
- sequences that start with an a and achieve a tie at some point

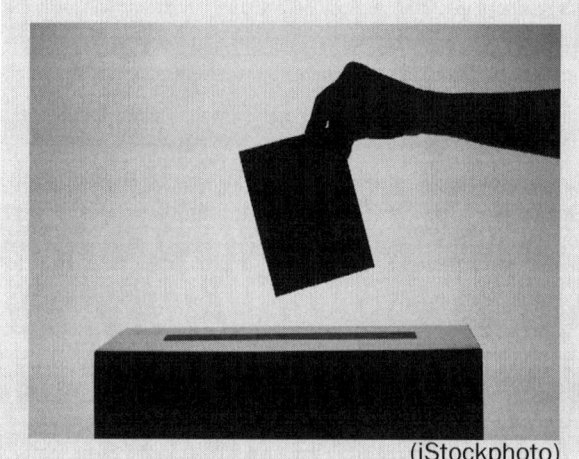
(iStockphoto)

Survivor!

The popular television series Survivor nicely illustrates a ballot-box type of problem. In individual tribal councils, as well as the final vote for an overall winner, ballots are drawn from a ballot box and read aloud. It is easy to hypothesize that the ballots are not drawn in a random order but instead are selected so as to maximize the suspense of the election outcome. Another interesting question, related to information theory, is that ballots are read only until the election outcome is certain; unread ballots are not presented to the remaining tribe members, thereby depriving them of strategic information about the voting behavior of their fellow competitors.

- sequences that start with a *b* and achieve a tie at some point

The probability that any sequence is found starting with a *b* is

$$\frac{B}{A+B}$$

since there are *B* ballots out of *A*+*B* total ballots where a *b* can be the first vote drawn. There are exactly as many sequences that start with an *a* and also achieve a tie because each one is matched with exactly one *b*-starting sequence. Therefore, the probability of reading the votes and achieving a tie along the way is exactly

$$\frac{2B}{A+B}.$$

This problem has spawned a number of related problems with interesting ties to Catalan numbers.

Further Reading

Freeman, Steven F. *The Unexplained Exit Poll Discrepancy.* Philadelphia: Center for Organizational Dynamics, University of Pennsylvania. 2004.

Hodge, Jonathan K., and Richard E. Kilma. *The Mathematics of Voting and Elections.* Providence, RI: The American Mathematical Society, 2005.

Sudman, Seymour. "Do Exit Polls Influence Voting Behavior?" *Public Opinion Quarterly* 50, no. 3 (1986).

Taylor, Alan D. *Mathematics and Politics: Strategy, Voting Power, and Proof.* New York: Springer-Verlag. 1995.

Matt Kretchmar

Europe, Eastern

Category: Mathematics Around the World.
Fields of Study: All.
Summary: Eastern Europe has a long tradition of both mathematics research and education.

Throughout history, the countries of Europe have had shifting political and social boundaries. Eastern European mathematics evolved within the context of many mathematics traditions, including Soviet Union mathematics, over the past centuries. Historically, gifted young scholars from regions around the world completed their mathematical studies at Europe's well-known and respected universities. Studies of mathematicians' letters and scientific papers show that they often maintained connections with people in other countries who shared their fields of interest. The Soviet Union exercised broad social and political influence over most of eastern Europe and also impacted U.S. mathematics in the twentieth century. Within the Soviet Union, students from the far reaches of the nations within its boundaries were often brought to Russia for work or education, as well as sent to other parts of the Soviet Union to teach or to establish research centers. In the twenty-first century, students in the United States and around the worked attend study abroad programs, such as the Budapest Semesters in Mathematics. In the twenty-first century, the United Nations Statistics Division classified the following countries belonging to eastern Europe: Belarus, Bulgaria, Czech Republic, Hungary, Moldova, Poland, Romania, Russia, Slovakia, and Ukraine. The *CIA World Factbook* adds Estonia, Latvia, and Lithuania, which were among the member nations of the Soviet Union, though the United Nations classifies them as belonging to northern Europe. Geographical boundaries continued to change in the twentieth century because of post–World War II structures and, later, the breakup of the "Eastern Bloc" nations, which were once under the Soviet Union's political influence. Therefore, mathematics contributions of some people from eastern Europe may be included within the histories of other regions or countries.

History of Russian and Soviet Mathematics Education

When examining past and present states of mathematics in Belarus, Moldova, Russia, Ukraine, Estonia, Latvia, and Lithuania, it is pertinent to acknowledge that they share a common sociopolitical root: they are all former member states of the Soviet Union. Further, the broader Eastern Bloc of Soviet Union allies included Bulgaria, Romania, Hungary, East Germany, Poland, Albania (until the early 1960s), and Czechoslovakia (which later split into the Czech Republic and Slovakia). The Eastern Bloc is sometimes known historically as "eastern Europe," versus the "western Europe" countries allied with the United States, a rival of the Soviet Union. During its several decades of existence in the twentieth century, the Soviet Union included many mathematicians who made significant contributions to the body of modern mathematical knowledge. Further, Russian and Soviet mathematicians were influential on many other countries.

One important landmark in mathematics education in Russia is the creation in 1701 of the School of Mathematical and Navigational Sciences in Moscow. Peter the Great, who had traveled widely in other parts of Europe to study the state of mathematics and science as part of his effort to modernize Russia and expand the empire, founded this school. It educated students in basic mathematics as well as more specialized subjects, such as astronomy and navigation. Notably, students from all social classes except serfs were admitted, and financial assistance was available. Graduates worked in the navy, as engineers, and as teachers in a variety of settings, so the school had a multiplier effect in terms of spreading mathematics education throughout Russia. Peter the Great also founded the Saint Petersburg Academy of Sciences in 1724, influenced in part by correspondence with mathematician Gottfried Leibniz, who also purportedly recommended a three-tiered educational system of schools, universities, and academies. Many eminent foreign mathematicians, such as Leonard Euler, Christian Goldbach, and Daniel Bernoulli, worked at the Saint Petersburg Academy.

As part of her goal of modernizing Russia in the European style, Empress Catherine the Great, who was born in Germany, established the first gymnasiums in Russia. These gymnasiums were schools meant to prepare students for higher education and were created in most major Russian cities in the nineteenth century. Nicolai Ivanovich Lobachevsky, one of the first Russian mathematicians to achieve international recognition, was a beneficiary of this expanded educational opportunity. He graduated from Kazan Gymnasium and Kazan University (in Tatarstan) and is most noted for his work in hyperbolic geometry, a form of non-Euclidean geometry. However, despite this considerable expansion, access to education was far from universal until the Soviet era. The Soviet Union was founded by revolution in 1917, when the monarchy of the Russian Empire was overthrown, but was not made official until 1922. The Saint Petersburg Academy of the Sciences evolved into the Russian and then Union of Soviet Socialist Republics (USSR) Academy of the Sciences. It reverted to the Russian Academy of Sciences following the dissolution of the Soviet Union, and remains an influential organization in the twenty-first century. Academies of sciences were also founded in most of the states of the Soviet Union. Universal compulsory education was established in 1919. Soviet schools had both political and educational goals but the expectation that all children would attend school rapidly increased literacy and played a key role in modernizing and industrializing the country.

In the Soviet Union, the study of mathematics and the sciences was emphasized, a choice that not only fostered rapid economic growth but also became a point of national pride, as by mid-century the Soviet Union was frequently seen to rival or even surpass the United States in scientific and applied research. When the Soviet Union successfully launched the satellite *Sputnik* in 1957, it raised concern in the United States not only because of the possibility that the Soviet Union was developing weapons for which the United States had no counter but also because it put into question the common assumption that the United States was the world leader in mathematics and science. One result of *Sputnik* in the United States was a substantial increase in federal funding for scientific education and research in the hope of catching up and surpassing the Soviet Union in the "space race."

As part of this concern that the Soviet Union was surpassing the United States, many studies were commissioned of the Soviet educational system and how it differed from the American system. Among the differences noted by researchers were the facts that in Soviet schools, specialists taught mathematics from the fourth grade onward, a uniform curriculum was used across the entire country, and much greater emphasis was placed on developing the talents of students who were identified as gifted in mathematics. The Soviet Union had "special schools," which were free boarding schools at high school level for gifted students and specialized in particular subjects. Four such schools were devoted to mathematics. Correspondence courses in advanced mathematics were also available to increase the number of students studying those subjects. American observers noted that the level of mathematics required for university admittance during the Soviet period was much higher than what would be expected for entering freshmen in the United States. At the same time, other authors have noted that English-language sources often do not reflect the full scope and influence of Russian and Soviet mathematics. These omissions may be because of Cold War influences and a period of Soviet isolationism from the United States and much of Europe, a policy that contrasts strongly with earlier Russian connections and the growing collaborations following the Soviet era.

Notable Soviet and Russian Mathematicians

Andrey Kolmogorov (1903–1987) is known for his work in the fields of probability theory and topology, including the Kolmogorov axioms, Kolmogorov's zero-one law, and Kolmogorov space.

Stefan E. Warschawski (1904–1989) studied at the University of Königsberg and Göttingen. His doctoral thesis was on the boundary behavior of conformal mappings.

Sergei Lvovich Sobolev (1908–1989) worked in mathematical analysis and partial differential equations. Sobolev spaces (named after him) can be defined by growth conditions on Fourier transforms.

Israel Moiseevich Gelfand (1913–2009) worked in the field of functional analysis. He is known for the Gelfand representation in Banach algebra theory; the representation theory of the complex classical Lie groups; contributions to distribution theory and measures on infinite-dimensional spaces; integral geometry; and generalized hypergeometric series. His name is linked to the development of mathematical education.

Igor Shafarevich (1923–) is the founder of the major school of algebraic number theory and algebraic geometry in the Soviet Union. He has also written well-known textbooks.

Grigori Perelman (1966–) declined the Fields medal, a prestigious award in mathematics often equated to the Nobel Prize, for his work on the Poincaré conjecture, named for Henri Poincaré. He cited inequities and reportedly noted, "If the proof is correct then no other recognition is needed."

Other well-known Soviet or Russian twentieth-century mathematicians include Boris Pavlovich Demidovich, who worked on problems in mathematical analysis, and Yakov Isidorovich Perelman, who was

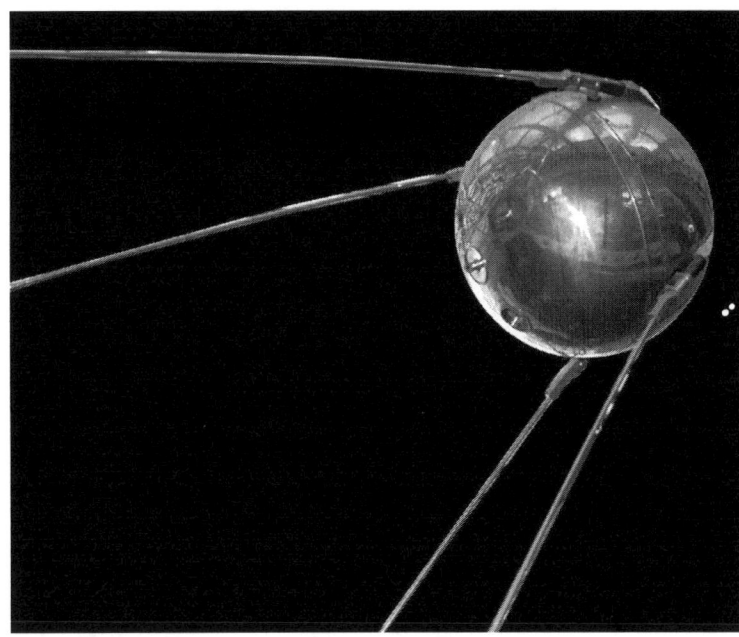

The launch of the first artificial satellite, Sputnik 1, *by the Soviet Union on October 4, 1957, started the race to the moon. (National Aeeronautics and Space Administration)*

a science writer and author of many popular science books.

Czech Republic and Slovakian Mathematicians

Kurt Gödel (1906–1978) proved fundamental results about axiomatic systems. Gödel's Incompleteness Theorems are named for him.

Stefan Schwarz (1914–1996) studied semigroups, number theory, and finite fields and founded the Mathematico-Physical Journal of the Slovak Academy of Sciences in 1950.

Hungarian Mathematicians

Hungarian mathematicians of the twentieth century are well known in the mathematical world. Many of them immigrated to the United States after World War II.

Frigyes Riesz (1880–1956) was a founder of functional analysis. He produced representation theorems for functional on quadratic Lebesgue integrable functions, named for Henri Lebesgue, then introduced the space of q-fold Lebesgue integrable functions. He also studied orthonormal series and topology.

George Pólya (1887–1985) worked in probability, analysis, number theory, geometry, combinatorics, and mathematical physics. He wrote books about problem-solving methods, complex analysis, mathematical physics, probability theory, geometry, and combinatorics. He was regarded by many as a great teacher and influenced many mathematicians.

Cornelius Lanczos (1893–1974) worked on relativity and mathematical physics. He invented what is now called the Fast Fourier Transform, named for Joseph Fourier. He published more than 120 papers and books.

John von Neumann (1903–1957) worked in quantum mechanics, game theory, and applied mathematics, as well as helping pioneer computer science. His doctoral thesis was on set theory. His definition of ordinal numbers is the one commonly used in the early twenty-first century.

Rózsa Péter (1905–1977) is known for teaching, for her books on the history of mathematics, and for her series of theorems about primitive recursive functions.

Paul Erdos (1913–1996) is well known among mathematicians for his insatiable ability to pose and solve problems. It is often said that he lived on mathematics and coffee, touring the circle of his friends and pupils and giving lectures on combinatorics, graph theory, and number theory. He advocated for elegant and elementary proof. One of the most prolific mathematicians in history, he wrote more than 1500 papers.

Paul Richard Halmos (1916–2006) is known for his contributions to operator theory, ergodic theory, functional analysis (in particular Hilbert spaces, named for David Hillbert), and for his textbooks.

Alfréd Rényi (1921–1970) worked on probability theory, statistics, information theory, combinatorics, graph theory, number theory, and analysis.

László Lovász (1948–) published his first paper called *On graphs not containing independent circuits* when he was only 17 years old. He is a prominent figure of post–World War II mathematicians.

Notable Polish Mathematicians

Stefan Banach (1892–1945) worked on the theory of topological vector spaces, measure theory, integration, and orthogonal series. His doctoral thesis "On Operations on Abstract Sets and their Application to Integral Equations" (1920) marks the birth of modern functional analysis. He defined the "Banach space."

Benoit Mandelbrot (1924–2010) is known as the father of fractal geometry. The Mandelbrot set, a connected set of points in the complex plane, is named after him.

Mathematicians From Romania

János Bolyai (1802–1860) is perhaps the most famous Romanian mathematician because of his treatise on a complete system of non-Euclidean geometry in his book *Appendix*. In his own words, he created a new world out of nothing.

Caius Iacob (1912–1992) worked in the fields of analytic geometry, descriptive geometry, analysis, and complex functions.

Grigore C. Moisil (1906–1973) worked on differential equations, the theory of functions, and mechanics. He set up the first Romanian computer science course. Moisil was appreciated for his philosophy and humor.

Other important Romanian mathematicians include Dimitrie Pompeiu, Ferenc Radó, Isaac Jacob Schoenberg, Simion Stoilow, Gheorghe Titeica, Gheorghe Vranceanu, Octav Onicescu, Ion Colojoara, and Dan Barbilian.

Competitions and Contests

Building on eastern Europe's strong mathematics traditions, many mathematical contests are hosted frequently or entirely within the region, such as International Mathematical Olympiad, Romanian Master of Sciences (formerly called the Romanian Masters in Mathematics—it was expanded to include physics), Czech-Polish-Slovak Match, Bulgarian Competition in Mathematics and Informatics, Romanian National Olympiad, and the International Kangaroo Mathematics Contest (often called "Math Kangaroo") among others. Individuals from all over the world participate regularly in these competitions. There are also several winners of the Fields Medal who were born or worked in eastern Europe.

Further Reading

Davis, Robert B. "An Analysis of Mathematics Education in the Union of Soviet Socialist Republics." Report for the National Institute of Education. December 1979. http://www.eric.ed.gov/PDFS/ED182141.pdf.

Dickson, Paul. *Sputnik: The Shock of the Century*. New York: Walker Publishing, 2001.

Sinai, Iakov. *Russian Mathematicians in the 20th Century*. Singapore: World Scientific Publishing, 2003.

Vogeli, Bruce R. *Soviet Secondary Schools for the Mathematically Inclined.* Washington, DC: National Council of Teachers of Mathematics, 1968.

SIMONE GYORFI
SARAH BOSLAUGH

Europe, Northern

Category: Mathematics Around the World.
Fields of Study: All.
Summary: Since the Enlightenment, Northern Europe has made considerable contributions to mathematics research and continues to do so.

Northern Europe has produced many outstanding mathematicians and scholars in related fields, from the development of calculus by Isaac Newton in the seventeenth century to the cosmological models developed by Stephen William Hawking in the twentieth and twenty-first centuries.

Northern Europe also led the way in developing many practical applications of mathematics and later statistics, including taking a national census like the Domesday Book undertaken in England in 1183 and developing mathematical ways to measure the influence of personal habits on health as in the studies of Richard Doll and Austin Bradford Hill on the relationship between smoking and disease. In the twenty-first century, the United Nations category of northern Europe includes the Åland Islands, the Channel Islands, Denmark, Estonia, Faeroe Islands, Finland, Guernsey, Iceland, Ireland, the Isle of Man, Jersey, Latvia, Lithuania, Norway, Svalbard and Jan Mayen Islands, Sweden, and the United Kingdom of Great Britain and Northern Ireland. However, the changing political boundaries in many of these countries throughout history, as well as the rise and fall of the Soviet Union, which included countries like Estonia, Latvia, and Lithuania, mean that mathematical contributions of some individuals may be included within the histories of other regions.

Sir Isaac Newton was one of the most influential mathematicians of the modern era. He shares credit with Gottfried Leibniz for developing integral and differential calculus, and he also made major contributions in the fields of physics and astronomy. Newton's 1687 book *Philosophiae Naturalis Principia Mathematica* laid the groundwork for classical mechanics including a description of the three laws of motion and remains one of the most influential books in the history of science. He also built the first reflecting telescope and developed a theory of color based on the visible spectrum displayed when visible light is refracted through a prism. Through his work with the laws of gravity and Kepler's laws of planetary motion, named for Johannes Kepler, Newton was able to demonstrate mathematically the validity of heliocentrism, which is the scientific principle that Earth and other planets revolve around the sun.

The nineteenth century saw several major breakthroughs in mathematics by scholars from northern Europe. In England, philosopher and mathematician George Boole developed the system now known as "Boolean logic," which has many practical applications and was instrumental in the development of modern digital computers. His most famous works are *The Mathematical Analysis of Logic* (1847) and *The Laws of Thought* (1854). His slightly younger contemporary, Norwegian Niels Henrik Abel, invented the field of group theory (contemporaneously with Frenchman Evariste Galois), which has many applications in mathematics and physics. Abel is well known for a proof he wrote at age 19 that there can be no general algebraic solution of an equation greater than degree four. In Ireland, Sir William Rowan Hamilton provided an important reformulation of Newtonian mechanics and invented an extension of the number system called "quaternions."

In the period 1910–1913, the British scholars Bertrand Russell and Alfred North Whitehead wrote the influential *Principia Mathematica* in which they attempted to derive the foundations of mathematics from a set of axioms and inference rules. Russell was also a prominent writer and political activist who won the Nobel Prize for Literature in 1950, while Whitehead was also noted as a philosopher. More recently, Andrew Wiles, who was born and educated in the United Kingdom but immigrated to the United States, achieved fame for proving Fermat's Last Theorem (named for Pierre de Fermat), one of the most famous previously unsolved problems in mathematics.

Honors

There is no Nobel Prize for mathematics but several different international awards are offered that have been termed the "Mathematics Nobel Prize" because of their prestige. The Fields Medal is awarded every four years to one or more mathematicians of age 40 or younger by the International Mathematical Union. Winners from the United Kingdom have included Klaus Roth (1958), Michael Atiya (1966), Alan Baker (1970), Simon Donaldson (1986), Richard Borcherds (1988), and Timothy Gowers (1998). Lars Ahlfors of Norway won in 1936, the first year the medal was given; Atle Selberg of Norway won in 1950; and Lars Hormander of Sweden won in 1962. Another major mathematical prize, the Abel Prize, is named after Norwegian mathematician Niels Henrik Abel and is awarded annually by the Norwegian Academy of Science and Letters. The Abel Prize has been awarded since 2003. Northern European winners include Michael F. Atiyah of the United Kingdom and Lebanon in 2004 and Lennart Carleson of Sweden in 2006.

The Wolf Prize in Mathematics has been awarded almost annually by the Wolf Foundation since 1978 and more than one prize may be given per year. Northern European winners include Lars Ahlfors of Finland (1981), Atle Selberg of Norway (1986), Lars Hormander of Sweden (1988), Lennart Carleson of Sweden (1992), Andrew Wiles of the United Kingdom (1995/1996), and David B. Mumford of the United Kingdom (2008).

Northern European countries have been regular competitors in the International Mathematical Olympiad, an annual competition held since 1959 for high school students. Each competing country sends a team of six students who are assigned six questions to solve. Individual students are awarded medals based on their scores, and countries are also compared based on the total score for their team.

There have been many medal winners from northern European countries. The United Kingdom began participating in 1967 and even hosted the 1976 and 2002 competitions. Ireland first participated in 1988. The northern Europe countries from the former Soviet Union—Estonia, Latvia, and Lithuania—first participated in 1993, which coincided with the removal of Russian troops from the area and other political reorganization throughout the former Soviet Union. Among the Scandinavian countries, Sweden first participated in 1967, Norway in 1984, Finland in 1965, Denmark in 1991, and Iceland in 1985. Sweden hosted the 1991 competition, and Finland hosted it in 1985.

Further Reading

Knox, Kevin C., and Richard Noakes. *From Newton to Hawking: A History of Cambridge University's Lucasian Professors of Mathematics*. Cambridge, England: Cambridge University Press, 2003.

Krantz, Steven G. *An Episodic History of Mathematics: Mathematical Culture Through Problem Solving*. Washington, DC: Mathematical Association of America, 2010.

School of Mathematics and Statistics, St. Andrews University. "The MacTutor History of Mathematics." http://www-groups.dcs.st-andrews.ac.uk/~history/.

Westfall, Richard S. *Isaac Newton*. Cambridge, England: Cambridge University Press, 2007.

Sarah Boslaugh

Europe, Southern

Category: Mathematics Around the World.
Fields of Study: All.
Summary: Modern Western mathematics was developed in southern Europe and continues to thrive there.

The system of modern mathematics originated in southern Europe, with the ancient Greeks undoubtedly building on traditions already used in Egypt and by the Phoenicians. Like many areas of the world, the nations of southern Europe have had many different boundaries, names, and political alliances throughout history, and so the mathematical contributions of some individuals may be included within the histories of other regions. For example, many nations were member states of the former Soviet Union. The United Nations now includes Albania, Andorra, Bosnia and Herzegovina, Croatia, Gibraltar, Greece, Holy See, Italy, Malta, Montenegro, Portugal, San Marino, Serbia, Slovenia, Spain, and the former Yugoslav Republic of Macedonia in Southern Europe.

Ancient Greeks and Romans

The earliest Greek school of mathematics is ascribed to Thales (c. 640–550 b.c.e.), who came from Miletus, in present-day Turkey, and Pythagoras (c. 569–500 B.C.E.) who hailed from the Mediterranean island of Samos and later moved to Sicily. Archytas, who subscribed to the Pythagorean philosophy and worked on the harmonic mean, was from Tarentum in modern-day Italy. One of the most well-known Greek mathematicians of the ancient world, Euclid of Alexandria (c. 330–260 B.C.E.), was also not from the Greek mainland. He lived in Alexandria, in modern-day Egypt, and his work proved hugely influential to subsequent mathematicians with his detailed hypotheses and proofs. The great mathematician Archimedes of Syracuse (c. 285–212 B.C.E.) also studied in Alexandria but was from Sicily, where he spent most of his life.

These early Greek mathematicians were undoubtedly an influence on the Romans but the Romans themselves were seemingly more interested in applied mathematics—especially how it related to engineering and building—than in the pure mathematics that was favored by the Greeks. Mathematics was certainly taught in Roman schools and historians have long pondered why Roman mathematicians did not have more influence. This dearth of mathematical advancement has generally been ascribed to the Romans' lack of a designation for "zero" and their awkward system of numbers, which may have prevented any great advances in theory. The Roman Empire did, however, see a continual flourishing of mathematics in Greece and the Greek diaspora, in particular the city of Alexandria. Anicius Manlius Severinus Boethius (c. 475–525) was a well-known Roman mathematician who worked during the declining years of the Roman Empire.

The Renaissance

The Bishop of Seville, Isidorus Hispalensis (570–636), helped develop mathematics in Spain and there were great advances made in arithmetic with the Moorish invasions of Spain and the incorporation of many of the advances made in the Muslim world. The great trading cities of Genoa and Venice soon established themselves as important centers of finance, as did Florence during the Renaissance. Venice, in particular, because of its geographical position and its connections with the Arab world, saw the importation of many books and manuscripts on Arab mathematics—at that stage well advanced in pure mathematics theories compared to Europe. This Arab influence saw Leonardo Pisano Bigollo (c. 1170–1250), the son of an Italian merchant in North Africa, develop theories—the most well-known being the Fibonacci numbers, which were termed after his assumed name.

Several centuries later, the advent of the printing press also led to a republication of the works of Greek mathematicians such as Euclid, albeit in Latin translation. Cardinal Bessarion, the former Archbishop of Nicaea, helped bridge the link between Byzantium and Rome, helping to preserve some of the Greek learning that was lost when the city of Constantinople was captured and sacked in 1453. Leonardo da Vinci (1452–1519) developed mathematics theories, testing out some of them in siege machines designed for Cesare Borgia and others. Girolamo Maggi (c. 1523–1572), another Italian mathematician, was involved in designing military defenses in Cyprus. He was captured by the Ottoman Turks and executed in Constantinople but not before writing two major treatises from memory while in prison there.

The Renaissance saw a new interest in mathematics in Italy, with Galileo Galilei (1564–1642) being a well-known mathematician and scientist. He was a great influence on many subsequent mathematicians, including Alessandro Marchetti (1633–1714). Evangelista Torricelli (1608–1647) invented a barometer; Giovanni Ceva (1647–1734) proved Ceva's theorem in elementary geometry; and the Jesuit Franceso Cetti (1726–1778) helped connect mathematics to other scientific discoveries. Later Italian mathematicians include Giulio Ascoli (1843–1896) who taught in Milan, and Carlo Emilio Bonferroni (1892–1960) who developed the theory of Bonferroni inequalities. The Italian Mathematical Union was established in 1922 by Salvatore Pincherle and others, and its journal, the *Bollettino dell'Unione Matematica Italiana,* is widely respected around the world.

Professional Associations

Professional associations in the region other than the Italian Mathematical Union include the Bosnian Mathematical Society; the Croatian Mathematical Society; the Cyprus Mathematical Society; the Montenegro Mathematical Society; the Portuguese Society of Mathematics; the Mathematical Society of Serbia; the Mathematics, Physics, and Astronomy Society of Slovenia; and the Royal Spanish Mathematical Society. Mathematicians also gather from all over Europe in the European Mathematical Society. The International Mathematical

Olympiad is a competition for high school students that originated in 1959. Albania first participated in 1993, Bosnia and Herzegovina in 1993, Croatia in 1993, Greece in 1975, Italy in 1967, Montenegro in 2007, Portugal in 1989, Serbia in 2006, Slovenia in 1993, Spain in 1983, Yugoslavia in 1963, and the former Yugoslav Republic of Macedonia in 1993. Greece was a host of the competition in 2004, Slovenia in 2006, Spain in 2008, and Yugoslavia in 1967 and 1977.

Further Reading

Field, Judith Veronica. *The Invention of Infinity: Mathematics and Art in the Renaissance.* New York: Oxford University Press, 1997.

Hodgkin, Luke. *A History of Mathematics: From Mesopotamia to Modernity.* New York: Oxford University Press, 2005.

Manaresi, Mirella. *Mathematics and Culture in Europe: Mathematics in Art, Technology, Cinema, and Theatre.* New York: Springer, 2007.

JUSTIN CORFIELD

Europe, Western

Category: Mathematics Around the World.
Fields of Study: All.
Summary: Western Europe has been home to many of the important astronomical and mathematical discoveries of the early modern age.

Historically, the term "western Europe" has had cultural and political definitions. For example, during the Cold War it was often used to designate a collection of noncommunist countries allied in some way with the United States. In the early twenty-first century, the United Nations Statistics Division for western Europe contains Austria, Belgium, France, Germany, Liechtenstein, Luxembourg, Monaco, the Netherlands, and Switzerland. There is a rich history of mathematics scholarship, education, and achievement in western Europe. Important work in a diverse array of mathematical areas like calculus, number theory, analytical geometry, probability, statistics, functional analysis, graph theory, logic, and number theory was produced by people from this geographic region, as well as many mathematical contributions to related disciplines like physics, astronomy, optics, engineering, and surveying.

Historical Contributions

Western European mathematicians have made major contributions to the development of mathematics and the application of mathematical theory to practical problems, from German mathematician and astronomer Johannes Kepler, who worked with Danish astronomer Tycho Brahe and helped established the laws of planetary motion, to French mathematician René Thom, who founded the study of catastrophe theory.

Much of modern science and mathematics has its roots in work done in Europe in the seventeenth century. Johannes Kepler studied at the University of Tubingen, where he learned both the geocentric model of astronomy (the view that Earth is the center of the universe, with the other planets revolving around it) and the heliocentric model of German astronomer Nicolaus Copernicus (the view that the sun is the center of the universe and the planets, including Earth, revolve around it). He later worked with Brahe and established the laws of planetary motion in several influential publications: *Astronomia Nova, Harmonices Mundi,* and *The Epitome of Copernican Astronomy.* Also in Germany, mathematician Gottfried Leibniz developed the field of calculus independent of Sir Isaac Newton in England.

In France, mathematician and philosopher René Descartes developed analytical geometry, including the development of Cartesian coordinates, did important work in optics, and was also one of the fathers of modern Western philosophy with influential books such as *Meditations on First Philosophy, Discourse on the Method* (which contains the oft-quoted statement cogito ergo sum, or "I think, therefore I am"), and *Principles of Philosophy.* Also in France, the basics of probability theory were developed by mathematicians Pierre de Fermat and Blaise Pascal, while Fermat also did important work in number theory, analytic geometry, and optics. Fermat's Last Theorem, mentioned but not proved by Fermat in 1637 in the margin of a book, was among the unsolved problems in mathematics until British mathematician Andrew Wiles proved it in 1994. Pascal invented the mechanical calculator and the hydraulic press and is well known among middle school students for Pascal's Triangle, a presentation of binomial coefficients.

In the eighteenth century, Swiss mathematician and physicist Leonhard Euler spent much of his adult life working at the Russian Academy of the Sciences in St. Petersburg. He developed the concept of the function and the notation $f(x)$, one of several notation conventions he developed that are still used in the early twenty-first century (others include using the letter *e* for the natural logarithm, *i* for an imaginary unit, and the Greek letter *sigma* (Σ) for summation). He also made important contributions to calculus, number theory, graph theory (he solved the famous Seven Bridges of Konigsberg problem), and applied mathematics. French and Italian astronomer and mathematician Joseph-Louis Lagrange, who was born in Italy but worked primarily in France and Prussia, created the calculus of variations, developed a method of solving differential equations and transformed Newtonian mechanics into a branch of analysis, which facilitated the development of mathematical physics. He was also the first professor of analysis at the École Polytechnique, an elite engineering school founded in France in 1794. Also in France, mathematician and astronomer Pierre-Simon LaPlace played a key role in the development of Bayesian statistics, named for English minister and mathematician Thomas Bayes, and mathematical astronomy. He also posited the existence of black holes and gravitational collapse in the solar system.

In the nineteenth century, mathematician German Carl Friedrich Gauss made important contributions to several mathematical and physics fields including statistics, number theory, astronomy, surveying (he invented the heliotrope), and optics. The well-known normal distribution is sometimes referred to as the "Gaussian distribution" because he is often credited with discovering it. In France, Augustin-Louis Cauchy not only worked as an engineer but also pursued mathematical studies in his spare time and was appointed to the Académie des Sciences in 1816. He made numerous contributions to mathematics and physics, including his development of complex function theory, clarification of the principle of calculus, and development of the argument principle. In France, mathematician Evariste Galois proved, in parallel with the work of Norwegian mathematician Niels Henrik Abel, that there was no general method for solving polynomial equations of degree of greater than degree four.

In 1900, German mathematician David Hilbert gave an influential talk at the International Congress of Mathematicians in which he identified 23 unsolved problems in mathematics, which served as a spur for other mathematicians to focus on those problems (10 have been solved as of 2010). Hilbert is also well known for formulating the theory of Hilbert spaces, which are key to functional analysis, and did important work in mathematical logic and proof theory. Austrian mathematician Kurt Gödel, best known for his two incompleteness theorems, immigrated to the United States to escape World War II and spent his later years at Princeton University. A group of primarily French mathematicians, including Jean Dieudonne and André Weil, began publishing anonymously under the pseudonym "Nicolas Bourbaki." They are now known as the "Bourbaki Group" or "Association des collaborateurs de Nicolas Bourbaki" and have published several books in which they attempt to ground different areas of mathematics in set theory.

Awards and Honors

There is no Nobel Prize for mathematics but several different international awards are offered that have been termed the "Mathematics Nobel Prize" because of their prestige. The Fields Medal is awarded every four years to one or more mathematicians of age 40 or younger by the International Mathematical Union. Winners of the Fields Medal from western Europe include Laurent Schwartz of France (1950), Jean-Pierre Serre of France (1954), Rene Thom of France (1958), Pierre Deligne of Belgium (1978), Alain Connes of France (1982), Gerd Faltings of Germany (1986), Jean Bourgainof Belgium (1994), Pierre-Louis Lions of France (1994), Jean-Christophe Yoccoz of France (1994), Laurent Lafforgue of France (2002), Wendelin Werner of France (2006), Ngo Bao Chau of Vietnam and France (2010), and Cedric Villani of France (2010).

The Abel Prize, named after Norwegian mathematician Niels Henrik Abel, is awarded annually by the Norwegian Academy of Science and Letters. Western European winners include Jean-Pierre Serre of France (2003), Jacques Tits of Belgium and France (2008), and Mikhail Gromov of Russia and France (2009).

The Wolf Prize is awarded in several fields, including mathematics, by the Wolf Foundation. The first prizes were given in 1978 and it is awarded almost annually, with the possibility of more than one winner in a field in a given year. Western European winners include Carl L. Siegel of Germany (1978), Jean Leray of France (197), André Weil of France and the United States (1979), Henri Cartan of France (1980), Friedrich Hirzebruch of Germany (1988), Mikhail Gromov of Russia and France (1993), Jacques Tits of Belgium and France (1993), Jurgen Moser of Germany and the United States (1994/1995), Jean-Pierre Serre of France (2000), and Pierre Deligne of Belgium (2008).

Western European countries have been regular competitors in the International Mathematical Olympiad, held annually for students younger than 20 who have not yet begun tertiary education. There is both an individual and a team competition. Each country sends six students who are assigned six questions to solve. Countries are compared based on the total score for their team, while individual students may be awarded gold, silver, and bronze medals depending on how many problems they solve correctly. Germany has twice hosted the International Mathematical Olympiad and has participated since 1977.

East Germany also twice hosted the Olympiad and first participated in 1959, the year the Olympiad began. France began competing in 1967 and hosted the competition once. Belgium began participating in 1969. Austria began competing in 1970 and has served once as host. The Netherlands hosted the Olympiad in 2011 and has been competing since 1969. Luxembourg began competing in 1970, Switzerland began competing in 1991, and Liechtenstein began competing in 2005.

Further Reading

Bradley, Robert E., Lawrence A. D'Antonio, and C. Edward Sandifer, eds. *Euler at 300: An Appreciation*. Washington, DC: Mathematical Association of America, 2007.

Hahn, Robert. *Pierre Simon Laplace, 1749–1827: A Determined Scientist*. Cambridge, MA: Harvard University Press, 2005.

Joyce, David E. "The Mathematical Problems of David Hilbert." http://aleph0.clarku.edu/~djoyce/hilbert.

Krantz, Steven G. *An Episodic History of Mathematics: Mathematical Culture Through Problem Solving*. Washington, DC: Mathematical Association of America, 2010.

Mashaal, Maurice. *Bourbaki: A Secret Society of Mathematicians*. Translated by Anna Pierrehumbert. Providence, RI: American Mathematical Society, 2006.

Repcheck, Jack. *Copernicus' Secret: How the Scientific Revolution Began*. New York: Simon & Schuster, 2008.

School of Mathematics and Statistics, St. Andrews University. "The MacTutor History of Mathematics." http://www-groups.dcs.st-andrews.ac.uk/~history.

Segal, Sanford L. *Mathematicians Under the Nazis*. Princeton, NJ: Princeton University Press, 2003.

Sarah Boslaugh

Gerrymandering

Category: Government, Politics, and History.
Fields of Study: Data Analysis and Probability; Geometry; Number and Operations.
Summary: Mathematical algorithms are used in the process of redistricting and to help evaluate whether or not gerrymandering has occurred.

Gerrymandering is a form of political boundary delimitation, or redistricting, in which the boundaries are selected to produce an outcome that is improperly favorable to some group. The name "gerrymander" was first used by the *Boston Gazette* in 1812 to describe the shape of Massachusetts Governor Elbridge Gerry's redistricting plan, in which one district was said to have resembled a salamander. In the United States, congressional and legislative redistricting occurs every 10 years, following the decennial census. The aim of redistricting is to assign voters to equipopulous geographical districts from which they will elect representatives, in order to reflect communities of interest and to improve representation.

Both redistricting and gerrymandering can be characterized as mathematical optimization functions. For good-government redistricting, the optimization function is based on measures of representation and fair political outcomes. These measures may include the number of expected majority-minority districts and the number of competitive districts as well as bias and responsiveness of the expected seats-votes response curve. In contrast, a gerrymander may aim to minimize the number of districts in which a racial or ethnic minority can elect a representative, maximize the number of partisan seats, protect incumbents by creating districts that are not competitive, or obtain some other improper advantage.

Forms of Redistricting

Redistricting is the process of dividing a larger geographical unit into a fixed number of regions (known as districts). The formal aim of redistricting is to create the set of districts that yields the optimal results—as measured by some cost/benefit criteria—while at the same time meeting a set of constraints. The generalized redistricting problem applies to a variety of fields, including the assignment of sales territory; the site selection for warehouses, fire stations, and schools; and the division of political territories into election districts.

In political redistricting, a larger political unit, such as a state, is divided into a number of districts containing roughly equal numbers of people (or, in some jurisdictions, voters). The voters in each district will have the right to elect a fixed number of candidates to represent that district. In addition, the district plan must satisfy various legal requirements such that districts are geographically contiguous; composed from undivided subunits, such as counties, or census blocks; be nonempty; and do not overlap.

Mathematical Representation

In mathematical terms, both redistricting and gerrymandering are readily represented as a type of combinatoric partitioning problem. (The optimization problem is combinatoric because the rules for redistricting typically require that districts be constructed only from whole census blocks.) There are many specific formulations of this problem—all equivalent—including set-partitioning, integer-programming, polygon-dividing, and graph-partitioning.

The law typically requires that each district is contiguous and has roughly equal population. For legislative districts, equal population may be within 10% of the "ideal" population; for congressional districts, only minimal differences are permitted. Thus, a common characterization of the redistricting problem is the weighted graph partition problem: find a partition of the entire graph (for example, state to be redistricted) that induces connected subgraphs (guaranteeing contiguity) of equal node-weight (guaranteeing equal population) and that maximizes some goal function.

The choice of goal function depends on the objectives of the redistricter. For example, a redistricter intent on creating a partisan gerrymander might use a goal function that estimates the expected probability of one party controlling the legislature under a given plan, or alternatively, the expected number of party-controlled seats—a crude estimate of this is the number of districts with party registration over 55%. In contrast, the goal function for a more fair-minded redistricter might be the number of expected competitive seats, the expected bias of the expected seats-vote curve, or another measure of political representation.

Computational Issues

The behavior and characteristics of a district can be readily predicted based on the properties of the units it contains. For example, it is relatively straightforward—using modern statistical modeling and computational methods—to predict the number of seats each party is likely to capture in the next election, given a particular districting plan.

However, although each plan may be easy to evaluate, and the problem of choosing the "best" plan is easy to formulate, actually finding the best plan is extremely difficult. In fact, it is provably "NP-complete." NP-complete problems are generally considered by computer scientists and mathematicians to be computationally intractable. Surprisingly common forms of redistricting to optimize neutral, good-government, or partisan objectives (including compact, contiguous, equipopulous plans, proportionally representative plans, and partisan-seat maximizing plans) are all computationally intractable.

While algorithms exist to solve these problems precisely, reliably, and with certainty, the time required to obtain such a solution grows exponentially as the number of problems grow. Thus, it is impossible to use reliable solution methods for practical problems. Redistricting problems are instead solved computationally using heuristics (problem-solving procedures that provide no guarantees of yielding "good" solutions, although they may produce acceptable solutions in certain circumstances). In other words, when districts are created manually or with a computer, one usually cannot know whether these are the best districts possible.

Distinguishing Gerrymandering and Redistricting

In theory, and in U.S. law, a gerrymander is distinguished from a legitimate redistricting through its effect and the intent of the redistricter. If the intent of the redistricter is to produce an improper outcome and is effective in achieving that outcome, a gerrymander has occurred. In practice—except in more extreme cases—distinguishing gerrymanders from ordinary redistricting is challenging for three reasons. First, although it may seem easy to identify gerrymanders by district shape alone (and many measures of shape "compactness" have been proposed), in fact, none of these measures is related strongly either theoretically or empirically to improper political intent or effect. Politically relevant groups are not uniformly distributed in space. Further, partisanship and demographics are often strongly correlated. For example, members of some parties tend to live in cities, and the poor are often clustered in neighborhoods. As a result, geographical compactness measures that may seem neutral on their face have predictable political biases when applied. Thus both scholars and the courts have declined to accept measures of geographical compactness for gerrymander detection.

Second, it is not feasible to determine the optimal plan for a given objective, or the statistical distribution of possible redistricting plans, because the problem is too difficult to compute. This makes it challenging to determine whether a redistricter intended or achieved maximization of a particular goal, or not. Third, there is generally a lack of consensus on how to measure the various dimensions of political representation. Thus even good-government redistricters may disagree as to the best "goal function" to use when creating a plan. These three issues makes it challenging to use statistical and quantitative methods to determine whether the properties of a proposed plan are extreme, to determine the intent of the redistricter, and to determine whether a particular plan has unambiguously violated representational values.

Further Reading

Altman, M. "A Bayesian Approach to Detecting Electoral Manipulation." *Political Geography* 22, no.1 (2002).
———. "Is Automation the Answer? The Computational Complexity of Automated Redistricting." *Rutgers Computer and Technology Law Journal* 23, no. 1 (1997).
Altman, M., and M. McDonald. "The Promises and Perils of Computers in Redistricting." *Duke Constitutional Law and Policy Journal* 5 (2010).
Butler, D., and B. Cain. *Congressional Redistricting: Comparative and Theoretical Perspectives*. New York: Macmillan Publishing, 1992.
Cortona, P. G. di, C. Manzi, A. Pennisi, F. Ricca, and B. Simeone. *Evaluation and Optimization of Electoral Systems*. New York: Society for Industrial & Applied Math Press, 1999.
Puppe, Clemens and Attlia Tasnadi. "A Computational Approach to Unbiased Districting." *Mathematical and Computer Modeling* 48, nos. 9–10 (November 2008).
———. "Optimal Redistricting Under Geographical Constraints: Why 'Pack and Crack' Does Not Work." *Economics Letter* 105 (2009).

Micah Altman

Greek Mathematics

Category: Government, Politics, and History.
Fields of Study: Algebra; Connections; Geometry; Reasoning and Proof; Representations.
Summary: Greece provided the deductive foundation for many mathematical concepts.

Historians of mathematics and ethnomathematicians have noted that we do not know what all early civilizations did in mathematics. From the evidence that is available, however, it seems that ancient Greece in the late half of the first millennium B.C.E. was the first known civilization to specifically study pure mathematics—mathematics for its own sake, mathematics as aesthetically beautiful. There are occasional examples of pure mathematics in earlier civilizations, notably mathematical proportions in art and design in Egypt and elsewhere, but the earlier peoples used mathematics mostly for practical applications, even if those applications related to religion and art.

Most of the earlier civilizations had subsistence economies, where successful life depended on success in producing food and shelter, so mathematical thinking was used to contribute to these ends. Life was difficult for most people and required full-time concentration, so there was little time for the relaxation that would allow contemplation of mathematical relationships as beauty. However, by 600 and 500 B.C.E., Greece had become prosperous, with strong markets and trade ties around the eastern Mediterranean. There was subsistence work to be done, but the upper-class elite did not have these responsibilities and could devote time to philosophy and learning for its own sake. The trade also brought ideas from other areas, and the open marketplaces encouraged the exchange of ideas and the defense of one's own. These encounters set the stage for studying mathematics beyond the everyday uses and also for the idea of deduction to prove statements.

Early Greek Mathematicians
One of the earliest mathematicians known by name was Thales (624–547 B.C.E.) of Miletus (in modern Turkey). He was an early user of formal deduction in geometry and was known for demonstrating several basic geometric properties: that a diameter bisects a circle, that base angles of an isosceles triangle are equal, and that vertical angles formed by the intersection of lines are equal. He also used angle-side-angle and angle-angle-side triangle congruences and showed that an angle inscribed in a semicircle is always a right angle. In practical geometry, he recognized that the North Star (Polaris) could be used for navigation, and, most impressively, he is said to have predicted a solar eclipse in 585 B.C.E. (though some doubt this). He was also a businessman and bought oil-press mills when his predictions showed a good year for olives.

Pythagoras (572–497 B.C.E.) is more famous, and, for many, more interesting. After traveling as a young man, he settled in Crotona (in what is now southeastern Italy) and gathered followers in a secretive cultlike organization of number worshippers. They believed that whole numbers and ratios of whole numbers are central to everything—numbers rule the universe! They studied geometry, astronomy, and music, but linked all to numbers (including noticing how a plucked string sounds an octave higher when it is half as long, and that other common fractions of the length also make harmonic tones). Their worship led them to the beginnings of number theory as they studied odds and evens, prime numbers, and figurate numbers (numbers of objects arranged into squares, triangles, or other shapes). Some of the questions of number theory that they investigated remain as unsolved problems even in the early twenty-first century.

The most famous mathematics connected with Pythagoras and his group is the theorem of the relationship of the lengths of the sides and hypotenuse of right triangles. Others, notably the Egyptians and the Babylonians, also recognized this relationship, at least in simple cases such as the 3-4-5 triangle for the Egyptians, more such triples for the Babylonians, and, independently, the Chinese. However, the Pythagoreans were probably the first to prove the relationship in general, and hence, in Western mathematics, it is called the Pythagorean Theorem, $a^2 + b^2 = h^2$, where a and b are the lengths of the right triangle legs with the right angle between them, and h is the length of the hypotenuse across from the right angle. This theorem has been described as the first nonobvious theorem of mathematics.

The simplest example of the Pythagorean Theorem is a right triangle with each leg one unit long. This triangle has a hypotenuse of the square root of 2. Unfortunately for the whole-number-worshipping Pythagoreans, the square root of 2 can never be expressed as the ratio of any two whole numbers. Today, it is called an "irrational number," with an infinite, nonrepeating decimal expansion. An irrational number is contrary to the beliefs of the Pythagoreans—such a serious discrepancy that they kept this result secret. More broadly, the issue of irrational numbers caused a crisis in Greek mathematics. Some have even credited this problem to the general shift of Greek mathematics from numbers to a basic geometry that does not use measurement. The geometry of the Greeks became one that allowed figures to be constructed using only a compass and an unmarked straightedge.

Three Construction Problems
Three construction problems challenged the Greeks and many others in later centuries. One was the task of constructing a square with exactly the same area as a given circle—the hope was that this would aid in finding areas of round shapes. This would require finding a way to construct a line $\sqrt{\pi}$ units long. Another was to construct a cube of volume double that of a given cube, which would need a line of length the cube root of 2. The third problem asked for a trisection of a given angle—bisecting an angle was easy, but this asked for the angle to be cut into thirds. The problems were never solved by the Greeks, but their efforts led to interesting insights in geometry. The Greek mathematicians were redeemed in the nineteenth century when all three constructions were proved to be impossible, but there are still some skeptics who erroneously claim to have produced proofs for these constructions.

Deductive Reasoning and Euclid

This geometry and the use of deductive arguments became the standard not only of mathematics but also of clear thinking and logic. Plato's Academy posted a sign that said only those with a knowledge of geometry could enter—deductive geometry was the prerequisite knowledge for philosophy, government, and critical thinking in all areas. Greek civilization greatly expanded under Alexander the Great late in the fourth century B.C.E., reaching as far east as modern Afghanistan and south into Egypt. The city of Alexandria was established at the mouth of the Nile and became a center of trade—and a scholarly center with the construction of the library (also called museum) of Alexandria.

One of the early leaders of the library was Euclid (c. 300 B.C.E.), a mathematician whose life is little known, but his work is one of the most published works in all of mathematics. Probably drawing on the work of earlier scholars, he set up an axiomatic, deductive structure of geometry that became the basis for much future mathematical research. He began with five postulates that mostly drew upon the rules of geometric construction, plus some fundamental obvious truths and some basic definitions. From these, he developed deductive proofs of more geometric properties.

From these early theorems, further deductions eventually led to a "tree" of proven statements, each traceable back to the original theorems. His book, *The Elements*, is said to have been published more than any book except the Bible, and remains the framework for the introductory study of formal geometry even today. His fifth postulate did not come from constructions and defined parallel lines, leading to the difficult use of infinity—noting that parallel lines would not even meet no matter how far they were extended. It seems Euclid himself was worried about the issue of infinity and hesitated using this postulate as long as possible. Two thousand years later, challenges and changes to the fifth postulate would lead to the development of non-Euclidean geometries in the nineteenth century.

Archimedes

Archimedes (287–212 B.C.E.) is often considered the greatest of the ancient Greek mathematicians and one of the greatest in all of history. Unlike many mathematicians, he was recognized even in his lifetime. His achievements are especially notable in that he worked in both pure and applied areas of mathematics. In pure mathematics, Archimedes came close to developing integral calculus more than 1800 years before Newton and Leibniz. He wanted to find ways to calculate areas and volumes of round shapes and used the idea of dividing the shapes into very small slices, much like the similar slices used to integrate areas and volumes in calculus. He found volumes of spheres, cones, and cylinders and discovered an interesting relationship when these shapes have the same diameter and height: the volumes of these special cones, spheres, and cylinders form a 1:2:3 ratio.

Also using calculus-like techniques, he found the value of π by inscribing and circumscribing regular polygons inside and outside a circle and then increasing the number of sides on the polygons so they would close in and estimate the circumference of the circle. He calculated the value of π to be between $3\ 1/7$ and $3\ 10/71$. To help handle large numbers, he greatly expanded the numeration system.

Archimedes lived in Syracuse on the island of Sicily, and his applied work often was related to his life there. He studied the mechanics of simple machines such as levers, pulleys, and screws. He was reputed to have used some of this knowledge to help the king repulse an invasion from the Romans. Once the king asked him to check the authenticity of gold in a crown. He knew he could compare densities of pure gold and an alloy, but to do so, he needed to know the volume of the very irregularly shaped crown. As he entered his bath, he noticed the water level rise to compensate for his own volume; from that he recognized that he could measure the volume of the crown from the amount of water it would displace. The story says he jumped out of the bath and ran through town naked shouting "Eureka!" (I have found it!) in his excitement at the discovery.

Although Archimedes had helped fight off the Romans, they returned when he was an old man. Legend says he refused to leave the geometry he was writing in the sand when a Roman soldier told him to go. At the refusal, the soldier killed him. In some sense, this is symbolic, in that not only did Archimedes die at the hand of a Roman soldier but much of the Greek civilization fell to the expanding Roman Empire. The Romans were good engineers and built a network of roads and aqueducts, but they mostly used existing mathematics and contributed little beyond the work of the Greeks.

Other Greek Mathematicians

However, across the Mediterranean Sea, Alexandria and its library did not fall. Following from Euclid, the Alexandria library continued to be a center for Greek mathematics that would continue even several centuries after the decline of the overall Greek civilization. Some of the work was in astronomy. As early as 200 B.C.E., Eratosthenes calculated the circumference of Earth fairly accurately (incidentally, also indicating that he knew the Earth was round) by comparing the angle of the sun at noon in Alexandria and at Cyrene and using geometrical comparisons to do the calculation.

Later, other Greek astronomers, notably Ptolemy (100–178 C.E.), found more measurements of the movements of the planets. Some of their work led to the erroneous belief that Earth was the center of the solar system, but other studies provided a sound mathematical basis for early astronomical research.

Three other names of mathematicians bring the story of ancient Greek mathematics to a close in the early centuries of the Common Era. Hero (also called Heron) in the first century designed a device that, if constructed, could have been the first steam engine, but it did not get built. He also found a remarkable formula for the area of a random triangle when only the lengths of the three sides (a, b, and c) are given:

$$\text{Area} = s\sqrt{s(s-a)(s-b)(s-c)}$$

$$\text{where } s = \frac{a+b+c}{2}$$

the semiperimeter. Like the Pythagorean Theorem, this formula is considered one of the important early non-obvious theorems and is also useful in practical applications.

Diophantes, who lived in the mid-third century, has sometimes been called the "Father of Algebra." He broke from the Greek interest in geometry and studied numerical problems with techniques that resemble later algebraic methods. He was especially interested in problems whose statements and results were all whole numbers, thus restricting the range of solutions but offering challenges that led to creative work.

Hypatia (370–415) was famous as a mathematics researcher and teacher in Alexandria. Notably, Hypatia is one of the earliest important women mathematicians known in history. Originally taught by her father, who was also a mathematician, Hypatia wrote commentaries and expansions on earlier Greek work, a common type of mathematical research of the time. She was also especially noted as a teacher. However, she inadvertently was caught up in the religious politics of her time and was captured and killed by a mob. Thus, two phases of Greek mathematics ended in tragic deaths: Archimedes at the hands of Roman soldiers approximately marked the end of Greece's Golden Age in mathematics, while the mob killing of Hypatia came near the very end of Greek mathematical work.

Overall, Greek mathematics had continued for nearly 1000 years, providing an unequaled example for future mathematical work. The Greeks did important work in the applied areas but are especially recognized for laying the foundations for pure mathematics.

Further Reading
Boyer, Carl. *A History of Mathematics*. Hoboken, NJ: Wiley, 1991.
Burton, David M. *The History of Mathematics: An Introduction*. New York: McGraw-Hill, 2005.
Eves, Howard. *An Introduction to the History of Mathematics*. New York: Saunders College Publishing, 1990.
Katz, Victor. *A History of Mathematics: An Introduction*. New York: Addison-Wesley, 2008.

Lawrence H. Shirley

Gross Domestic Product (GDP)

Category: Government, Politics, and History.
Fields of Study: Data Analysis and Probability.
Summary: Gross domestic product is a figure used frequently in economics to discuss a country's complete economic output.

Until the Industrial Revolution, population size was a dominant factor in economic output. With the coming of technology, notions of productivity changed. The gross domestic product (GDP) is currently the most widely accepted and broadest indicator of aggregate economic activity.

The GDP represents a country's overall economic output, the dollar value of all final goods and services produced over a period of time within a nation's domestic boundaries. Many assert that the concept of quantifying a nation's economic output can be traced back to newspaper articles written in 1939 by British economist John Maynard Keynes, who was concerned about how Britain would manage its very limited resources at the start of World War II. The Keynesian formula for GDP was the sum of a country's consumption, investment, government spending, and exports, minus its imports.

In the United States, the GDP is calculated and released quarterly by the Department of Commerce. In general, the GDP is used to define emerging economic trends, devise appropriate policies, and gauge the effectiveness of current economic policies. More specifically, corporations use the data to forecast sales and adjust production and investment accordingly. Social scientists monitor the GDP as an indicator of well-being and as a proxy for individuals' voting and investment decisions. In 2003, economists Sir Clive Granger and Robert Engle won a Nobel Prize for their innovative, sophisticated methods of statistical time series analyses that enhance the understanding of market movements and economic trends. In 2010, mathematicians developed an objective quality of life index that uses linear functions and dimensionality reduction to combine four well-studied and widely used indices, including per capita GDP, to produce a relative ranking of countries.

Economists have devised three distinct methods of calculating a nation's GDP. While these approaches derive the same value, each views the GDP differently. The "product method" represents the market value of final goods and services newly produced within a nation during a particular time frame. The "expenditure method" is the national expenditure on goods and services within a specific time frame. The "income method" is the total of wages, rents, dividends, interest, and profits received by producers during a specified time frame. Regardless which method is used, the outcome is referred to as the "nominal" gross domestic product. When the nominal GDP is adjusted for inflation, it is called the "real" gross domestic product. The real GDP is used to measure the growth of a country's economy and real GDP per capita is often used as an indicator of aggregate standard of living.

The Product Approach to Measuring the GDP

The simplest and most direct way to calculate the GDP is the product approach. The product approach calculates the GDP as the market value of final goods and services newly produced within a specific nation. Goods and services produced throughout the year may be classified as either intermediate or final goods. Intermediate goods and services are those that are consumed during the production of other goods and services and are not counted when calculating the GDP; only the final value of a good or service is included in total output. This avoids an issue often called "double counting," in which the total value of a good is included multiple times in national output. The following equation is used to solve for the GDP using the product approach:

$$GDP = P - C$$

where P is the market price of final goods and services and C is intermediate consumption.

The Expenditure Approach to Measuring the GDP

The expenditure approach works on the principle that all of the products must be consumed, therefore the value of the total product must equal the people's total expenditures. The four main components in calculating the GDP via the expenditure method are consumption expenditures by households (C), gross private investment spending (I), government purchases of goods and services (G), and net exports (exports minus imports, $EX - IM$).

The expenditure approach can be represented in the following equation:

$$GDP = C + I + G + (EX - IM).$$

The Income Approach to Measuring the GDP

The income approach to measuring the GDP assumes that expenditures on final goods and services are eventually received by households and corporations as income. A key to calculating the GDP using this method is a concept known as "national income." The national income consists of five types of income: compensation of employees (W), proprietor's income (P), rental income (R), corporate profits (C), and net interest (I). Thus, national income = $W + R + I + P + C$.

Once the national income is calculated, several adjustments must be made before arriving at the GDP.

The first is an adjustment for the taxes paid by businesses to the government (indirect business taxes). Next, depreciation, or the consumption of fixed capital, is taken into account. Finally, the net foreign factor income (*NFI*) is included as an adjustment. The *NFI* is the difference between payments received from the foreign sector and payments made to the foreign sector for domestic production. The *NFI* represents the key difference between gross domestic product and gross national product. The following equation is used to solve for the GDP using the income approach:

GDP = Compensation of Employees + Rent + Interest + Proprietor's Income + Corporate Profits + Indirect business taxes + Depreciation + *NFI*.

Nominal Versus Real GDP

If using GDP to examine production over time, the effects of price increases and inflation must be taken into account. The real GDP is the total value of all goods and services adjusted to eliminate the effects of changing prices. The nominal GDP is calculated by using current market prices. Hence, the real GDP is the value of all goods and services produced by an economy in a given year in dollars of constant purchasing power.

Figure 1. Top 10 Countries by GDP in 2009 (**International Monetary Fund, World Economic Outlook Database).

Country	Year	Units	Scale	GDP
United States	2009	U.S. $	Billions	14,256.28
Japan	2009	U.S. $	Billions	5,068.06
People's Rep. of China	2009	U.S.$	Billions	4,909.28
Germany	2009	U.S. $	Billions	3,352.74
France	2009	U.S. $	Billions	2,675.92
United Kingdom	2009	U.S. $	Billions	2,183.61
Italy	2009	U.S. $	Billions	2,118.26
Brazil	2009	U.S. $	Billions	1,574.04
Spain	2009	U.S. $	Billions	1,464.04
Canada	2009	U.S. $	Billions	1,336.43

Mathematics concepts have also been used in recent years to debate related economic concepts that are rooted in mathematics, such as the principle of comparative advantage. Taken in its simplest form, it states that if two or more countries have already expanded their respective GDPs as far as possible under some set of international trade restraints, they can expand them further by relaxing those restraints. This has been formulated and proven mathematically, using techniques like convex analysis. However, one June 2000 letter to the editor of *SIAM News* (the monthly magazine of the Society for Industrial and Applied Mathematics) argues against such practices for "soft" social science concepts. Motivated by a then-recent protest of global free trade policy, the author stated, ". . . you can criticize the application of a theorem not only by questioning the validity of the hypotheses, but also by questioning the interpretation of the conclusion. … international trade in addictive drugs and guns being only the most glaring and brutal counterexamples to the 'goodness' of increasing GDP."

Further Reading

Abel, Andrew, and Ben Bernanke. *Macroeconomics.* 5th ed. Upper Saddle River, NJ: Pearson Addison Wesley, 2005.

Baumohl, Bernard. *The Secrets of Economic Indicators: Hidden Clues to Future Economic Trends and Investment Opportunities.* 2nd ed. Upper Saddle River, NJ: Pearson Education, 2008.

Eisen, Peter. *Economics: Barron's Business Review Series.* Hauppauge, NY: Barron's Educational Series, 2000.

Frumkin, Norman. *Guide to Economic Indicators.* 4th ed. Armonk, NY: M. E. Sharpe, 2000.

Moss, David. *A Concise Guide to Macroeconomics.* Cambridge, MA: Harvard Business School Press, 2007.

Kristi L. Stringer
Casey Borch

Home Buying

Category: Business, Economics, and Marketing.
Fields of Study: Algebra; Data Analysis and Probability; Number and Operations; Problem Solving.
Summary: Interest rates on mortgages are set using sophisticated mathematical techniques.

Homes are the largest single purchase most people make in their lifetimes. Buyers usually take out a home loan, called a "mortgage," rather than pay cash. When the desired property is identified and funds for a down payment (typically 20%) are acquired, home buyers work with a bank or other lender to finance the purchase. When determining what a person can afford to borrow, lenders consider several variables. In the past, these judgments were often highly subjective decisions made by individual lender agents, but the increased popularity and availability of credit cards in the latter half of the twentieth century, as well as federal legislation designed to combat discrimination in lending, required more objective methods of assessment.

For example, FICO scores mathematically measure risk of nonpayment. Created by Fair Isaac Company (FICO), founded by engineer Bill Fair and mathematician Earl Isaac, an individual's FICO score is a weighted combination of variables such as previous credit performance, current debt, and length of credit history. Also, lenders may use debt-to-income ratios to indicate the size of payment a borrower can afford. Comparing home loans can be challenging because different lenders may use this information differently. Also, interest rates, closing costs, and additions to the base payments need to be considered in the comparison. The process of buying a house may involve additional expenditures beyond the mortgage. Buyers routinely hire a home inspector to independently assess the condition of the home, and many such inspectors charge a fee based on the square footage. In some areas, radon tests or soil analyses might be required. If problems are found, either the buyer or the seller may have to hire a structural engineer or other professional to rectify the problem before an agreement is reached or the loan approved. Property taxes, based on the assessed value of the home and land, and homeowners insurance, which is also a function of the assessed value and replacement cost of the home and its contents, are also part of almost all home-buying transactions. Home buying tax credits or reduced interest rates may offer the buyer additional options and are designed to stimulate the economy.

FICO Scores and Credit Ratings

Each of the three major credit bureaus (Experian, TransUnion, and Equifax) calculate a credit score based upon advice from the Fair Isaac Company, an independent company that specializes in business analysis, including risk assessment. However, not all three companies use identical inputs, and each may yield a different result. A FICO score is between 300 and 850, with higher scores indicating better risks. The exact formula used for FICO score calculation is proprietary and changes periodically, but the personal data incorporated in the FICO formula include, in order of importance, payment history; amounts owed; length of credit history; new credit applied for; and types of credit used. The FICO score influences requests for credit, and many banks charge higher interest rates to people with lower scores.

Debt-to-Income Ratio

Most standard loan applications request information on both income (annual income, bank account balances, investments such as stocks and bonds) and debt (amount owed on loans, credit cards, and standard monthly bills like car payments, utilities, insurance). This information can be used to determine the percentage of income already committed to be spent each year: debt-to-income ratio = total debt ÷ total income.

If a prospective borrower's expected debt-to-income ratio is higher than the bank's cutoff (most banks have a limit between 32% and 40%), a loan may be denied, particularly if the prospective borrower's FICO score is low. A high FICO score might result in the prospective borrower being granted the loan even with a higher debt-to-income ratio.

Calculating a Mortgage Payment

The principal (amount to be borrowed), the annual interest rate, the payment schedule (for example, monthly or bimonthly), and the length of the loan (10, 15, 30 years, for instance) all factor into the payment. The formula for the payment (R) is given below, where P is the principal, r is the adjusted interest rate (the annual rate divided by the number of payments in one year), and n is the number of payments to be made over the life of the loan

$$R = \frac{rP}{\left(1-\left(1+r\right)^{-n}\right)}.$$

Additions to this base payment include the following:

- PMI: Borrowers paying less than 20% down may be required to purchase private

Table 1.

Payment #	Interest Owed	Payment	Principal Paid	Balance
Closing				$100,000.00
1	$500.00	$599.55	$99.55	$99,900.45
2	$499.50	$599.55	$100.05	$99,800.40
3	$497.99	$599.55	$101.56	$99,497.24
...	...	...	...	...
358	$8.90	$599.55	$590.65	$1,190.17
359	$5.95	$599.55	$593.60	$596.57
360	$2.98	$599.55	$596.57	$(0.00)

mortgage insurance (PMI) to protect the lender's investment. The cost of this insurance is added to the loan payment.
- Escrow: Many banks require that payments be made into an escrow account to accrue funds to pay property tax and insurance.

Finalizing a Mortgage Loan

Transaction fees for processing a mortgage are more commonly called "closing costs"; hence finalizing a home loan is called "closing." Charges to be paid when closing on a loan typically include an origination fee for the lender, appraisal fee for the appraiser, title search and recording fees for the attorney, and points. Points are up-front interest fees charged by the lender, with one point costing 1% of the principal. Banks often give borrowers the option of purchasing additional points at closing to decrease the interest rate on a mortgage.

The reduction of interest for purchasing a point can vary from bank to bank; one point may reduce the rate by as little as 0.1% or as much as 0.25%. The cost of the points purchased at closing is tax deductible, as is any interest paid over the life of the mortgage.

Amortization of Loans and Fixed Rates

Loan payments include a portion that reduces the principal balance and a portion that the lender keeps—the interest. The amount of interest included in a payment varies over the life of the loan, but can be determined by remembering that each payment includes "simple interest payable on the balance." Calculating the schedule of payments, including the split between principal and interest for each payment, is referred to as "amortizing." The Latin roots of the term mean "death pledge," indicating linguistically the willingness to forfeit something of great value if the debt is not paid. In this case, failure to pay the debt results in foreclosure by the bank and the loss of the property.

To illustrate, suppose a home buyer borrows $100,000 at 6% fixed annual interest (the interest rate does not change over the life of the loan) payable monthly for 30 years (360 payments). Using the loan formula, the monthly payment is $599.55, assuming no PMI or escrow. Since the homebuyer is paying 6% annual interest and 12 payments a year, the adjusted (monthly) interest rate is 0.5% for all payments.

Thus, the homebuyer owes the lender 0.5% of $100,000 in the very first payment; $500 will be kept by the lender as interest and $99.55 will be used to reduce the principal. For the next payment the balance is $99,900.45; the interest will be 0.5% of that balance or $499.50. These calculations can be summarized in an amortization table, which is usually provided to the buyer as part of the mortgage agreement. The first and last several rows for this example are presented in Table 1. As the loan progresses, the interest portion decreases and the remaining amount applied to the principal increases.

Variable Rates

The example uses a "fixed" interest rate, but many lenders also offer variable rate loans, meaning that the interest rate may be changed according to some economic indicator (called the "index"), such as the prime rate. There are legal restrictions on this practice: the lender must inform the borrower of the size and frequency of such changes (called the "interval"), and the maximum (called the "cap") for the rate. For example, a two-year adjustable rate mortgage (ARM) payable monthly over 30 years might have the following particulars: 6% to start indexed on the 6-month U.S. Treasury Bill, and adjustments of at most 2% are allowed every two years with a cap of 10%.

The initial calculations, such as down payments or the payment amount, do not change in this case. For this two-year ARM, the first two years of the amortization table above does not change. One important question to ask the lender when considering an adjustable rate loan is what happens to the payment when the interest changes? Most redo the amortization calculations starting at the next payment, so a new rate would mean a new payment amount.

The housing crisis of 2007–2009 resulted in many homeowners finding themselves with homes whose values had decreased to the point that they were worth less than the amount owners still owed on their mortgages. There were many factors that influenced this outcome. Prior to the Great Depression, home ownership was much more rare than in the early twenty-first century when homes were often financed with balloon-payment mortgages in which a loan is amortized over only part of its lifetime, leaving a large principal payment due at the end. The federal push to open the housing market using fully amortized, fixed-interest mortgages required lenders to assume much greater financial risk, which can be mathematically modeled but not perfectly predicted. To manage that risk, mortgages became financial commodities in the larger financial marketplace. Housing prices, interest rates, and other aspects of financial markets are highly variable, and some people blamed the housing crisis on too much reliance on sophisticated mathematics.

In general, it was probably not the models themselves but the sometimes-incorrect ways in which the models were often used. In addition, many lenders ignored reliable risk predictors, such as FICO scores and debt-to-income ratios, resulting in more people taking on higher loan payments than they could afford. Home prices rose from demand to the point where properties were extremely overvalued. They later decreased in value, so the property was worth much less than the balance on the loan, leading to a large increase in foreclosures. Homeowners defaulted on loans, ruining their credit ratings, and banks paid large foreclosure fees. There were also more short sales, where banks agreed to accept less than the mortgage balances when homes sold to avoid foreclosure charges and poor credit ratings for the homeowners.

Further Reading

Johnson, Tim. "Paying the Price." *Plus Magazine* (July 14, 2009). http://plus.maths.org/content/paying-price.

Perry, Timothy, and Daniel Prouty. *The Book of Home Purchase: Make Quick, Simple On-Site Room-to-Room Calculations of Repair and Replacement Costs.* Bergamo, Italy: Bergamo Publications, 1998.

Shestopaloff, Yuri. *Mortgages and Annuities: Mathematical Foundations and Computational Algorithms.* Toronto: AKVY Press, 2009.

Holly Hirst

Incan and Mayan Mathematics

Category: Government, Politics, and History.
Fields of Study: Connections; Measurement; Number and Operations; Representations.
Summary: The Incan and Mayan civilizations had a variety of mathematical achievements, including number systems and calendars.

The Inca Empire existed from 1438 until 1533 c.e., when it was conquered by the Spanish and the last Inca emperor, Atahualpa, was murdered. At its height, the Inca Empire comprised most of present-day Peru, Bolivia, and Ecuador, as well as parts of Colombia, Chile, and Argentina. It was a culturally diverse but politically centralized empire, based in the capital of Cuzco. Having no written words, the Incas invented a clever method of recording numbers, usually for administrative purposes, using knotted cords called a *quipu*.

The Maya civilization flourished between 250 and 900 c.e. The homeland of the Mayans was the Greater Yucatan Peninsula, including present-day Guatemala and Belize, as well as parts of Mexico, Honduras, and El Salvador. In contrast to the Inca Empire, the Maya civilization was never a political entity but consisted of a multitude of independent city-states. Among the many remarkable accomplishments of Mayan culture were hieroglyphic writing, a vigesimal and duodevigesimal number system, the invention of a symbol for zero, an elaborate system of calendars, and highly accurate astronomical observations.

Incan Quipus

A quipu is a bundle of colored, knotted cords. Every quipu has a main cord that is thicker than the others. Pendant cords are tied to the main cord, and subsidiary cords are tied to pendant cords or other subsidiaries. Quipus have been found with as many as 2000 pendants and six levels of subsidiaries. The pendant and subsidiary cords carry knots. Three types of knots are used: simple knots, figure-eight knots, and long knots with two to nine turns. To record numbers, the Incas used a decimal number system. Each digit other than the units is represented by a cluster of the appropriate number of simple knots. The Incas did not have a special knot for zero but simply left an empty space on the cord.

Units are represented by a long knot with the appropriate number of turns. If the unit is one, however, a figure-eight knot is used, since a long knot with only one turn is identical to a simple knot. For example, the number 701 is represented by a cluster of seven simple knots, an empty space, and a figure-eight knot. The digits are ordered with the units away from the main cord. Since the units are distinguished from the other digits, the same cord can carry several numbers. The colors of the cords and the topology of pendants and subsidiaries do not contribute to the numerical information but signify the item that is being counted. There are about 800 quipus in museums today. The largest number found on a quipu is 97,357.

The Mayas were known for building complex and highly decorated ceremonial structures, including temple-pyramids, palaces, and observatories, all contructed without the use of metal tools. (Photos.com)

Quipus are not suitable for performing arithmetic. In 1590, Spanish Jesuit missionary José de Acosta described how the Incas carried out difficult computations by moving around maize kernels. A Peruvian drawing from about 1615 shows a tablet, called a *yupana*, that might have been used for this purpose. This yupana is divided into smaller squares, each containing 1, 2, 3, or 5 dots, which could be maize kernels. Acosta explicitly mentioned the numbers 1, 3, and 8. This has led to speculations that the Incas used so-called Fibonacci numbers in their calculations since 1, 2, 3, 5, and 8 are the first such numbers.

Mayan Numbers and the Invention of Zero

The Mayan number system is neither a pure grouping system, like Roman or Aztec numbers, nor a pure positional system, like Hindu–Arabic numbers, but a mixture of the two, like Babylonian or Incan numbers. Numbers from 0 to 19 are written with dots representing 1, lines representing 5, and a symbol for 0 resembling an eye. Thus, 17 is written as two dots and three lines. For numbers larger than 19, a base-20 and, at one place, a base-18 positional system is used. The first place represents units, and the second place represents multiples of 20.

The third place, however, does not represent multiples of $20 \times 20 = 400$ but multiples of $18 \times 20 = 360$. From then on, the fourth place represents multiples of $20 \times 360 = 7200$, the fifth place multiples of $20 \times 7200 = 144,000$, and so on. Mayan numbers were originally written vertically with the units at the bottom. For convenience, Mayanists write them horizontally with the units to the right. Thus, the Mayan number 9.12.11.5.18 means the following:

$$9 \times 144,000 + 12 \times 7200 + 11 \times 360 + 5 \times 20 + 18$$
$$= 1,386,478.$$

After the Babylonians, the Mayas or possibly their Olmec predecessors were the first culture in the world to invent a symbol for zero. The earliest known occurrence of this zero symbol is found on a stela in Uaxactun, Guatemala (357 C.E.). The earliest indisputable inscription using the Hindu–Arabic decimal system including a symbol for zero is from Cambodia (683 C.E.).

Mayan Calendars

The Mayas used three different calendars: the Tzolkin, the Haab, and the Long Count. A typical Mayan date looks like the following:

9.12.11.5.18 6 Etznab 11 Yax.

Here, "9.12.11.5.18" is the Long Count date, "6 Etznab" is the Tzolkin date, and "11 Yax" is the Haab date. This was the day of death of the great ruler, Pacal, of the city-state, Palenque, corresponding to August 29, 683 C.E.

The Tzolkin calendar is based on two independent cycles of 13 and 20 days, respectively. A Tzolkin date consists of a number from 1 to 13 followed by one of the following 20 names of days:

Ahau	*Kan*	*Lamat*	*Eb*	*Cib*
Imix	*Chicchan*	*Muluc*	*Ben*	*Caban*
Ik	*Cimi*	*Oc*	*Ix*	*Etznab*
Akbal	*Manik*	*Chuen*	*Men*	*Cauac*

Both the number and the day name change daily such that the calendar runs as follows: 1 Ahau, 2 Imix, 3 Ik, and so forth. Every possible Tzolkin date occurs once during the Tzolkin year of $13 \times 20 = 230$ days. This follows from the so-called Chinese Remainder Theorem, which the Mayas must have known at least in some special cases, and the fact that 13 and 20 have no common divisors.

The Haab calendar consists of 18 months of 20 days, followed by five extra days. The length of the Haab year is thus $18 \times 20 + 5 = 365$ days. The names of the months are the following:

Pop	*Tzec*	*Chen*	*Mac*	*Kayab*
Uo	*Xul*	*Yax*	*Kankin*	*Cumku*
Zip	*Yaxkin*	*Zac*	*Muan*	
Zotz	*Mol*	*Ceh*	*Pax*	

The days of each Haab month are numbered from 0 to 19. The Haab calendar thus runs as follows: 0 Pop, 1 Pop, 2 Pop, and so forth. The final five days, called *Uayeb*, are numbered from 0 to 4; these days were considered unlucky.

The least common multiple of 260 and 365 is $73 \times 260 = 52 \times 365 = 18,980$, which means that the combined Tzolkin–Haab calendar repeats itself after 73 Tzolkin years, or 52 Haab years, or 18,980 days.

The Mayas believed in a cycle of eras of $13 \times 144{,}000$ days or approximately 5125 years, each era ending with a time of great change. A Long Count date is a five-digit Mayan number recording how many days have elapsed since the last transition of cycles. There is a unique correspondence between the last digit of the Long Count date and the Tzolkin day name. If the last digit is 0, the day name is Ahau; if the last digit is 1, the day name is Imix, and so forth. According to various Mayan sources, the previous era ended on the following date:

13.0.0.0.0 4 Ahau 8 Cumku.

The problem of translating Long Count dates into dates in the Gregorian calendar is known as the Correlation Problem and has been a topic of considerable controversy. Today, most Mayanists believe that 13.0.0.0.0 4 Ahau 8 Cumku corresponds to August 11, 3114 B.C.E. The Mayans thus expected the next cycle change upheaval to occur on 13.0.0.0.0 4 Ahau 3 Kankin, corresponding to December 21, 2012 C.E., when the present Long Count cycle ends.

Mayan Astronomy and the Dresden Codex

The Dresden Codex is one of only four original Mayan books that have survived to the present day. It contains astronomical tables in which the number 584 figures prominently; this is the best integer approximation to the average period of Venus, as seen from the Earth, of 583.92 days. In the Codex, 584 is divided into parts of 236, 90, 250, and 8, reflecting the phases of Venus.

First Venus appears as the Morning Star for 236 days, then it disappears on the far side of the sun for 90 days, then it reappears as the Evening Star for 250 days, and finally it disappears again for eight days while it is between the Earth and the sun. The difference between 90 and 8 is explained by the fact that, as seen from the Earth, Venus moves more slowly relative to the sun when it is on the far side of the sun. The difference between 236 and 250 is thought be because of a local difference between the eastern and western horizons.

It is a strange coincidence that $584 = 8 \times 73$ and $365 = 5 \times 73$ have the large common prime factor of 73. This implies that five Venus periods correspond very closely to eight Haab years, and indeed the Codex contains a Venus table of this length of time. The Mayas knew, however, that this correspondence was not exact. To compensate, they subtracted either four days after days, giving a period of 583.93 days, or eight days after days, giving a period of 583.86 days.

It has been suggested that the Mayas used the first correction four times and the second correction once, thus subtracting a total of 24 days after 301×584 days, which gives a Venus period of exactly 583.92 days. This explanation, however, was questioned by the famous physicist, Nobel laureate, and amateur Mayanist Richard Feynman.

Further Reading

Coe, Michael D. *Breaking the Maya Code*. New York: Thames & Hudson, 1992.

Feynman, Richard. *Surely You're Joking, Mr. Feynman!* New York: Vintage, 1992.

O'Connor, J. J., and E. F. Robertson. "Mactutor History of Mathematics Archive: Mayan Mathematics." http://www-history.mcs.st-and.ac.uk/HistTopics/Mayan_mathematics.html

Teresi, Dick. *Lost Discoveries: The Ancient Roots of Modern Science—From the Babylonians to the Maya*. New York: Simon & Schuster, 2002.

David Brink

Income Tax

Category: Government, Politics, and History.
Fields of Study: Measurement; Number and Operations.
Summary: Mathematics is used to compute income tax returns and analyze income-tax fraud.

Albert Einstein once quipped that preparing a tax return was an activity too difficult for a mathematician and was better suited for a philosopher. Many would point to the complex and ever-changing laws regarding taxation, rather than the underlying mathematical concepts, as being the problematic part of understanding income taxes.

Numbers and their operations, along with algebra, are very useful in the calculation of the taxes owed by individuals and corporations. In addition, probability, statistics, and geometry are among the fields used by those interested in the analysis of the process and outcomes of taxation, such as tax irregularities and evasions, tax burden, and the effects of taxation on overall economic welfare.

History

In 1861, the U.S. Congress imposed a tax on personal incomes to help finance the Civil War. Prior to that time, it had depended mainly on excise taxes and customs duties. The first income tax was a proportional (or flat) tax: anyone who made an income of more than $800 per year had to pay a fixed 3% of that income in taxes. The next year, a two-tiered progressive rate structure was put into place. Taxable incomes up to $10,000 were still taxed at 3%, while higher incomes paid 5%, though people were allowed to take various deductions from their incomes before calculating the tax. Taxes were also withheld by employers for the first time. This tiered taxation method became the standard for income tax, although some countries in the early twenty-first century use a dual income tax system in which individuals and corporations are taxed at a low rate while labor income is taxed at a high rate.

Income taxes were abolished in 1872; but after a great deal of legal debate, they returned permanently with the passage of the Sixteenth Amendment in 1913. Everyone who earns income in the United States is subject to federal individual income tax and, in most cases, state income tax as well. Some municipalities also charge local income tax. Employers are required to withhold money from taxpayers' paychecks and to remit the funds to the appropriate government agencies. Self-employed taxpayers are required to submit quarterly payments.

Calculating the Income Tax

As can be seen from the Internal Revenue Service 1040 individual income tax form, a series of calculations are required to determine the amount of income tax owed.

Total Income: All sources of taxable income must be added to calculate total income, including not just wages but also funds accrued from sources such as tips, interest earned, alimony, capital gains, retirement withdrawals, royalties, and business income.

Adjusted Gross Income: Certain types of expenses can be subtracted from the total income, including some expenses related to moving, business, education, alimony paid, self-employment, and student loans. After subtracting the allowable expenses, the result is the adjusted gross income.

Taxable Income: Additional deductions and exemptions are subtracted from the adjusted gross income to arrive at the taxable income.

- *Deductions*: Taxpayers can elect to take the standard deduction, which is a set amount depending on filing status, or they can itemize their deductions to see if a tally of the allowable deductions results in more than the standard amount. People who paid mortgage interest, significant medical costs, large charitable donations, and/or business expenses will often find that itemizing produces a larger deduction than the standard.
- *Exemptions*: The federal government allows taxpayers to deduct a fixed amount for each dependent in the household; in 2009 that amount was $3,650 per dependent.

Tax owed: The tax is then looked up in the tax table, reading the appropriate column depending on filing status (single, married filing jointly, married filing separately, head of household), unless the taxable income is over $100,000, in which case a tax computation worksheet is used. In general, single people pay more taxes than married couples filing jointly with the same income.

Understanding the Federal Tax Tables and the Tax Computation Worksheet

The government defines a series of tax brackets, which are percentages linked to income ranges. The income ranges for a specific tax bracket vary depending on the filing status of the taxpayer.

The federal government sets different ranges for the following categories: single, married filing jointly, married filing separately, and head of household. In 2009, for example, the tax brackets were 10%, 15%, 25%, 28%, 33%, and 35%. The range for a single tax payer in the 10% bracket was $0 to $8,350 in taxable income. For a married couple filing a joint return, the income range for the 10% bracket was $0 to $16,700.

The tax table and tax computation worksheet values do not correspond directly to the tax brackets. For example, a single person earning $62,025 in 2009 would appear to fall into the 25% tax bracket ($33,950 to $82,250). However the tax shown in the tax table is $11,694, which is less than $17,250 (25% of $69,000). The tax table value was determined by applying the tax brackets to the taxable income in stages. In 2009, the tax brackets for a single taxpayer were

- 10% bracket: $0 to $8,350
- 15% bracket: $8,350 to $33,950
- 25% bracket: $33,950 to $82,250

The first $8,350 of the taxable income earned falls into the 10% bracket, yielding $835 in taxes. The next $25,600 ($33,980–$8,350) of the taxable income falls into the 15% bracket, yielding $3,840. The last $28,075 ($62,025–$33,950) falls into the 25% tax bracket, yielding $7,018.75.

The total tax is $835 + $3,840 + $7,018.75 = $11,694 (rounded to the nearest dollar).

The tax tables are provided in $50 increments, so anyone earning between $62,000 and $62,050 would pay the same amount of tax.

The tax computation worksheet calculations work the same way. For a single person with a taxable income of $130,000 (28% tax bracket), the worksheet calculation is to multiply by 0.28 and then subtract $6,280. The $6,280 figure is subtracted to compensate for the lower taxes paid on the portions of the $130,000 income that fall into the lower tax brackets.

Other Methods of Calculating Taxes

Some groups are concerned that the federal tax code is too complicated, confusing, and unfair. There are those who advocate simplifying the tax code and leaving the graduated tax bracket structure, and others who advocate a flat tax—one percentage rate for all with no exemptions or deductions.

Most states follow the federal government's lead and have a series of tax brackets. In 2009, Colorado, Illinois, Indiana, Massachusetts, Michigan, Pennsylvania, and Utah all had flat taxes ranging from 3 percent to 6 percent. Alaska, Florida, Nevada, South Dakota, Texas, Washington, and Wyoming did not collect any individual income tax.

Mathematical Modeling

The impact of taxation is of great personal and political concern. Income taxes, in particular, can generate a great deal of debate, and many people feel personally and directly affected by changes in these taxes. Mathematical methods are used to model a variety of phenomena related to taxes. For example, equilibrium modeling seeks to explain and predict the broad economic repercussions of different market factors, including taxes.

These complex models take into account the flow of cash, commodities, and other goods between various people and businesses, which have different motivations and constraints. Other potential variables can include prices, interest rates, and taxes. A system (like the U.S. economy) is in equilibrium when the inflows and outflows, or supply and demand, are balanced. These models are computationally intense and generally solved using numerical methods, graph theory, geometry, and stochastic simulation.

Several countries and U.S. states, as well as companies and accounting firms, use software based on Benford's Law to check income tax returns for fraud. Benford's Law is named for engineer and physicist Frank Benford. According to stories about Benford, he was inspired by the fact that pages of logarithm books associated with numbers starting with the digit 1 were dirtier and more worn than other pages. Thinking that it was unlikely that scientists had some special preference for these numbers, he analyzed over 20,000 sets of data from a wide variety of sources, such as baseball statistics, numbers he found in magazine articles, and atomic weights.

All of these data sets followed a similar pattern in terms of the first digits of the numbers. About 30% of the time, the first digit of the numbers was a 1. Each subsequent numeral 2 through 9 occurred less and less often as the initial digit, such that the probability of any number n from 1 through 9 being the first digit is the following:

$$\log\left(1+\frac{1}{n}\right).$$

One simple way that data can be tested is by comparing the observed first-digit counts to Benford's Law. For example, accountant Mark Nigrini examined 169,662 IRS files and found that they follow Benford's Law, with an allowable statistical margin of error. Former president Bill Clinton and (as of 2010) Secretary of State Hillary Rodham Clinton's tax returns for several years were also analyzed.

Nigrini concluded that the Clintons may have used some rounded-off dollar estimates rather than exact numbers, but his test did not uncover any fraud. Generally, studies show that fraudulent data contain too few numbers starting with 1 and too many starting with 6.

Further Reading

Fu, Michael, Robert Jarrow, Ju-Yi Yen, and Robert Elliott. *Advances in Mathematical Finance*. Basel, Switzerland: Birkhäuser, 2007.

Nievergelt, Yves. *A Graphic Introduction to Functions: The Federal Income-Tax Law*. Bedford, MA: COMAP, 1989.

Radulescu, Doina Maria. *CGE Models and Capital Income Tax Reforms: The Case of a Dual Income Tax for Germany* (Lecture Notes in Economics and Mathematical Systems). Berlin: Springer, 2007.

Woytek, Steve. *Mathematics in Living, Credit, Loans and Taxes Book IV*. Boulder, CO: Pruett Publishing, 1976.

Holly Hirst

Industrial Revolution

Category: Business, Economics, and Marketing.
Fields of Study: Measurement; Number and Operations; Problem Solving.
Summary: New energy sources, management styles, and more intensely divided labor revolutionized manufacturing and technology.

The term "Industrial Revolution" refers to the great social transformation, beginning in the mid-eighteenth century, during which manufacturing replaced agriculture as the center of productive activity. This transition had profound implications for economic and political institutions and international relations, as well as for the landscape and environment, family, education, and culture. Its two main dimensions were technological innovation and the social organization of production. The Industrial Revolution was facilitated by the increased use of realistic perspectives in painting and drawing that flourished in the Renaissance, as well as by the invention of the printing press in the fifteenth century, which spurred intellectual growth in many fields, including mathematics. These developments allowed for better visual representation and distribution of mathematical ideas and inventions to a much broader audience than the older master-apprentice models.

Characteristics

Some historians question the use of the term "revolution," since these developments indisputably occurred incrementally over a period of a century or more. Nonetheless, their cumulative impact dramatically changed virtually every aspect of life, first in Great Britain and eventually worldwide. New technologies both drew on existing mathematics and prompted its further development. New institutions of intellectual life also fostered the emergence of increasingly abstract mathematics.

The key technological feature of the Industrial Revolution was the application of new sources of power: first the steam engine (late eighteenth century), and later electricity and the internal combustion engine (late nineteenth century). As the Industrial Revolution spread in the late twentieth century, nuclear energy and emerging "green energy" sources have been developed. A crucial problem of the early Industrial Revolution was the means of transmitting power from the steam engine to the machines used in production itself. This problem gave rise to the mathematical theory of linkages.

Equally important to the Industrial Revolution was the large-scale organization of labor. In England, the Enclosure Acts (1760–1845) forced small farmers into urban areas, while vagrancy laws, poor laws, and workhouses (places where those who were not able to support themselves could seek shelter and employment) instilled labor discipline. A large labor pool was thus created for the new factories. Market competition impelled factory owners to use the cheapest possible labor—children as young as 5 as well as adult women and men—and to maximize profits by extending the working day to 14 hours or more per day, seven days per week.

The vastly larger scale of production made possible by mechanization and the steam engine created a qualitatively distinct industrial organization of labor. It intensified the division of labor, de-skilling some jobs and creating new forms of specialization.

The Industrial Revolution therefore meant profound changes in work, residence patterns, family relations, and urban life. This in turn sparked interest in social statistics. Edwin Chadwick (1800–1890) and Friedrich Engels (1820–1895) pioneered the use of quantitative measures to describe social problems. Belgian mathematician Adolphe Quetelet applied the statistical techniques previously used in astronomy to social problems, further developing them and helping to institutionalize the discipline of statistics.

James Watt and the Steam Engine

James Watt (1736–1819), the grandson of a mathematics teacher, possessed the combination of manual dexterity and an aptitude for mathematics. He trained as a maker of mathematical instruments, securing a position at the University of Glasgow, a major center of the British Industrial Revolution, where he first encountered the inventive yet inefficient Newcomen steam engine. While the Newcomen engine served to pump water from coal mines, Watt's improvements turned the steam engine into a practical means of supplying power to factories and of transporting manufactured goods to market.

James Watt's parallel motion mechanism (1804), in particular, allowed the force of an engine to act in both push and pull directions, converting rotary motion to linear motion. This provided an empirical, though imprecise, solution to the geometrical problem of constructing a straight line without tracing a straight line. In Euclidean geometry, it is axiomatic that a straight line can be produced, but—in contrast to the circle—no method existed to do so.

Following Watt, a spatial linkage that traced exact straight lines was created by mathematician Pierre-Frederic Sarrus in 1853 and proved geometrically by Charles-Nicolas Peaucellier in 1864. The mathematical theory of linkages was further developed by Pafnuty Chebyshev, James Joseph Sylvester, Alfred Kempe, and Arthur Cayley.

Mathematics and the Industrial Revolution

The late eighteenth and early nineteenth centuries were extremely fruitful in the development of modern mathematics. However, the connections between this work and the Industrial Revolution are mainly indirect.

A notable exception was Charles Babbage (1791–1891) and his work on some of the earliest computing machines. Numerical tables used in applied mathematics were calculated by hand and often contained many errors. Babbage sought to replace these human "computers" with machines, as so many manufacturing jobs were being mechanized. He began work on his first "difference engine" in 1822, moved on to a programmable "analytical engine," and continued experimenting with steam-powered computing machines for much of the rest of his life. Ada Lovelace, generally credited as the first computer programmer, created a program that could have run on Babbage's machine, had it been built.

Some technical problems that arose in connection with the Industrial Revolution proved amenable to solution via abstract mathematics developed in other contexts. For example, analysis of electrical circuits, waves, and oscillations is simplified by using complex numbers, originally explored in relation to the solution of algebraic equations.

In France, the École Polytechnique, founded by mathematicians Lazare Carnot and Gaspard Monge in 1794 to train military engineers, supplied technical training and expertise for emerging French industries. Its faculty, students, and examiners included many of the most influential French mathematicians of the nineteenth century, and its textbooks, such as the calculus texts of Adrien-Marie Legendre and Sylvestre-Francois Lecroix, influenced mathematics instruction internationally.

Watt and Horsepower

As Watt marketed his steam engines, he developed the standard unit of "horsepower" to demonstrate the superiority of his product over the horses traditionally used to power a mill wheel. Based on his observations, he calculated that one horsepower was equal to approximately 33,000 ft-lb/min. The "watt," which came into use as a unit of power in the late nineteenth century, was named for James Watt.

Further Reading

Musson, A. E., and Eric Robinson. *Science and Technology in the Industrial Revolution*. New York: Gordon & Breach, 1989.

Sangwin, Christopher. "Revisiting James Watt's Linkage with Implicit Functions and Modern Techniques." *Mathematics Magazine* 81, no. 2 (2008).

Weightman, Gavin. *The Industrial Revolutionaries: The Making of the Modern World 1776–1914*. New York: Grove Press, 2010.

Bonnie Ellen Blustein

Infantry (Aerial and Ground Movements)

Category: Government, Politics, and History.
Fields of Study: Data Analysis and Probability; Geometry.
Summary: Mathematics has long played a significant role in infantry operations, including influencing cryptography, logistics, and military strategy.

The oldest military unit and still the backbone of most modern armies, infantry units consist of soldiers who engage the enemy face-to-face. Historically, infantry units marched from one location to another. In modern times, infantry units may be deployed in a variety of ways, including overland in trucks; by sea, such as the troops landing on Omaha Beach on D-Day; or by air, either from planes or helicopters. Paratroopers are often considered elite among infantry units. In general, infantry are distinct from other land-based mobile units, such as cavalry, employing different tactics and strategies.

Mathematics has always played a major role in warfare, including infantry movements. Early Babylonian clay tablets show evidence of sophisticated mathematical calculations of the volume of dirt that would be needed for siege ramps and what sort of minimum manpower would be required to accomplish the task. The sophistication of mathematics in ancient Greece was no doubt in part because of its usefulness to war—the Greeks may have left a legacy of philosophy and art but spent much of their time and resources at war among themselves and with their neighbors.

Napoleon Bonaparte is widely considered to be a military genius who revolutionized the use of light infantry and artillery. He was also an avid mathematics student and was often accompanied in the battlefield by mathematicians, including Joseph Fourier. He discussed his own solutions to mathematics problems with notable mathematicians, such as Lorenzo Mascheroni, Pierre Laplace, and Joseph Lagrange, including what is known as Napoleon's Theorem. He was quoted as saying, "The advancement and perfection of mathematics are intimately connected to the prosperity of the state." Many modern officers have been educated at the U.S. Military Academy at West Point and other military academies, which emphasize mathematics and engineering in their curriculums, and both military and civilian mathematicians continue to play critical roles in infantry tactics and deployment, especially in the modeling and simulation of twenty-first-century combat strategies.

History

Archimedes, one of the most famous ancient mathematicians, applied his knowledge of geometry, the estimation of weights and volumes, and three-dimensional rotations to defending the city of Syracuse from siege by Roman forces (214–212 B.C.E.). In addition to the standard trick of cutting holes into the walls for archers to fire arrows through, Archimedes helped to design the catapults used by the Syracuse artillery units, and he called for traps to be built in the walls to drop heavy stones on approaching ships. Cranes were even used to drop grappling hooks onto ships and capsize them. The siege took much longer than it otherwise would have, and the Roman commander reportedly ordered that Archimedes's life be spared out of respect for his intellect—an order that was ignored, and Archimedes was killed when the siege finally succeeded.

The Renaissance was a time of flourishing mathematics, with applications in a wide variety of sciences, including cartography. While the Age of Discovery certainly was one cause for the demand for increasingly more precise maps, so too was the desire to accurately direct the movement of troops and ships while at war. Accurate chronometers were developed at the order of the military, which also called for more precise ways of determining latitude in order to increase the usefulness and accuracy of maps.

Modern Warfare

Eventually, mathematics would be used to more accurately determine the velocities and paths of projectiles, which in turn influenced not only the behavior of artillery units but also the design of infantry firearms, which became increasingly critical in conflicts like the U.S. Civil War and World War I.

World War II, because of its extraordinary size and resource consumption, put mathematicians to use in all areas of the military, a close relationship that has continued and been further assisted by the development of modern-day computers. The advent of paratroopers in World War II added a new level of complexity to the deployment of infantry troops, taking into account not only point-to-point movement on the ground but also precision insertion via parachute. Humans leap-

Scientists such as Luis Alvarez helped create the Identification Friend or Foe (IFF) radar camera, shown above, and improved antenna systems to identify friendly aircraft without using visual confirmation. (U.S. Air Force)

ing from a moving plane do not fall straight down, so calculations had to be made to take altitude, speed, and other factors into account in order to determine when, where, at what altitude, and at what intervals paratroops should deploy to successfully land on a predetermined spot. A hybrid transportation algorithm that first mathematically computes an ideal solution, which is then used for stochastic simulations, has been successfully used to model deployment of troops and equipment.

Other investigations into this problem often use numerical methods, fluid dynamic equations, 3-dimensional flows, mesh resolution techniques, and simulation methods. The use of aircraft for combat reconnaissance was also largely pioneered during World War II, though it was hampered by their limited speed and at times by unreliable radio communications, which did not facilitate the rapid decisions infantry commanders in the field were required to make.

Modern communication methods allow for rapid computer modeling and real-time decision making, virtually as soon as the data are collected. Military radar was also in its infancy in World War II, though work by mathematicians and scientists such as physicist Luis Alvarez would improve its utility. For example, Alvarez helped create transponders, then known as Identification Friend or Foe (IFF) radar beacons, and improved antenna systems, which identified friendly aircraft without visual confirmation and facilitated precision delivery of troops and bombs even in poor weather.

Mathematics at War

The quantification of troops, inventory, and distances as well as the order of battle and the estimations of travel speeds and damage to fortifications have likely always played a role in warfare. The term "order of battle" originally referred to the order in which troops were positioned relative to the position of the commander but has come to refer to the composition of the forces involved in a field operation, including their command structure, personnel, disposition (the geographical locations of the headquarters of units and subunits), and equipment.

In U.S. Army practice, an order of battle prepared for an intelligence report also includes information on personalities (known enemy personnel and relevant information pertaining to them), unit history relevant to the current situation, a logistics report on how units obtain supplies, and a combat effectiveness section that is prepared using combat modeling applications based on sophisticated algorithms. Orders of battle are fundamental to a military commander's situational awareness. Commanders depend more on combat effectiveness projections as modeling techniques have become more sophisticated and data from field operations have been applied in order to continually evaluate them.

In essence, the same mathematics responsible for governing the artificial intelligence of enemy forces in video games like Call of Duty is used—albeit with a great deal more data and more powerful processing—to evaluate enemy forces in real life. These models draw on a diverse array of mathematical methods. Game theory in general is concerned with modeling strategy. Statistical analysis, Andrey Markov chains, business logistics, and fluid dynamics have all played significant roles. During World War I, mathematician Frederick Lanchester devised Lanchester's Laws, which use systems of ordinary differential equations to determine which of two sides will remain at the end of a battle, as functions of the defenders' strengths and time, assuming neither side breaks off combat. They continue to be the basis for many modern simulations. Some models simplify problems or address only small portions of a vastly complex problem, including trying to quantify "soft" or qualitative aspects of combat, though hybrid modeling with both discrete and continuous components is a growing way to reliably model critical subsystems and also their interactions with one another. Mathematical analysis of satellite data and images is also used for detecting landmines and improvised explosive devices, which are some of the greatest threats to troops on the ground.

Perhaps the biggest impact of mathematics on the infantry is that the use of combat modeling means the ability to predict—if not always accurately, at least with a greater degree of accuracy than in the past—the outcome of various combat scenarios and, thus, to manage risk and reward when allocating troops. Military effectiveness can be maximized at multiple levels, from the allocation of funds at the budget stage to recruitment techniques to the command structure of the armed forces to troop movements.

Further Reading

Biddle, Stephen. *Military Power: Explaining Victory and Defeat in Modern Battle.* Princeton, NJ: Princeton University Press, 2004.

Booß-Bavnbek, B., and J. Høyrup. *Mathematics and War.* Basel, Switzerland: Birkhäuser, 2003.

BILL KTE'PI

Insurance

Category: Business, Economics, and Marketing.
Fields of Study: Algebra; Data Analysis and Probability; Number and Operations.
Summary: Society has long used mathematical methods to quantify risk and protect against loss, and professionals like actuaries help make these decisions.

Insurance involves the exchange of a fixed amount of money or sequence of payments (called premiums) by the insured to an entity or group for indemnification of the insured from specified losses. Thus, insurance involves trading a small but certain cost (the premium) for payment of a potentially large but uncertain loss in the future.

It is used to manage risk of loss in uncertain situations by hedging the risk (for example, by pooling money with others and sharing losses) or transferring it to some entity, like an insurer, for a price. Because the price paid today must cover future costs and future uncertain indemnification payments, the insurance industry employs many mathematicians to calculate and predict expected future costs and payments.

Importance

Risk transfer and risk pooling via insurance are very important. Following the government, insurance is probably the second most important mechanism available to alleviate social upheaval and to reduce risks to citizens. Social upheaval is reduced by supplying a financial safety net in times of loss. Risk reduction is achieved since insurance establishes risk reduction incentives, such as lowering the cost of insurance, for those who undertake risk reduction behaviors. Examples of risk reduction behavior include premium reduction in automobile insurance for defensive driving classes or having air bags; lowering premiums and providing loss control consulting to business firms concerning risk exposures; and lobbying governments for stronger safety standards.

Insurance allows entrepreneurs to create new products, explore new energy alternatives, and engage in selective risk-taking beneficial to society, such as creating new pharmaceuticals, which might be too uncertain or create potential liability exposure consequences too great to be undertaken if not insured. Through insurance, cash flows of firms are stabilized, bankruptcy likelihood is reduced, and the cost of capital to firms is lowered.

History

Because of the individual and societal benefits of insurance, it is no wonder that the rudiments of insurance date back millennia—although the modern approach to insurance awaited the development of mathematical tools to create the logical underpinning of the industry. The Code of Hammurabi (c. 1750 B.C.E.) details how early Babylonian merchants who had a loan on cargos or vessels could pay a little extra so that if the ship were lost at sea, the loan would be forgiven—an early example of risk transfer. Early civilizations also had arrangements wherein members pooled resources, and if one suffered a loss, such as a building burning down, others would pitch in and furnish materials and labor to rebuild the member's lost structure—an example of risk pooling. Before formal life insurance companies were developed, people in England in the seventeenth century would band together in groups called friendly societies, each contributing a small sum such that if an emergency or death occurred, the group would pay medical expenses, funeral costs, and sometimes give a stipend to the widow. Some of these friendly societies later developed into insurance companies.

Mathematics of Premiums

A crucial element in insurance is determining the insurance premium. The premium is the amount of money to be paid by the insured whose risk of loss is being indemnified, but needs to be an amount sufficient for the insurer selling the insurance to both cover potential loss costs and make a profit. Indeed, many early insurance-type organizations failed from the lack of correct assessments of risk and potential exposures to financial loss by the group furnishing the insurance—an incorrect quantification of risk. Without quantification of risk, the expected lost costs cannot be formalized and monitored. It is in this area of risk quantification that mathematics of insurance arises, mostly in the area of probability and statistics, which deal with the quantification of uncertainty.

The mathematics of insurance, known as "actuarial science," had its birth amid the incredible growth in mathematics in the seventeenth century. Most major mathematicians of the seventeenth and eighteenth centuries contributed to insurance mathematics in a variety of ways, such as calculating annuity tables based on interest rates and tables listing the probability of death at each age (called "life tables"). Some, such as Abraham DeMoivre, made a living, in part, by consulting on the calculation of annuity values. The first life table was constructed in 1694 by mathematician and astronomer Edmund Halley, now most famous for identifying Halley's Comet.

The development of modern probability theory—an essential element of the quantification of risk needed to price insurance—is usually attributed to French mathematicians Blaise Pascal and Pierre Fermat from a series of letters from 1654 concerning games of chance left unfinished. Using this new mathematical theory, the fair price of insurance could be rationally developed for the first time. For example, if, in the case of the occurrence of an event having a probability p, a benefit B is to be paid at some future time T, then the fair price today is pBv^T where v is the "discount rate" accounting for interest available on money invested today and paid at time T, expressed algebraically as the following:

$$v = \frac{1}{1+i}.$$

In this formula, i denotes the annual interest rate on invested money. Subsequent developments in mathematics have allowed for uncertainty in B, v, and T, enabling one to obtain the fair value of the insurance in more-complex risk transfer situations.

A mathematical foundation for insurance lies in the Law of Large Numbers (LLN), developed by mathematician Jacob Bernouli, and the Central Limit Theorem (CLT), developed by Abraham de Moivre and extended by mathematician Pierre Simon Laplace. The LLN is fundamental to insurance since it proves that the empirical relative frequency with which an event occurs in a risk pool will, as the size of the sample increases, approach the "true" probability of the event.

This allows insurance companies to objectively obtain the likelihood of loss-producing events from their experience in large collections of policyholders. The CLT proves that the average of a sample of homogeneous independent observations, such as losses within a pool of risks, will be well approximated by the bell-shaped Gaussian distribution as the number in the pool increases. From this idea, the setting of premiums for insurers who are appropriately confident of remaining solvent can be calculated.

Further Reading
Baranoff, Etti G., Patrick Brockett, and Yehuda Kahane. "Risk Management for the Enterprise and Individuals." *Flatworld Knowledge* (2009). http://www.flatworldknowledge.com/printed-book/1635.
Pearson, Egon. *The History of Statistics in the Seventeenth and Eighteenth Centuries Against the Changing Background of Intellectual, Scientific and Religious Thought*. New York: Macmillan, 1978.
Trieschmann, James S., Robert Hoyt, and David Sommer. *Risk Management and Insurance*. Cincinatti, OH: South-Western College Publishing, 2004.

Patrick L. Brockett

Intelligence and Counterintelligence

Category: Government, Politics, and History.
Fields of Study: Algebra, Number and Operations; Problem Solving.
Summary: Quantitative data, mathematical models, cryptography, data analysis, and social network analysis have proved powerful tools in intelligence.

The intelligence industry is tasked with gathering information and predicting or inferring past, present, or future behavior based on that information. While code-breaking is the most popularly known intersection of intelligence and mathematics, lattice theory is at least as relevant and various forms of data analysis are constantly relied upon. Math can "connect the dots" to maximize the usefulness of a small set of data.

"Mathematicians Won the War"
During World War II, the mathematics underlying cryptography played an important role in military planning. Winston Churchill admired Alan Turing, the Cambridge University mathematician who had mastered the Nazi codes, recognizing him as the man who had perhaps made the single greatest individual contribution to defeating Germany. After the first frosts of the Cold War descended in the Soviet East, approximately $2 billion was spent in the development of game theory.

After the Cold War came the "war on terror." The adversary uses rational strategies to attack, so rational strategies are needed for defense.

The "War on Terror"
The National Security Agency (NSA) is a riddle wrapped in a mystery inside a code—a black palace of glass located in Fort Meade, Maryland. It dwarfs the location of the Central Intelligence Agency (CIA). Its budget is unknown, and it is the world's largest employer of mathematicians, primarily number theorists, whose work depends integrally on the presumed complexity of factoring large numbers.

In May 2006, one of the NSA's secrets escaped. *USA Today* reported that the phone companies AT&T, Verizon, and Bell South had handed customer records over to the agency—not transcripts of calls, they said, just who was calling whom. Technically, only telephone numbers were being recorded, but one could easily obtain a name from a phone number. This information was being used to determine who might be a terrorist. With the NSA data, one can draw a picture or a graph with "nodes" (or dots) representing individuals and lines between nodes if one person has called another. The field of social network analysis (SNA) deals with trying to determine information about a group from such graphs, such as who the key players are or who the cell leaders might be.

Even if everyone in the graph is a known terrorist, graphs do not directly portray information about the order or hierarchy of the cell. SNA researchers look instead for graph features like centrality—they try to identify nodes that are connected to many other nodes, like spokes around the hub of a bicycle wheel. Indeed, Monterey Naval Postgraduate School researcher Ted Lewis, in his textbook *Critical Infrastructure Protection*, defines a critical node to be such a central hub.

There are two problems in creating such a graph. First, the "central player" might not be as important as the hub metaphor suggests. For example, Jafar Adibi of the University of Southern California looked at e-mail traffic between employees of the company Enron before its famous collapse and drew a graph. He found that if you naively analyzed the graph, you could mistakenly conclude that one of the "central players" was CEO Kenneth Lay's secretary. Second, as the journal *Studies in Conflict and Terrorism* reported in 2003, one can capture all the central players in a terrorist cell and leave the cell with a complete chain of command still capable of carrying out a devastating terrorist attack.

Lattice Theory Applied to the "War on Terror"

While it is true that NSA expert Kathleen Carley of Carnegie Mellon University was twice able to correctly predict who would take over Hamas when its leaders were assassinated (Hamas, the Palestinian Islamic Resistance Movement, is considered a terrorist organization by the U.S. government), her analysis uses detailed information about the individuals in the organization, not just which anonymous nodes were linked with which. Since terrorist cells are composed of leaders and followers, it is important to utilize lattice theory, which takes into account order and hierarchy.

Formal concept analysis (FCA), a branch of applied lattice theory, helps identify persons of interest. Individuals who share many of the same characteristics are grouped together as one node, and links between nodes in this picture, called a "concept lattice," indicate that all the members of a certain subgroup with certain attributes must also have other attributes. For instance, one might group together people based on what cafés, bookstores, and houses of worship they attend and then find out that all the people who go to a certain café also attend the same church, but maybe not vice versa. At Los Alamos National Laboratory, the laboratory that helped build the first atomic bomb, formal concept analysis has been used to mine data drawn from hundreds of reports of terrorist-related activity and to discover patterns and relationships that were previously in shadow—connections that human analysts could not have easily found without something like FCA.

> ### Critical Infrastructure Protection
>
> The U.S. government tried to prevent the publication of a study showing how the U.S. milk supply could be poisoned by terrorists, an analysis that uses queuing theory. Similar mathematics has been used to study the threat of dirty bombs.
>
> Which border do you guard? Which border do you want the terrorist to think is weak? You want to funnel him toward your snare, thinking the field is open. Reflexive theory—a branch of mathematical psychology developed by the Soviet military and funded by the U.S. State Department—gives a quantitative method to address these questions. The same mathematical analysis could potentially be used to alleviate the problem of improvised explosive devices in Iraq. Phoenix Mathematics, Inc. is developing software tools to help border patrols allocate personnel and spread disinformation to the adversary.

Tools from lattice theory can be applied to help intelligence agencies determine whether they have disrupted a terrorist cell. In early June 2005, the Pentagon announced plans to revise its strategy in the "war on terror." While then U.S. president George W. Bush repeatedly cited that 75 percent of Al Qaeda's leadership had been killed or captured, Al Qaeda remained active. The Pentagon shifted its target to mid-level captains and foot soldiers. Lattice theory, along with some extramathematical analysis, will help law enforcement agencies determine which individuals in a terrorist cell should be captured first, in order to maximize the chances of disrupting a cell by expending as few resources as possible. Lattice theoretical methods tell us the probability that a terrorist cell has been disabled based on how many terrorists have been captured and what rank they held in the organization.

Social choice theory has been applied to the hierarchical relationships within terrorist cells, determined from the direction of communications traffic, to model network formation. Researchers at New York

University have identified two types of coalitions. They have found that the detection of one type of cell is more effective in disrupting networks, whereas the detection of the other type of cell is more effective in identifying all the members of the cell. They have also used the lattice theory to try to determine the leaders from the graph of a terrorist network. Lattice theory and graph theory can even account for gaps in one's knowledge of the structure of a terrorist cell by making assumptions about how the "perfect" terrorist cell must be organized. The knowledge of the structure of the perfect terrorist cell could also be used by terrorists to counter intelligence efforts.

Winning the Battle for Hearts and Minds

Former U.S. defense secretary Donald Rumsfeld stated in a *USA Today* article on October 22, 2003, "Today, we lack metrics to know if we are winning or losing the global war on terror. Are we capturing, killing, or deterring and dissuading more terrorists every day than the madrassas and the radical clerics are recruiting, training, and deploying against us?" To model the growth of a terrorist network, one could use the same differential equations that govern the spread of an infection, like severe acute respiratory syndrome (SARS). Such models could be used to help the government understand, and eventually contain, the spread of a terrorist insurgency.

On March 16, 2003, then U.S. vice president Dick Cheney predicted on *Meet the Press* that Americans would be "greeted as liberators" in Iraq. Ideas from statistical physics have been used to model the battle for the hearts and minds of the people of Iraq. Just as a magnetic pole may be north or south, a person could be either for the occupation or against it. The model shows that there can be a tipping point in the evolution of public opinion. It may seem as if much of the population is with one side (for example, the United States) but then, dramatically, a wave of hostility sweeps down, and one witnesses the birth of an insurgency.

Terrorism of the Futures Market

When bombs explode, the stock market drops. Mathematician Stefan Schmidt of the Technical University in Dresden, Germany, has attempted to quantify the impact on the market of a terrorist incident. The only people who know when a bomb will explode are, of course, the terrorists. By playing the market, they may already have obtained as much money as they need, thus stifling U.S. Treasury Department efforts to cut off their funding. The terrorism of the futures market may be the terrorism of the future.

Further Reading

Argamon, Shlomo. *Computational Methods for Counterterrorism*. Berlin: Springer, 2009.

Associated Press. "Mathematicians Offer Help in the War on Terror." *USA Today* (October 9, 2004).

Brams, Steven J., et al. "Influence in Terrorist Networks: From Undirected to Directed Graphs." *Studies in Conflict and Terrorism* 29 (2006).

Farley, Jonathan David. "Breaking Al Qaeda Cells: A Mathematical Analysis of Counterterrorism Operations (A Guide for Risk Assessment and Decision Making)." *Studies in Conflict and Terrorism* 26 (2003).

———. "Evolutionary Dynamics of the Insurgency in Iraq: A Mathematical Model of the Battle for Hearts and Minds." *Studies in Conflict and Terrorism* 30 (2007).

———. *Toward a Mathematical Theory of Counterterrorism: Building the Perfect Terrorist Cell*. Carlisle Barracks, PA: U.S. Army War College, 2007.

Krebs, Valdis. "Mapping Networks of Terrorist Cells." *Connections* 24 (2002).

Lefebvre, Vladimir A., and Jonathan David Farley. "The Torturer's Dilemma: A Theoretical Analysis of the Societal Consequences of Torturing Terrorist Suspects." *Studies in Conflict and Terrorism* 30 (2007).

Lewis, Ted G. *Critical Infrastructure Protection in Homeland Security: Defending a Networked Nation*. Hoboken, NJ: Wiley, 2006.

Lindelauf, R., P. E. M. Borm, and H. J. M. Hamers. "The Influence of Secrecy on the Communication Structure of Covert Networks." *Social Networks* 31 (2009).

Memon, Nasrullah, Jonathan David Farley, David L. Hicks, and Torben Rosenørn, eds. *Mathematical Methods in Counterterrorism*. Norderstedt, Germany: Springer Verlag, 2009.

Rosoff, H., and D. von Winterfeldt. "A Risk and Economic Analysis of Dirty Bomb Attacks on the Ports of Los Angeles and Long Beach." *Risk Analysis* 27 (2007).

Wein, Lawrence, and Yifan Liu, "Analyzing a Bioterror Attack on the Food Supply: The Case of Botulinum Toxin in Milk." *Proceedings of the National Academy of Sciences of the United States of America* 102 (July 12, 2005).

Woo, Gordon. "Quantifying Insurance Terrorism Risk." *Risk Management Solutions* (2002).

Zhao Guomin, Liu Mao, Zhang Qingsong, Wang Li, and Yang Yang. "Risk Control of Terrorism Attack Based on Order Theory." *Proceedings of the 2006 International Symposium on Safety Science and Technology.* Changsha, China (2006).

Jonathan David Farley

Inventory Models

Category: Business, Economics and Marketing.
Fields of Study: Data Analysis and Probability; Measurement; Number and Operations.
Summary: Mathematical inventory control models help businesses make decisions, and they are widely studied in the discipline of operations research.

In an ideal world, retail stores would stock all of the products that customers are interested in buying and stock these in sufficient quantity to cater to all customers. In reality, store area is limited and a company would not benefit by stocking an excess of each product. The problem, then, is to calculate the optimal amount of supply.

These decisions take into consideration how many units should be kept so that most, if not all, customers can be served on a particular day, because if customers do not find what they want, they will shop elsewhere. At the same time, a store does not want too many units on hand, as there are costs attached to storing excess units, and they may remain unsold, which also reduces profit.

The problem can be considered in manufacturing, where a product consists of many small components, and a business has to decide how many components it must order and store so that fabrication runs smoothly. Similar examples exist in service industries and military ships. Mathematical inventory control models help businesses make decisions, and they are widely studied in the mathematical discipline of operations research.

Mathematics of Inventory

Computational logistics is a mathematical and business field concerned with planning the flow and storage of goods, services, or information from the point of origin to the point of use. One key planning consideration is the trade-off between transport and inventory costs, a factor recognized at least as early as the mid-1880s. Mathematicians, computer scientists, and others continue to develop new inventory management and optimization models as well as the algorithms and software necessary to implement them. Mathematician Samuel Karlin was awarded the John von Neumann Theory Prize in 1987, as well as the National Medal of Science in 1989, for diverse mathematical contributions, including inventory theory.

Inventory models used to calculate optimal order quantities and reorder points, often broadly called economic order quantity (EOQ) models, existed long before the arrival of the computer. Advances in mathematical methods and computer technology have facilitated more realistic models that account for more variables. Optimizing inventory depends on factors such as storage space, storage cost, demand rate, time between demands, cost of ordering, time for retrieving stored item or receiving an ordered item, discounts for bulk orders, and many other real-world costs.

Just-in-time models are based on the idealized principle that items are available exactly when they are needed, with zero storage time or delay. Just-in-time inventory management and lean manufacturing ideas existed as far back as Henry Ford's Model T factories but became widely feasible in the late twentieth century with advances in technology that affected variables, like the lead time required to place an order for more stock. Reduction of process variability, using better monitoring, waste reduction, or inventory buffers, are typically seen as key to achieving optimal models under this system. A just-in-time model can save money by reducing inventory, but tighter constraints make them consequently more vulnerable to disruptions that violate the constraints.

Many basic EOQ models are simplified by assuming that variables such as demand are fixed or uniform across some period of time. These deterministic models are easy to solve analytically but may produce unrealistic results. They are often useful for theoretical study or businesses with greater variability tolerances. Many variables that influence inventories, such as demand and delay times for orders of new goods, are more realistically modeled as random variables. As a result, inventory models are often probabilistic or stochastic. Constraints tend to be operationalized as costs.

For example, the physical area available for storage, such as square footage of shelf space or warehouse volume, can be reformulated as a cost constraint by calculating a cost per unit area or volume. Cost may also be parameterized into components like procurement and maintenance costs. Markov chains and linear programming techniques are useful for formulating and solving various types of inventory models. Statistical methods are used to obtain valid data for modeling and simulations.

Further Reading

Cachon, Gerard, and Christian Terwiesch. *Matching Supply With Demand.* New York: McGraw-Hill, 2008.

Luenberger, David G., and Yinyu Ye. *Linear and Nonlinear Programming.* 3rd ed. New York: Springer, 2010.

Porteus, Evan. *Foundations of Stochastic Inventory Theory.* Stanford, CA: Stanford Business Books, 2002.

Sethi, Suresh P., et al. *Inventory and Supply Chain Management With Forecast Updates.* New York: Springer, 2010.

Sherbrooke, Craig. *Optimal Inventory Modeling of Systems.* 2nd ed. New York: Springer, 2004.

Ravi Sreenivasan

Loans

Category: Business, Economics, and Marketing.
Fields of Study: Algebra; Number and Operations.
Summary: Determining the terms of a loan so that they are fair but compensate for risk is a challenge of algebra.

Most people have personal experience with one or more types of loans, such as home mortgages, car loans, or home equity loans. In each case, the general format of the loan is the same: the lender provides temporary funds to a borrower, and the borrower repays these funds over a prespecified period of time, according to a prespecified pattern. As it is for any financial asset or liability, mathematics is a critical tool for determining the appropriate parameters of loans, including the periodic payment necessary for the borrower to completely pay off the loan by the end of the loan's life.

Mathematicians work on many problems related to loans. For example, individuals who take out large loans, like mortgages, are often required to purchase insurance for those loans. Actuaries use mathematical and statistical methods to assess lending risk to decide whether insurance is needed and how much. They also work on more complex problems related to interest rates and credit, such as deciding what constitutes usury (unreasonably high interest rates) for loans whose yield rate is not fixed or determining the reliable predictors of credit risk.

History

Loans appear to have been a part of economic activity ever since economies began to become sophisticated. In response to certain historical unfair lending practices, a number of proscriptions against usury were recorded in ancient sources, such as the Old Testament, and works by Aristotle and Tacitus. More generally, an active lending market is important to an economy, as it facilitates the availability of funds for investment.

Loans, like other financial instruments, are two-sided transactions. There is a lender and there is a borrower, and cash flows are made between them—what one party pays, the other receives. Algebraically, this process is usually reflected by identifying the cash flows as either positive or negative; a positive cash flow for the lender would be a negative cash flow of the same magnitude for the borrower, and vice versa. For the lender, the loan transaction is essentially an investment, and thus an asset. For the borrower, the loan represents a liability and ultimately needs to be paid back.

The most common method in the twenty-first century of paying off a loan is via amortization, in which interest and a portion of the original borrowed principal are paid back in each of the periodic payments. There are a number of parameters associated with the typical amortization loan, including the following:

- The original amount borrowed (B)
- The length or term (n) of the loan (for personal loans, such as mortgages and auto loans, the length of the loan is typically measured as the number of monthly payments to be made by the borrower to the lender; theoretically, however, payments can be made according to any schedule, such as weekly, annually, or uneven periods of time)

- The periodic (for example, monthly) interest rate (i) on the loan, which determines the amount of interest paid by the borrower to the lender
- The periodic (for example, monthly) payment (R) made by the borrower to the lender

In the most common type of amortized loan, the payment made by the borrower each period is constant over time. Each payment consists of two components: an interest payment and a partial principal repayment. Across the life of the loan, the sum of all of the n partial principal repayments is equal to the total original amount borrowed, B. As each payment is made, the outstanding balance of the loan is lessened by the amount of the partial principal repayment in that payment.

For the lender, a loan is basically an investment and an asset. For the borrower, the loan is a liability and needs to be paid back. (Photos.com)

The effect of this approach is that, while each payment R is of the same size, the split between the interest component and the principal component of each payment changes over time. More specifically, as time moves on, the principal component increases and the interest component decreases. This is because the indebtedness (the outstanding balance) of the loan decreases over time, and thus the periodic interest charged on the loan (which is equal to the interest rate multiplied by the loan's outstanding balance) also decreases over time.

To illustrate, suppose that $1,000 is borrowed, and this four-year loan is to be paid off with four equal annual payments of R, one at the end of each of the four years during the life of the loan. Suppose that the effective annual interest rate $i = 0.10$, or 10%. In this situation, the annual payment R can be determined by the formula

$$R = \frac{iB}{1 - \left(\frac{1}{1} + i\right)^n}$$

where $i = 0.10$, $B = \$1,000$, and $n = 4$. Thus, $R = \$315.47$.

This value for R can be verified by considering the impact of each annual payment separately. For example, consider the first payment of R. During the first year, the borrower incurs interest charges of 10% of the outstanding balance at the beginning of the year, or $100. Thus, $100 of the $315.47 first payment covers the interest for borrowing the original $1,000 during the first year; the remaining $215.47 of the first payment then serves to partially pay off the loan, leaving an outstanding loan balance, or indebtedness, of $1,000 − $215.47 = $784.53. During the second year, the borrower incurs interest of 10% of that new outstanding balance, or $78.45. That portion of the second payment of R covers this interest, and the remainder ($315.47 − $78.45 = $237.02) serves to further pay down the loan. Thus, after the second payment, the borrower has loan indebtedness of $784.53 − $237.02 = $547.51. Continuing this process through the fourth and final payment will reveal that, after that final payment, the original $1,000 loan has been completely and precisely paid off.

Occasionally, people will pay off installment loans before their final due date by making early payments or paying slightly more than is due at each installment. In this case, they may be entitled to a rebate on some of the originally computed interest. Rebates can be figured using several methods, including variables such as how the interest was originally computed and the way in which the regular and extra payments were divided between principal and interest. The actuarial method of calculation is generally more favorable to the borrower than rebates calculated under other methods, such as the Rule of 78s.

There are other ways of paying off loans; for example, paying the interest regularly and then paying off the entire principal at the end of the loan term. In fact, this process is essentially how a specific type of financial instrument, a bond, works. When corporate or governmental entities issue bonds, they are borrowing money. More precisely, they are borrowing an amount equal to the price of the bond from the investor or investors who purchase the bond. The issuing organization pays periodic interest to the investors (in the form of coupons) and at the expiration date of the bond pays back to the investors a lump sum, known as the "redemption value."

Further Reading

Broverman, Samuel A. *Mathematics of Investment and Credit*. Winsted, CT: ACTEX Publications, 2008.

Kellison, Stephen. *Theory of Interest*. New York: McGraw-Hill, 2008.

Rick Gorvett

Market Research

Category: Business, Economics, and Marketing.
Fields of Study: Communication; Data Analysis and Probability; Problem Solving.
Summary: Quantitative and qualitative methods are used to analyze data and guide business decisions.

Market research is a field of study and practice focused on gathering information about markets and customers for the purpose of improving sales or other business outcomes, though similar techniques have been applied to public awareness campaigns designed to change behavior such as smoking and weight loss. Market research draws from a variety of disciplines, with mathematics, statistics, actuarial science, psychology, and business being particularly influential. Careers in market research require strong quantitative skills and market researchers may be required to use concepts from algebra, trigonometry, geometry, calculus, economics, or statistics. Statistical data collection using surveys, experiments, and focus groups is widespread. Both quantitative and qualitative methods are used to analyze these data and guide decisions. Mathematical and statistical models are also developed to try to explain consumer behavior, predict future sales and trends, direct the optimal placement of advertising media or allocation of advertising funds, make consumer recommendations, and simulate market behavior. The availability of enormous consumer databases accumulated from credit cards, store discount cards, and many other sources has spurred the use of data mining techniques, like data fusion and clustering, to merge sometimes-incomplete data sources and then classify subgroups of consumers according to selected criteria.

Types of Market Research

Market research is a broad field and it is important to understand several distinctions about how and why such research is conducted. The first distinction is between marketing intelligence and market research projects: the former is an ongoing, broad-based process of gathering and analyzing information; the latter are focused on a particular question or product and generally have a defined budget and time for completion. A second distinction is between exploratory and confirmatory research: exploratory research is usually conducted early in the decision cycle, and its goal is to discover what options exist; confirmatory research comes into play later in the cycle when the goal is to narrow options and decide which course of action to follow. These distinctions are crucial because the same technique can be used for different purposes; for instance, surveys or focus groups can be part of an ongoing and broad-based data collection effort or may be a one-time effort focused on a particular product or some aspect of a product. Both research techniques may be used either to gather a broad array of data whose purpose may not be known (which might be conceptualized as "seeing what's out there") or as a tightly focused effort at making distinctions to guide decision making among a small set of already-known options.

Another distinction is whether the research will be focused on sales to consumers or to other businesses. The former is sometimes called "business-to-consumer" (B2C), and the latter is called "business-to-business" (B2B) marketing. Most people are familiar with consumer market research and may have taken part in it, whether they were aware of it or not. Consumer market research is focused on the goal of selling a product to a large number of people (or, in a more general sense, of discovering their preferences). For instance, an entrepreneur might want to design a sports sneaker that will appeal to urban young men of high school age. Because of this focus on describing the preferences and judgments of groups, consumer market research often incorporates knowledge and techniques from social sciences, such as psychology, sociology, and anthropology. Techniques include surveys, focus groups, and ethnographic observation (observation of how people make choices or use products without interfering in that process).

B2B refers to commercial transactions between businesses. For instance, a wholesaler may sell goods to a retailer (who will then sell them to the public), or a supplier may provide goods necessary for business operations, such as paper, computers, and other office supplies for business. Although B2B accounts for a high volume of sales, the process of market research is different because the consumers may be assumed to have a high degree of knowledge about the product they will be buying, and usually a single individual or small department can make the decision for large purchases of goods. For these reasons, B2B market research may be focused differently, for instance, on discovering how a corporation views its own brand and how a product may be allied with that effort. However, as with consumer marketing, the goal is still to gain information that will allow businesses to develop and market products that meet the needs and desires of potential purchasers.

Another distinction is between qualitative research, which generally collects verbal data, and quantitative research, which collects information that may be translated into numbers. Qualitative research is often used for exploratory research and to gather information very early in the research process; for example, focus groups and unstructured interviews may be used to gather reactions to a new idea or product. When the research effort has progressed sufficiently that a few questions have been selected for further investigation, more structured quantitative research (for instance, a questionnaire-based survey) may be used to gather precise information relating to these questions.

The Research Process

The process of market research proceeds in a manner similar to much social science research, with the main difference being the ultimate goal. In the social sciences, it is generally to add to human knowledge, while in market research, it is generally to make an optimal business decision. In either case, the first step is to iden-

History of Market Research

Formal consumer marketing research got its start in the 1920s with the founding of ACNielsen Corporation in Chicago by engineer Arthur C. Nielsen. Nielsen pioneered many concepts now common in market research, including market share and combined consumer surveys with quantitative audits of sales (both from account books and by observing what was on store shelves) to track sales patterns. Nielsen was also involved in early radio marketing research and later applied the same methods to measure the audiences for different television programs (forerunners of the well-known "Nielsen ratings," which are still used today). In the early days of radio and television it was common for advertisers to sponsor an entire program, rather than to buy a short segment of time to deliver a commercial message, and so the issue of how many people and which particular demographic groups were listening to specific radio programs (or watching specific television programs) became crucial because the sponsor wanted to deliver their message to the right market and be associated with programming that would appeal to that market.

A famous example is the development of "soap operas" on radio and television. These were serial programs about domestic life and were sponsored by soap companies because the programming was developed to appeal to female audiences who presumably were the primary purchasers of household soap products. Marketing research was largely limited to internal departments of mainstream packaged-goods companies until the 1980s but since then has become a major industry as more companies became interested in using market research, and independent consulting firms were developed to answer this need.

tify the question to be answered or the problem to be solved, a process that is particularly important when the research will be conducted by a separate department or a consulting group. The next step is to elaborate on the problem—exactly what information is required or what questions much be answered in order for a decision to be made? The third step is to identify which research techniques are most appropriate for answering the questions, including consideration of the time and other resources available. Once these steps have been completed, a study can be designed, including specification of a time frame and the data sources to be used.

In research, the distinction is often made between primary and secondary data sources. Primary sources are data that are collected by an individual or organization for its own use, for instance, conducting focus groups to see how people react to several versions of a new product a business is planning to introduce to the market. Secondary sources are those collected by someone else and then made available to others. Examples include government data sets such as the U.S. Census and data collected by private or university researchers for specific projects that are later made available for use by others.

Both primary and secondary data have their advantages and disadvantages. Collecting primary data allows the research team to specify exactly what data they want, for instance, color and design preferences among housewives in a specific urban area. They are generally more expensive because the researchers must collect the data themselves and they are necessarily more limited in scope. It is generally cheaper to use secondary data, and the scope is often much broader (for example, it may have been collected on a national or international basis) than could be collected by a small research team. However, secondary data may be several years out of date by the time it is available and may not focus specifically on the questions of interest for a particular marketing research project. Often, both types of data are combined in the same research project; for instance, U.S. Census data about neighborhoods (racial composition, median household income, etc.) can easily be combined with information from a primary, purpose-designed survey of individuals.

Further Reading

Burns, Alvin C. and Ronald F. Bush. *Marketing Research.* Upper Saddle River, NJ: Prentice Hall, 1998.

Mariampolski, Hy. *Qualitative Market Research: A Comprehensive Guide.* Thousand Oaks, CA: Sage, 2001.

McDaniel, Carl D. *Marketing Research: The Impact of the Internet.* Cincinnati, OH: South-Western, 2002.

McQuarrie, Edward F. *The Market Research Toolbox: A Concise Guide for Beginners.* Thousand Oaks, CA: Sage, 2006.

Percy, Larry, ed. *Marketing Research That Pays Off: Case Histories of Marketing Research Leading to Success in the Marketplace.* London: Haworth Press, 1997.

Swzwarc, Paul. *Researching Customer Satisfaction and Loyalty: How to Find Out What People Really Think.* Sterling, VA: Kogan Page, 2005.

Sarah Boslaugh

Mathematics: Discovery or Invention

Category: Mathematics Culture and Identity.
Fields of Study: Communication; Representation.
Summary: One of the central questions of the philosophy of mathematics is that of mathematical realism.

Mathematicians engage in a great many activities, including investigating and extending old and new concepts within the field, as well as developing new techniques to solve problems in mathematics and other disciplines. The question is, when they carry out this activity, do they discover existing laws or do they invent and create? If invention is involved, is it individual or is it social? This question is a polemical topic that has been subject to strong controversy and refers to ideas that have emanated everywhere from ancient Greek personages, such as Plato, up to modern advocates of artificial intelligence (AI).

Platonists

Those who subscribe to the discovery position are usually classified as Platonists. Plato expressed that mathematical ideas are discovered, existing independently of human observation or changes of a physical nature. However, the general trend known as "mathematical realism," which includes formalism and logicism, also catalogued within the discovery perspective. Mathematics is seen as the science of logic with its laws based on enduring truths, whether they have been discovered or not. Those who subscribe to this position cite, for example, the existence of universal constants, such as π, φ, Euler's e, or Feigenbaum's α and δ in bifurcation theory. It is put forth that the circumference of a circle has always measured π times diameter, whether or not that fact had been discovered by a particular society or culture.

It is also claimed that the discovery of mathematical laws, objects, and relations occurs simultaneously, or over time, in distant places. The most famous examples include the simultaneous, but independent, discovery of calculus by Isaac Newton and Gottfried Leibniz in the seventeenth century and the independent discovery of the universal constant π by the Babylonians, Greeks, Chinese, and others at different historical moments. Many of the structures from very abstract areas of mathematics are often found to model phenomena in the physical world, such as the case of Cantor's set, originally an abstract construct, which serves as a model for error distribution of the noise in transmission lines (for example, electric power lines or telephone wires). This case is also taken as evidence that mathematics is, apart from a consistent logical system when accepting the axioms, a language that describes the physical universe, whether or not that description was intended by the mathematician who discovered the pattern, technique, theorem, or other relevant mathematical object.

Criticisms of Platonists

This idea adds another element to the discussion. For the realists, it is important to distinguish between mathematics itself, as a timeless science of logic, together with the laws that govern its existence, and the practice of mathematics, which includes many aspects that are language-like and that, they agree, are created, such as particular symbolism, notation, formalization, and nomenclature. Often the Platonists are dismissed by arguments that ridicule or simplify Plato's allegory of the cave to an alleged discovery of an almost physical mathematical realm. This simplification seems because of a literal, instead of a metaphorical, interpretation of the way that many working mathematicians refer to their subject, a way of expression that reflects the actual feeling of "concreteness" that is provoked by daily contact, manipulation, and struggle with their abstract objects. Roger Penrose, for example, who identifies with the Platonist perspective, speaks of the Mandelbrot set as a structure whose constant surprises, within its self-similarity, are waiting to be explored.

Diversities of Non-Platonists

On the other hand, those that challenge Platonism and mathematical realism in general are not a homogeneous group.

One of these positions asserts that the existence of mathematics can be understood only as part of human culture. It is argued that the reality of mathematics is a sociocultural and historical phenomenon and that mathematics exists only because there are human beings who create it. Advocates of this position argue that mathematics is in the same category as law, religion, and money. It is only human consciousness and society with its conventions that makes them real.

Philosopher Ludwig Wittgenstein regarded mathematics as a type of ". . . communication; people play 'language-games' and 'sign-games' to invent, rather than discover, mathematics." The Social Constructivists, supporters of this position, argue that mathematical development is guided by fashions and trends in human societies. They claim that mathematical truth is invented and depends on the sociocultural context.

The term "quasi-empirism" is used for the type of modern mathematical research that relies on computers and other quasi-experimental methods that seem to contradict the deductive nature of mathematics and question the existence of absolute and eternal mathematical truth. The Social Constructivists assert that this activity demonstrates the fallibility of mathematical activity and removes it from the realm of any absolutes, thus supporting their claim that mathematics is "man-made."

The embodied theories consider mathematics as an exclusively human endeavor, invented according to the physical and cognitive human reality. Exponents of this position privilege the biological evolution of the human brain and consider mathematical objects as a reflection of human cognition. Hence, according to this perspective, mathematics is constructed by the human brain, and its apparent truths were created because they actually work efficiently in the universe in which we find ourselves.

Further Reading

Davis, Philip, and Reuben Hersh. *The Mathematical Experience*. Boston: Mariner Books, 1999.

Ernst, Paul. *Social Constructivism as a Philosophy of Mathematics.* Albany: State University of New York Press, 1998.

Lakoff, George, and Rafael Nuñez. *Where Mathematics Comes From: How the Embodied Mind Brings Mathematics Into Being.* New York: Basic Books, 2001.

Mazur, Barry. "Mathematical Platonism and its Opposites." http://www.math.harvard.edu/~mazur/papers/plato4.pdf.

Penrose, Roger. *The Emperor's New Mind.* Oxford, England: Oxford University Press, 1989.

Persson, Ulf. "Platonism, a Synopsis." http://www.math.chalmers.se/~ulfp/Platonism/platon.pdf.

Rehmeyer, Julie. "Still Debating with Plato." *Science News* (April 25, 2008). http://www.sciencenews.org/view/generic/id/31392/title/Math_Trek_Still_debating_with_Plato.

Wittgenstein, Ludwig. *Remarks on the Foundations of Mathematics.* Translated by G. H. von Wright, R. Rhees, and E. Anscombe. Oxford, England: Basil Blackwell, 1978.

Mariana Montiel

Middle Ages

Category: Government, Politics, and History.
Fields of Study: Algebra; Geometry; Measurement.
Summary: Medieval mathematics developments included Scholasticism and the emergence of secular universities.

The European Middle Ages, or the "medieval period," lasted from the fall of Rome to the Renaissance and was identified by Renaissance thinkers as separating their own period from that of classical civilization. The Middle Ages were construed as a time of backwardness, but in fact progressed in spite of economic, medical, and political difficulties. Mathematicians made original contributions to such areas as algebra and astronomy and commentaries on historic texts preserved Greek works. Mathematics historians have studied Arabic, Persian, Turkish, Indian, Islamic, and European contributions during the Middle Ages. For example, Adolf Yushkevich wrote a seminal work on the history of mathematics in the Middle Ages. He highlighted similar features of medieval mathematics based on the cultures in Europe and Asia and, along with Boris Rozenfeld, studied Arabic contributions.

Early Middle Ages

The transfer of western Europe from the Roman Empire to the Goths occurred gradually through the fourth and fifth centuries, partly by conquest and partly by migration and assimilation. The old travel and trade network decayed and scholarship retreated mostly into monasteries. The philosopher Boethius straddled the Roman and Goth eras. He valued mathematics highly, endeavoring to translate several important mathematical works from Greek to Latin and dividing the seven liberal arts into two tiers: a lower tier, the trivium—containing logic, grammar and rhetoric—and an upper tier, the quadrivium—containing the four mathematical arts of arithmetic, geometry, astronomy, and music theory. Boethius is remembered primarily for his work *Consolation of Philosophy,* written while he was imprisoned before execution. Christianity became a primary supporter of higher learning, music, and art in Europe, and also a strong participant in government owing to the high levels of literacy among Church officials. Monasticism also gained momentum during the early middle ages, inspired by the isolated communities in Syria and Egypt. Owing to the importance of study in religious life, many monasteries functioned also as schools and libraries.

Carolingian Renascence

Around the ninth century, Charlemagne and his successor, Louis the Pious, enacted various reforms to effect uniform standards in a renascence of the Roman Empire. Charlemagne had schools created to restore education across Europe, reunifying the dialectized Latin and creating a script for it, the Carolingian minuscule. The standard curriculum saw Boethius's trivium and quadrivium become the foundations for the bachelor and master of arts degrees. A standard currency facilitated reformation of the economy and long-distance trade and taxation. The Roman influence is evident in monumental architecture, which incorporates elements from classical styles in clear, relatively simple arrangements. Circles, squares, cubes, and cones feature prominently, as does symmetry. Carolingian architecture and painting became the basis for the more ornate Romanesque style and, ultimately, the Gothic.

Byzantium and the East
The Greek-speaking part of the Roman Empire, also called Byzantium, survived the Latin half's decline. In the sixth century, Byzantium extended around the eastern Mediterranean from Egypt to Greece, expanded across all of north Africa, and even took Carthage and Italy from the Goths. Then, severely weakened by epidemics thought to be the Black Death, the Byzantine Empire shrank to what is now Turkey and Greece, plus Carthage and some parts of Italy. Even after this decline, Byzantine culture stood as the standard for both western Europe and the Near East. Owing to increasing influence from Christianity, art and monumental architecture tended to manifest in churches (such as Hagia Sophia), and philosophy intertwined with Christianity on many topics, including ethics, existence, governance, and death.

Hellenistic knowledge percolated gradually eastward from Byzantium, first in translation into Syriac and then into Arabic, which fueled a philosophical community in Damascus. By the seventh century, Neoplatonism, which had been Christianized in late antiquity, had been accommodated into the Islamic framework. This set the backdrop against which Aristotelianism, and all of its disagreements with Platonism, had to be accommodated next.

In the eighth century, Baghdad became the cultural focus of the East. The scholarly community there attracted scholars of diverse races and religions. The Islamic Golden Age continued into the eleventh century, with many advances of significance to western Europe, including those by al-Khwarizmi in algebra, by Ibn al-Haytham (Alhazen) in optics and scientific method, by al-Battani (Albategnius) in astronomy, by Jabir ibn Hayyan (Geber) in alchemy, and by Ibn Sina (Avicenna) in medicine. A rich tradition of poetry and calligraphy also emerged.

Al-Andalus
In the eighth century, the Moors of north Africa took most of the Iberian Peninsula that ultimately became the Umayyad caliphate based at Córdoba after the Abbasids came to power in Baghdad. While the Abbasid caliphate suffered from political fragmentation, the Umayyad territories in the Iberian Peninsula thrived.

Astronomy and botany were especially active in al-Andalus, both for intellectual interest and for applications in timekeeping, astrology, and medicine. While the societal framework was predominantly Islamic, numerous Jews and Christians participated in high culture during extended periods of cosmopolitanism. Al Zarqali (Arzachel) discovered the ellipticity of planetary orbits in the eleventh century, and ibn Baija (Avempace) deduced that the Milky Way was not a continuous cloud but numerous stars. Studies of Aristotle by ibn Rushd (Averroës) shaped philosophy and religion for centuries later.

High Middle Ages
From the eleventh to the thirteenth centuries, western Europe was peaceful enough to entertain a high degree of cultural development. Windmill- and waterwheel-powered industries developed, economies flourished, and urban populations grew quickly, spreading into formerly Moorish Iberia, into southern Italy, and even into the Baltic and the Near East. The Arabic heritage was absorbed and then reacted against in a philosophical movement called "Scholasticism."

Scholasticism emerged from the works of Aristotle. They were translated from Arabic into Latin and provided a basis for a worldview based on empiricism and logic. Although the philosophy was secular, it was pursued largely for its power to support Christian doctrine. The Arabic writers had already weighed Platonist versus Aristotelian views and largely harmonized the philosophy with religious givens. Much of the result was hence incompatible with new movements in Christianity, and the Scholastics sought to rebuild it by returning to the original sources. The scientific content was developed notably by Robert Grosseteste and Roger Bacon in England and Albertus Magnus, Thomas Aquinas, and Duns Scotus in France. These five also ranked highly in the Church, illustrating the continuing need that religion had for higher education and the support for intellectuals that the Church provided.

Early in the Middle Ages, higher learning had been concentrated in monasteries and Church schools. With the new secular engagement, universities appeared, beginning with Bologna in 1088, then Paris in 1150, then Oxford in 1167, then others. Learning emerged from the monasteries into urban surroundings and engaged more with secular needs, such as commerce and industry. Gothic architecture replaced the hefty, solid Romanesque, with height and lightness built from thin stone ribs reaching up and out to become the ribs of vaulted ceilings. Acute arches and vaults replaced the Romanesque semicircle, and walls gave way to large glass windows. Gothic designs manifest Euclidean geometry problems, including constructing regular polygons, dividing arbitrary angles into equal parts, dividing lines into equal parts, fitting circles through points, tangent to lines or tangent to other circles.

In the fourteenth century, frequent plagues and crop failures decimated the population, undermining social structure, industry, and economies. From the turmoil sprang new outlooks on all fronts. Among the more famous literary achievements, Dante wrote his *Commedia* and other tracts (including some scientific ones), Chaucer wrote his *Canterbury Tales*, Bocaccio wrote the *Decameron*. Such fresh thoughts ultimately gave rise to the Renaissance in fifteenth-century Italy.

A number of European mathematicians were important in helping to introduce eastern mathematics into Europe. Many Greek works were unknown in Europe and were found only in Arabic. Adelard de Bada translated the Arabic texts of Arabic and Greek mathematicians into Latin. Leonardo Pisano Fibonacci was educated in north Africa and traveled extensively. In Pisa he introduced the Hindu–Arabic place-valued decimal system and the use of Arabic numerals into Europe, while also making fundamental contributions of his own.

Further Reading

Cantor, Norman F. *The Civilization of the Middle Ages: A Completely Revised and Expanded Edition of Medieval History, the Life and Death of a Civilization*. New York: HarperCollins, 1993.

Crosbie, Alfred F. *The Measure of Reality: Quantification and Western Society, 1250–1600*. Cambridge, England: Cambridge University Press, 1997.

Lindberg, David C. *The Beginnings of Western Science: The European Scientific Tradition in Philosophical, Religious, and Institutional Context, Prehistory to A.D. 1450*. 2nd ed. Chicago: University of Chicago Press, 2007.

Menocal, M. R. *The Ornament of the World: How Muslims, Jews and Christians Created a Culture of Tolerance in Medieval Spain*. Boston: Little, Brown, 2002.

Alistair Kwan

Military Draft

Category: Government, Politics, and History.
Fields of Study: Data Analysis and Probability; Number and Operations; Problem Solving.
Summary: Military drafts must make use of probabilities to ensure the draft is equitable.

The U.S. military is made up of volunteers. However, if more people are needed than the number of people who volunteer, there needs to be a method for procuring enlistment. The method used is called a "military draft." It is the law that all male citizens ages 18–25 are to register with the Selective Service. If a need arises, the U.S. Congress would have to pass legislation instituting a draft. The U.S. president would have to sign the bill into law.

When a draft occurs, there is a lottery of the registered men that is intended to be fair. Each registered man of the same age should be as likely as every other registered man to be selected. Once selections are made, some men are excused if they are not fit to serve. A military draft has not been used since 1973.

The Current Lottery
The current lottery method that would be employed if there were to be a draft is to place a capsule with dates for every possible day of the year (month and day) into a barrel. For example, December 1, January 27, and March 13 would be three such capsules. A second barrel will contain the numbers 1 through 365. These barrels are well mixed. In fact, one way to mix the barrels is to not place the capsules into the barrels in order. Rather, the capsules are placed into the barrels in a random manner. One capsule is drawn from each barrel, one at a time, and paired. For example, if November 4 is drawn from one barrel and 78 is drawn from the other barrel, then November 4 and 78 are paired. This continues until all 365 days have a number. The number becomes the day's rank. This process forms 365 ranked groups. Each group consists of those registered men whose birthday is the corresponding date pulled from the barrel and who will turn 20 in the current year.

For example, assume that each date is paired with the following number:

Future soldiers being sworn into the army. The U.S. military is made up of volunteers; however, if more people are needed, the U.S. government would have to pass legislation instituting a military draft. (Elizabeth M. Lorge/U.S. Army)

November 4 paired with 78,
December 28 paired with 1, and
January 12 paired with 25.

Then all men who turn 20 in the current year of the draft and have a birthday on November 4 will be the 78th group to be called to serve. Before they are drafted, groups 1–77 would be exhausted of possibilities (that is, all fit to serve in the previous 77 groups would be called to serve first). All registered men who turn 20 in the current year and have a birthday on December 28 are in the first group. All men who turn 20 in the current year and have a birthday on January 12 are in the 25th group. Again, these groups are made up of men who will turn 20 in the year of the draft. Once all 365 groups are used, then the rankings are followed again, calling all men turning 21, then 22, 23, 24, 25, 18, and 19.

What it Means to be Random

A selection process of this nature is random only if any person is as likely as any other person to be selected to serve. Thus, each of the 365 birthdays must be as likely as each other birthday to be ranked first. Each of the remaining birthdays must be as likely as any to be ranked second. A man's birthday should not allow one to predict the likelihood of his being drafted.

Vietnam Draft

The 1969 lottery drawing for the Vietnam War was demonstrated not to be random. A barrel with 366 plastic capsules was used, where each capsule had a birth date on it (month and day); one capsule was for those who were born on leap day. One at a time, the capsules were drawn by hand. The first to be drawn was ranked first. The second to be drawn ranked second. Thus, if September 21 was drawn first, then all men aged 18–26 with a birthday on September 21 would be the first group called to service.

The procedure that was followed to order the men with the shared birthday depended on each man's initials. A separate lottery was held in which the 26 letters of the alphabet were ranked. This followed the same process as the birthdays, in that 26 letters were placed in a barrel and one by one were drawn. Using the resulting ranking, each man within a shared birthday was ranked according to the permutation of the first letter of his last name, the first letter of his middle name, and the first letter of his first name. Overall, this should have been a fair method for selection, as it was based on randomized birthdays and letter permutations.

Why It Was Not Random
The above-mentioned method would be random if implemented properly. However, it turned out that men with birthdays later in the year (for example, December birthdays) were much more likely to be drafted than those with birthdays in the beginning of the year. What happened is quite simple. The capsules were placed in the barrel month-by-month beginning with January, and the barrel was not well mixed. The December capsules were on top and they had a higher probability of being pulled out first, resulting in lower draft numbers for those men.

Further Reading
Friedman, Lauri S. *Military Draft* (*Writing the Critical Essay: An Opposing Viewpoints Guide*). Farmington Hills, MI: Greenhaven Press, 2007.

Hay, Jack. *Military Draft* (*History of Issues*). Farmington Hills, MI: Greenhaven Press, 2007.

CARMEN M. LATTERELL

Missiles

Category: Government, Politics, and History.
Fields of Study: Algebra; Calculus.
Summary: Mathematicians have long worked on improving missile accuracy and performance.

Stone or arrow missiles have been used for thousands of years. Missiles with explosives can be traced back to China following the Song dynasty. Mathematical and technological advances have led to countless improvements in missile design, trajectory, range, and accuracy and have continually revolutionized warfare. Aristotle theorized on laws governing projectile motion, as did mathematicians like Leonhard Euler and Daniel Bernoulli, who derived or refined mathematical principles of projectile motion using geometry, calculus, and differential equations. In the nineteenth century, mathematicians Alfred Freenhill and Percy MacMahon worked on a missile trajectory model that related resistance to the cube of the velocity, suggested from experimental data. During World War I, mathematics took on an increasingly important role. John Littlewood created techniques to reduce the work required for accurate missile trajectory calculations, and Gilbert Bliss used the calculus of variations to account for variables like wind and the rotation of the Earth. During the 1950s, mathematician John von Neumann headed the committee that led to the development of U.S. intercontinental ballistic missiles. During the space age, mathematicians made a breadth of contributions, like Evelyn Boyd Granville, who worked on the development of missile fuses at the National Bureau of Standards.

Mathematician Peter Swerling, known for his theory of radar, also researched optimal estimation of satellite and missile orbits and trajectories. Missiles of the twenty-first century can be defined as weapons that follow a trajectory for the purpose of delivering explosive warheads to targets by means of lift and rocket propulsion. They may be launched from ground, submarines, and airplanes to nearly any target on the face of the Earth. Mathematicians working in government, industry, and academia continue to contribute to the development of all types of missiles and missile defense systems.

Trajectory and Guidance
The basic flight path of a missile is a parabolic arc. Sixteenth-century mathematician Niccolo Tartaglia described cannonball flight paths. Seventeenth-century mathematician Evangelista Torricelli published a geometric method for computing projectile range. Benjamin Robins, an eighteenth-century mathematician, invented the ballistic pendulum. His experiments, later expanded by Euler, demonstrated that air resistance could not be ignored in calculating trajectories. Scientist Heinrich Magnus showed that other forces could affect spinning spheres and cylinders; this effect is now known as the Magnus Effect. The importance of higher mathematics, like calculus, in computing trajectories contributed to the inclusion of these topics in many military school curricula in the nineteenth century.

In the early twenty-first century, mathematics continues to play a key role in missile accuracy. Most modern guidance systems use mathematical methods to determine the trajectory needed, such as angular coordinates between the missile and the target or the distance between the target and the missile. Sometimes computations are done ahead of time and the missile follows a predetermined path. Other times, the missile can make adjustments to the flight path in order to correct the trajectory as needed and may follow a path that is very different from the basic parabola. Some systems utilize astronomy—the accuracy of a missile is determined by examining the relationship of the missile to a fixed start position. Others employ altitude maps and compute the missile's distance from the ground to determine the path of the missile. These systems, however, are subject to error. Navigation systems that utilize a path calculated prior to launch may be influenced by instrument errors, while systems that utilize flight path data are more accurate but are subject to the effects of countermeasures such as radar decoys or infrared flares.

Advanced missiles are propelled by an internal combustion mechanism and guided by radiation, lasers, radio waves, or computers. Guidance often involves the use of mathematical techniques, like Kalman filtering, named after Rudolf Kalman, which allows a missile's course to be manipulated. Many of these latest-generation weapons come complete with cameras that record visual and spatial location information to aid human operators in their direction. Other missiles are guided by locations systems, such as INS, TERCOM, or GPS, which are programmed to recognize the weapon's global positioning at its origin and use it to calculate the distance, trajectory, and course to the target. These modern flight systems use positioning, targeting, and guidance data, along with thrust and aerodynamics, to maneuver missiles while they are in flight, even allowing them to seek and destroy moving targets.

Defensive Systems

With the development of more advanced missiles has come the need for more advanced defense systems. For example, satellites could measure the missile's trajectory and speed to determine a probable impact point and relay this information to an interceptor vehicle. The interceptor might initially utilize celestial guidance to track the incoming missile, and then use preset guidance to collide with the incoming missile. The U.S. Missile Defense Agency employs many engineers, scientists, and mathematicians to work collaboratively on defense solutions.

Further Reading

National Aeronautics and Space Administration. "Beginner's Guide to Rockets." http://exploration.grc.nasa.gov/education/rocket/bgmr.html.

Shneydor, N. A. *Missile Guidance and Pursuit: Kinematics, Dynamics and Control*. Cambridge, England: Woodhead Publishing, 1998.

Van Riper, Bowdoin. *Rockets and Missiles: The Life Story of a Technology*. Westport, CT: Greenwood Press, 2004.

Calli A. Holaway
Michael G. Lovorn

Mutual Funds

Category: Business, Economics, and Marketing.
Fields of Study: Algebra; Data Analysis and Probability; Measurement; Number and Operations.
Summary: Many mathematicians attempt to develop mathematical models that forecast the future direction of the stock market and thus to produce better investment results for mutual funds.

Mutual funds are a type of investment in which large numbers of people pool their money and a fund manager invests these funds in one or more types of security. Investors own shares in the fund, and the value of those shares is determined by the total value of all the securities owned by the fund. Mutual funds are a popular investment vehicle because they allow people to achieve a varied investment portfolio with a relatively small investment, thus limiting their risk in comparison to buying individual stocks, bonds, or other assets.

Many types of mutual funds are available, depending on the desires of the investor. For instance, are they more interested in a riskier fund that may produce a higher yield for their investment, or a safer fund that is more likely to preserve the value of their capital? Some mutual funds specialize in a single type of investment—for instance, international stocks, health sector stocks, U.S. government bonds, or real estate—while others invest in a variety of securities in order to achieve a desired balance between yield and risk. Although mutual funds are often perceived as a safe investment, they are not guaranteed by the Federal Deposit Insurance Corporation (FDIC) as are bank deposits, and it is possible to lose money by investing in mutual funds. Economists, statisticians, actuaries, and others frequently try to predict the stock market using time series analyses and other mathematical methods. Prediction has historically proven to be quite challenging because of the complexities of time series data and the different socioeconomic variables and human psychological factors that appear to influence the stock market.

History and Growth

Although the first mutual funds were offered in the United States in the 1920s, the modern mutual fund industry dates from 1940 when the Investment Company Act established a body of rules regarding financial investments. In 1949, less than $2 billion were invested in mutual funds, but they became a more popular investment vehicle in the 1960s. By 1973, $47 billion was invested in mutual funds. By 1987, this amount had grown to $4 trillion, and by 2000, to $6 trillion, representing the investments of over 83 million investors. One factor in the growth of individual investments in mutual funds is the shift in the United States from guaranteed pension plans to retirement savings plans like the 401(k) in which an individual worker is responsible for choosing how to invest his or her retirement funds.

In 2008, there were over 8000 mutual funds in the United States versus about 3000 stocks listed on the NASDAQ stock exchange and a similar number on the New York Stock Exchange. It may at first be counter intuitive that there should be more funds than stocks, but this fact is not surprising if one considers any mutual fund as a composite made up of individual stocks or a subset of the total number of stocks (although, of course, a mutual fund may also include bonds and other components). Any set of n elements has 2^n possible subsets, so a set of 10 elements has 1024 subsets and a set of 25 elements has over 33 million.

Risk Minimization

One appeal of mutual funds is that they allow people to reduce their risk through diversification. Modern portfolio theory attempts to select assets to minimize risk, maximize return, or some combination of those two (in general, higher risk is associated with higher return, although this does not hold absolutely). The basic concept behind the theory is that stocks or other assets, such as bonds, in the fund are evaluated in the context of other assets, and the goal is to maximize return or minimize risk for the total collection of assets, called a "portfolio." American economist Harry Markowitz developed portfolio theory beginning in the 1950s, and in 1990, was awarded the Nobel Prize in Economics for this achievement.

Management

Because the performance of a mutual fund is often related to that of the economy as a whole, the performance of specific mutual funds as well as mutual funds as a class is often evaluated against the performance of indices such as the Dow Jones Industrial Average (a scaled average of the stocks of 30 large, publicly owned companies) or the S&P 500 (a weighted index of 500 large-cap common stocks). There are always pitfalls in making these types of comparisons; for instance, the return of mutual funds as a whole appears larger than it really is because funds that do poorly often go out of existence and are thus dropped from the average (survivorship bias). Interestingly, over time most individual funds produce somewhat worse results than a large index, such as the S&P 500, suggesting that the talent of individual managers (who choose when to buy and sell the stocks or other investments that comprise a mutual fund) are less efficient than the stock market as a whole. For this reason there are mutual funds today that are not "actively managed" in the sense that an individual manager makes buying and selling decisions. Instead, such funds simply own the stocks that comprise some index, such as the S&P 500, with buying and selling decisions motivated by changes in the makeup of the index (for instance, because of mergers or to new stocks joining or leaving the index).

This method is not a criticism of mutual funds per se but simply an argument for the efficiency of the market. Studies of the stock picks of professional analysts also tend to perform only marginally better than those selected randomly—most famously by throwing darts at a dartboard. Despite this well-known result, many individuals and investment firms have developed complex mathematical models that attempt to forecast the future direction of the stock market and thus produce better investment results. In addition, people have tried to predict the movement of the stock market with other types of data; for instance, in 2010, two graduate students found that the emotional content of tweets (messages sent on Twitter, a social networking

Web site that can receive and send text messages from mobile devices, such as mobile phones) from the general public could be used to predict movement of the Dow Jones Industrial Average several days in advance.

Further Reading

Fink, Matthew P. *The Rise of Mutual Funds: An Insider's View*. New York: Oxford University Press, 2008.

Grossman, Lisa. "Twitter Can Predict the Stock Market." *Wire Science* (October 19, 2010). http://www.wired.com/wiredscience/2010/10/twitter-crystal-ball.

Malkiel, Burton G. *A Random Walk Down Wall Street*. Rev. ed. New York: W. W. Norton, 2007.

Paulos, John A. *A Mathematician Plays the Stock Market*. New York: Basic Books, 2003.

Wepsic, Eric, Sendehil Revuluri, and Chris Welty. "Mathematical Modeling in the World of Finance." *Math Horizons* 6 (September 1998).

Sarah Boslaugh

National Debt

Category: Government, Politics, and History.
Fields of Study: Data Analysis and Probability; Measurement; Number and Operations.
Summary: The accumulation of federal government budget deficits over time is the national debt, which is best considered relative to GDP or other factors.

Mathematician Richard Feynman once said, "There are 10^{11} stars in the galaxy. That used to be a huge number. But it's only a hundred billion. It's less than the national deficit! We used to call them astronomical numbers. Now we should call them economical numbers." In modern society, entities from individuals through governments need money to function. Most government funds are generated by taxing individuals, businesses, goods, and services.

At the same time, governments must spend money for various purposes. If a government has more income than expenditures in a given fiscal period, usually one year, the excess of income over expenditures is called a "surplus"; if a government has more expenditures than income, the excess of expenditures over income is called a "deficit." The sum of all of these single-year surpluses and deficits over the entire history of the federal government is called the "national debt." Mathematics has long been used to quantify expenditures, deficits, and debts. Taxation and deficiency problems were mentioned in the Chinese mathematical text *The Jiuzhang suanshu* (Nine Chapters on the Mathematical Art), and Indian mathematician Brahmagupta referred to debts to mean what are now called "negative numbers." William Playfair created some of the earliest graphical representations of social and economic data around the time of the American Revolution, such as trade balances between England and other countries and the English national debt. By the twentieth century, mathematical measurement, estimation, and modeling were increasingly used. Standard economic measures like gross domestic product (GDP) were common, and there were theories and research on principles like return on capital, interest rates, and exchange rates, many of which cannot be known with certainty. Stochastic modeling, random walks, particle theory, and Brownian motion, named for botanist Robert Brown,

Federal Reserve System

The Federal Reserve System (sometimes called the "Fed") is the national bank of the United States and is independent of other United States institutions, including the Treasury Department. While it is not directly related to administering the deficit or to making decisions on government spending, it helps to manage the money supply in the United States by facilitating the lending of money between banks and by lending money to banks directly, which in part determines the interest rates that banks charge for borrowing money. These interest rates in turn influence the rates on the Treasury bonds that finance deficits. Many mathematicians and actuaries work for the Federal Reserve. For example, mathematician and Federal Reserve board member (as of 2010) Gary Anderson and economist George Moore developed the Anderson–Moore algorithm for solving linear saddle point models, which are used in economic modeling.

have been used extensively in the mathematical modeling of financial processes. After events like the Wall Street crash of 1929, there was also interest in forecasting models that could warn of debt crises. Mathematicians continue to research and create models to address both historic and new financial concerns, and many people have created representations such as the national debt clock and deficit calculators to extrapolate trends. Others argue against too much aggregation or extrapolation in mathematical models, citing inherent data collection errors in large-scale indices, like the consumer price index and gross national product, as well as subjectivity in individual perception and often-complex interactions between variables such as debt, deficit, production of good and services, and allocation of consumer resources.

Inflation

National debt differs uniquely from individual debt in the fact that governments usually have the power to print more money to pay debts. However, doing so often leads to undesirable economic consequences. More money in circulation can lead to increased demand for goods and services, which in turn may lead to inflation. Mathematicians and economists study inflation trends and cycles, as well as the reciprocal impacts of inflation on factors such as labor costs. While most economies function reasonably well with some level of inflation, too high a level of inflation leads to a host of problems, including hoarding of goods, increases in interest rates for credit and loans, and trade deficits with other countries. For this reason, most governments borrow rather than print the money to finance debt. In the United States, borrowing is accomplished primarily by selling government bonds. The purchaser, who may be an individual or another country, pays for a bond at the time of sale, and in return is promised a future amount of money, sometimes with interest payments made before the end of the bond period. The United States pays interest on its national debt bonds, which can be significant. For example, in 2009, interest on the national debt was $260 billion, approximately 8.5% of that year's federal budget and the fourth-largest single expense.

Intentional Debt

Having a large national debt poses many risks to an economy. A large national debt can help contribute to inflation and can lead to tax increases. Economists have also determined that GDP tends to grow more in an economy with a moderate level of national debt than in one with a high level of debt. Many variables affect spending, deficit, and debt. For example, governments often run deficits during economic recessions or depressions, spending money to attempt to stimulate the economy, partly under the notion that future gains will compensate and yield a positive long-term average or expected value. In 1900, the national debt in the United States was $2.6 billion and experienced overall nonlinear growth approaching the twenty-first century. Mathematical analyses have shown that debt increased sharply during World War I, while in the 1920s national debt decreased due to surpluses. It increased sharply again during the 1930s because of the Great Depression. Another increase occurred with spending for World War II. By 1950, the U.S. national debt had grown to $256.8 billion. After several relatively small increases, the national debt grew quickly beginning in the mid-1970s. Using exponential regression, mathematicians and economists have estimated that the national debt was doubling approximately every six years during this latter period up to nearly the end of the twentieth century. Projective models extrapolate such trends to estimate debt, often based on other estimated values, like the future population.

Debt Compared to GDP

In the same way that individuals can afford to spend more money when they receive a raise in salary, it can be misleading to look at the dollar amount of the federal debt without considering the overall size of the economy and the time value of money. For this reason, economists often evaluate the economic health of governments by considering national debt as a percentage of the country's GDP. In the United States in 1940, the national debt was 52.4% of the GDP. This number increased during World War II to 121.7% in 1946, meaning that national debt was actually larger than the GDP, but fell below 100% again in subsequent years. Mathematicians and economists have created models to forecast this index, with some predicting that factors like the housing and financial crises will cause the United States to once again pass the 100% threshold in the twenty-first century.

Further Reading

Paulos, John. *A Mathematician Reads the Newspaper*. New York: Anchor Books, 1996.

Stein, Jerome. *Stochastic Optimal Control, International Finance, and Debt Crises*. Oxford, England: Oxford University Press, 2006.

U.S. Office of Management and Budget. "Historical Tables: Budget of the U.S. Government." http://www.federalbudget.com/HistoricalTables.pdf.

Pete Johnson

Native American Mathematics

Category: Government, Politics, and History.
Fields of Study: Connections; Geometry; Measurement; Number and Operations; Representations.
Summary: Native Americans developed numbering systems and had a clear sense of dimension, geometry, and probability.

The term "Native American mathematics" is deceptive because there is no single culture for all Native Americans. Rather, each of more than 400 Native American tribes has its own distinct culture, with each mathematical element being specific to that culture. Nonetheless, in an examination of mathematical aspects, it is possible to discuss some commonalities across the many tribes, producing evidence of multiple number systems, arithmetic operations, geometry, and probability.

Number Systems

Native American numbering systems often used a simple grouping system that corresponded to different parts of the human body. For example, the idea of "tens" is contained in the numbering system of the San Gabriel Indians in California, where "all my-hand finished" represented the number 10, "all my-hand finished and one my-foot" represented the number 15, and "another finished my-foot the side" represented the number 20. It is inferred that they used single fingers on each hand to represent any number less than 10. Often, a Native American tribe would have names for large numbers, but had little use for such in their daily lives. For example, Michael Closs, a cultural historian, describes a Copper Eskimo elder while relating a story about two men who, trying to settle an argument, begins to count the hairs on a wolf and a caribou. The story ends with the count unfinished, as both men die of starvation. And, the story concludes with the phrase: "That is what happens when one starts to do useless and idle things that can never lead to anything."

Though using groupings of 5 and 10 as the structure for their number systems, the idea of a number base is not always evident. Also, some evidence exists for the use of 2, 4, and 20 as the structuring element. For example, the Yukis tribe in northern California used a combination of the quaternary (base four) and octal (base eight) systems. In turn, their counting mechanism depended on referring to the four spaces between the fingers on both hands, not the fingers themselves.

In a study of North American Native Americans, researchers documented the use of 307 different number systems; 33% were base 10, 33% were base 5, 23% were base 2, 10% were base 20, and the remaining 1% were base 3.

In any discussion of Native American mathematics, it is necessary to include the Aztecs, Incas, and Mayans. For example, the Aztecs' number system was based on the number 20, with the numbers 400 and 8000 given special significance. In contrast, the Incas used a slight variation of the base 10 system, and even had specific words for the numbers 1–10, 100, 1000, and 1,000,000. Finally, the Mayans, the most mathematically sophisticated of the three, had a vigesimal system using the number 20 as its base. The Mayan system also included special notations for multiples of numbers and used a special symbol glyph to represent zero.

Arithmetic

The idea or use of arithmetic operations was not something needed by early Native Americans, who depended on a hunting-gathering culture. Historians suggest that any signs of significant arithmetic are due to a tribe's interactions and trade with the early fur traders or buffalo hunters. For example, the language of the Navajo does not include words for "multiply" or "divide," yet that should not imply their inability to perform either process computationally or using real items.

Evidence of addition is found in words used to denote different numbers, using a process of addition by juxtaposition. For example, Alaskan Natives living near the Yukon River essentially used the words "five one" and "five four" to denote the numbers 6 and 9, respectively.

In direct contrast, the Miluk Coos, an Oregon tribe, used subtraction by juxtaposition, where "four ten" and "one ten" denoted the numbers 6 and 9, respectively. Some historians claim that 40% of the Native American tribes used some version of this subtraction process, especially for numbers close to multiples of 10.

Evidence of multiplication is found among Pawnee tribes, in a very creative fashion. Their term "50 persons" represented the number 1000, based on their use of the word "man" for the number 20, knowing that "man" had 10 fingers and 10 toes. Thus, "50 persons" was equivalent to 50 sets of 20 fingers and toes, or a total of 1000.

Measurements

The measurements invoked by Native Americans were context-sensitive and personal in nature. No standard units were established and used widely either within a tribe or across tribes. In most instances, the measurements used were specific to the context and informal. The Ojibwa tribe is a good example. For short lengths, their units were finger widths, hand spans, forearm lengths, and arm spans, while their longer lengths might reflect a changing position of the sun or even mention the unit "number of sleeps" involved in traversing a long distance.

Geometry

Native American geometry is evident in the colorful decoration and intricate patterns found on knife cases, moccasins, blankets, pouches, baskets, and pottery. At first, many of these patterns were created using porcupine quills but eventually the shift was made to using glass beads.

When creating a pattern, the different Native American tribes differed in their use of geometrical structures. In some instances, a tribe's members created irregular floral patterns, while other tribes used a geometry based only on straight lines, allowing them to create blocks, crosses, and triangles. The types of triangles ranged from isosceles to equilateral to right, with common traits being tall isosceles triangles or pairs of reflecting congruent triangles. Occasionally, circles and spirals appear as part of a design.

Many studies have focused on Native Americans' use of symmetry in strip patterns using beads. Of the seven possible symmetry groups, the most popular pattern is labeled "pmm2" in standard transformational schemes, which means that the pattern has horizontal, vertical, and rotational symmetry.

Also, in the process, the creator of the visual pattern possibly used counting or even some computing skills (for example, skip counting by threes to form a border). It is possible that creators of some of the patterns included elements of measurement (perimeter or area), number theory (multiples and divisors), and fractions (common, decimal, and ratios).

Tiling patterns are evident in the creation of blankets, going beyond strip patterns. Some historians claim tiling elements are also found in some of the Native American petroglyphs carved on the surfaces of caves, cliffs, and large stones.

Finally, Native Americans had a clear sense of dimension, using objects to represent the three possibilities. A stick represented dimension one, an animal skin represented dimension two, and an apple or walnut represented dimension three. However, in their paintings on flat surfaces, the idea of dimensional perspective is not utilized.

Probability

Elements of probability are found in some of the children's games played by various Native American tribes. For example, consider the Apaches' "Throw Sticks" game involving two or more people. In one version, three sticks are decorated with colorful designs on one side only, called the "face." The sticks are held in one hand and then dropped on the ground. The scoring is as follows: 10 points for three faces up, 5 points for two faces up, 2 points for 1 face up, and 1 point for no faces up. The score is kept by moving small sticks or "horses" around a circle of 30 stones. Play continues until someone travels the full circle. Elements of probability, such as likelihood, events, and dice-like actions, are all evident in this game.

Native Americans also played dice games, using dice made from bone, peach stones, deer horn, beaver teeth, or walnut shells. As most of these dice were two-sided, one side was colored to distinguish the two sides. When sets of dice were thrown, the scoring was based on the number of a given side appearing. Because the "dice" were crudely made, the chances of each side appearing are not equal. This observation actually validates the claim that Native Americans had a good sense of probability, because the higher score values were assigned to the least probable events.

Further Reading

Barta, Jim. "Native American Beadwork and Mathematics." *Winds of Change* (Spring 1999).

Closs, Michael. *A Survey of Mathematics Development in the New World*. Ottawa, Canada: University of Ottawa, 1977.

——— ed. *Native American Mathematics*. Austin: University of Texas Press, 1986.

Culin, S. *Games of the North American Indians, Volume 1: Games of Chance*. Lincoln: University of Nebraska Press, 1992.

Hughes, Barnabas, and Kim Anderson. "American and Canadian Indians: Mathematical Connections." *Mathematics Teaching in the Middle School* 2, no. 2 (November–December 1996).

Jerry Johnson

Oceania, Australia and New Zealand

Category: Mathematics Around the World.
Fields of Study: All.
Summary: The indigenous cultures of Oceania are mathematically interesting.

The United Nations classification for Oceania includes Australia and New Zealand as well as the hundreds of Pacific Islands groups under the headings Melanesia, Micronesia, and Polynesia. The Australian Mathematical Society was founded in 1956 and promotes mathematics and its applications. The New Zealand Mathematical Society was found in 1974 and promotes research and the dissemination of mathematics. Mathematicians born in Australia and New Zealand include Field's Medal winners Terence Tao (2006) from Australia and Vaughan Jones (1990) from New Zealand. High school students participate in the International Mathematical Olympiad. Australia began its participation in 1981 and hosted the contest in 1988, while New Zealand first participated in 1988. Mathematics historians and ethnomathematicians have researched the mathematics of the indigenous inhabitants of Australia and New Zealand. For example, the structures of Australian Aboriginal kinship systems can be modeled by the algebraic theory of groups, while the wood carving and tattooing done by the Maori of New Zealand embody geometrical principles of symmetry. These cultural achievements interest mathematicians and teachers of mathematics and also have influenced the humanities, the social sciences, and popular culture.

Australia

Studying Aboriginal kinship systems has greatly influenced anthropology and can be mathematically modeled. To give just one important example, Claude Lévi-Strauss, in support of his ideas on structural anthropology, cited what he called "the Australian facts" to help argue that a system of exchange (as illustrated by marriage partners reciprocally chosen from paired sections) underlies the origin of marriage rules.

Many Aboriginal societies are divided into two halves, with four sections in each half, for the purpose of determining kinship. The example best known to mathematicians, because of the classic work of Marcia Ascher, is that of the Warlpiri of Australia's Northern Territory. A schematic diagram is shown below.

```
 ┌─ A ←─ = ─┐ W ←─┐
 │  B ←─ = ─│ X ←─│
 │  C  ─ = ─└ Y   │
 └→ D  ─ = ─→ Z ──┘
```

The equal sign designates allowed marriages. That is, for members of a section in either half, there is one section in the other half from which marriage partners come. For instance, women in section A marry men from section W, and men from section A marry women from section W. Children's sections are determined by their mother's; the directed arrows show how. For instance, if a mother is in section A, her children are in section C; mothers in C have children in B; mothers in B have children in D; and mothers in D have children in A—completing a cycle. Similarly, mothers in W have children in Z, and so on. Thus, the matrilineal cycle has a length of 4. For fathers, if a man is in A, for instance, following the arrow backward shows that his mother is in D, so his father is in Z. Then his father's mother is in W, so his father's father is in A again. Thus the complete patrilineal cycle has a length of 2.

If one writes I for one's own section, m for one's mother's section, m^2 for one's mother's mother's section, f for one's father's section, and so on, the cyclic relationships can be expressed by $m^4 = I$ and $f^2 = I$. Other algebraic relationships, like $(mf)(mf) = I$, can be verified from the diagram. The resulting algebraic structure is that of the dihedral group of order 8. The Warlpiri, of course, do not have the concept of group, but those learning the system are asked to solve word problems like, "If someone's mother is in a particular section, then in what section is such-and-such a relative?" The Warlpiri abstract from the personal relationships to conceptualize the system itself. General terms of address reflect the individual's place in the structure. Kin relationships determine a person's behavior, obligations, place to live, and relationships to plants, animals, and landscape; they also link past, present, and future generations.

The Aboriginal view of the origin of their kinship system in the journeys of their ancestors during the ancestral past (known as the "dreamtime") is reflected in Aboriginal paintings. Such paintings are noted for their symmetry, and particular geometric elements indicate individual places, ancestral beings, or clans. The current interest in Aboriginal art has brought these geometric forms to a worldwide audience.

New Zealand

Geometric art pervades Maori culture in dance, song, music, weaving, painting, latticework, carving, and tattooing. Wood carving is the most prominent, though facial and body tattoos also continue to be symbols of Maori identity. Traditional Maori carving uses a small number of design forms and motifs, combined according to well-established rules. Rafters and ridgepoles of the Maori meetinghouse are decorated with carvings that embody tribal history. These carvings employ all seven of the symmetry groups that characterize strip patterns. They are often colored in ways that complement, rather than echo, the symmetries. Maori art also uses bilateral symmetry, but the symmetry is often broken by the nonsymmetrical use of colors or by the addition of small figures that vary. Maori tattoos use many of the same themes and motifs as does carving.

Also, individuals' tattoos serve to identify family, tribe, community, birthplace, and inherited or achieved authority.

Maori symmetric forms are united by their near identity while differing in their asymmetries. This aspect reflects the way the Maori characterize reality by pairs of things existing in a tension between union and separation. Understanding the formal geometric patterns thus gives insight into Maori culture. Maori geometric art has become part of global culture. For example, Maori carved wooden bowls appear in Paul Gaugin's paintings. In Herman Melville's *Moby Dick*, the tattooed harpooner Queequeg possesses—and sells—Maori tattooed ancestral heads. Enlightenment philosopher Immanuel Kant felt that he had to discuss Maori tattoos in examining the nature of beauty, though he concluded that Maori tattoo designs could be beautiful only if they were not on a human face. Additionally, Maori tattooing plays a key role in the acclaimed 1994 film *Once Were Warriors*.

Further Reading

Blakers, A. L. "The Australian Mathematical Society: Foundation and Early Years. I: Events Leading Up to the Foundation of the Society." *Australian Mathematical Society Gazette* 3, no. 2 (1976).

Greer, Brian, et al. *Culturally Responsive Mathematics Education*. New York: Routledge, 2009.

Kaeppler, Adrienne Lois. *The Pacific Arts of Polynesia and Micronesia*. New York: Oxford University Press, 2008.

Munn, Nancy D. *Walbiri Iconography*. Ithaca, NY: Cornell University Press, 1973.

Starzecka, D. C., ed. *Maori Art and Culture*. London: British Museum, 1996.

Tee, G. J. "The First 25 Years of the New Zealand Mathematical Society." *New Zealand Mathematical Society Newsletter* 76 (1999).

Washburn, Dorothy, and Donald Crowe. *Symmetries of Culture: Theory and Practice of Plane Pattern Analysis*. Seattle: University of Washington Press, 1998.

Judith V. Grabiner

Oceania, Pacific Islands

Category: Mathematics Around the World.
Fields of Study: All.
Summary: The people of the Pacific Islands historically used sophisticated mathematics, including a unique method of navigation.

The Pacific Ocean covers more than one-fifth of the Earth's surface and includes hundreds of islands. In the nineteenth century, few visitors to the Pacific Islands were able to match the skill of Pacific Islanders in solving arithmetic and algebra problems. The people of the Marshall Islands, scattered over dozens of atolls across the central Pacific, were master navigators who tracked their way over huge expanses of ocean without mechanical aids. The compass, sextant, and chronometer, which their European contemporaries were reliant upon for safe and successful voyaging, were completely unknown to them. What they possessed instead were a set of aids that relied upon an extremely complex type of knowledge related to what they could observe and even feel about the ocean around them. These aids were called *Mattangs*, *Meddos*, and *Rebbelibs* by their users and are known today as "stick charts."

Some other instances of mathematics in Pacific Island culture and in the Pacific Islands region include the often-complex geometric patterns found in basket weaving, such as the design named "stars," which has a tessellation pattern that is mathematically sophisticated and is reminiscent of Dutch graphic artist M.C. Escher's drawings. These patterns can also be found in traditional tattoos, where the type of pattern had great cultural significance and represented the rank and bravery of the tattooed person. Scientists have also modeled the number of species on islands as a mathematical power function that depends on land area, and they continue to study island populations of birds and other species in this context. Researchers have explored barriers to success in mathematics in the Pacific Islands and recommended that teachers include culturally relevant content in their classrooms. Professional development programs and consortiums offer training for teachers and explore mathematics education for Pacific-region children.

Stick Charts

Stick charts were made from strips of the midrib of a coconut frond or pandanus root bound together with coconut sennit in geometric patterns meant to represent currents flowing around their low-lying atolls. Small shells or coral pebbles were attached to indicate the location of islands, and curved sticks were used to represent wave patterns.

The first of these charts, the Mattang, was a small square chart used to teach how waves reflect and refract, or bend, around a single island or atoll (see Figures 1 and 2). By detecting a change in the direction of the prevailing swell, a navigator could discern the presence of an island or atoll over the horizon. The Meddo was an actual chart covering a small set of atolls and used for voyages to nearby atolls. Meddo charts also showed the direction of the main ocean swell and how it curves around specific islands and the distance from a canoe at which an island could be detected. The Rebbelib was a more complex version of the Meddo and was used to represent an entire chain of islands or even the whole of the Marshall Islands. It showed the complex relationship between the islands and the major ocean swell.

Stick charts were not made and used by all Marshall Islanders. Only a select few knew the method for making and reading the charts, and the knowledge

Figures 1 and 2.

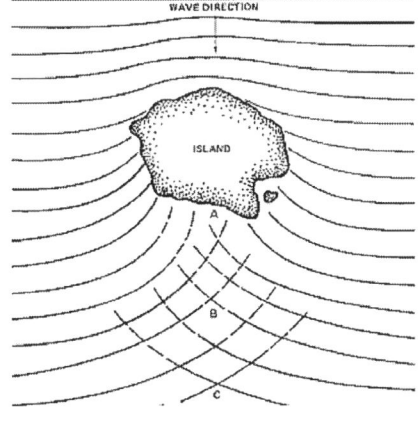
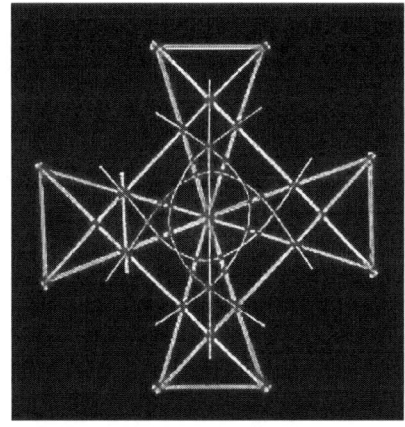

passed only from father to son. However, so that others could utilize the expertise of the navigator, 15 or more canoes sailed together in a squadron, accompanied by a lead navigator skilled in use of the charts. Because the knowledge contained in each chart was a closely guarded secret, they were not normally carried on a voyage. Instead, the navigator memorized the chart and gauged the wave patterns entirely by his sense of touch. Crouching in the bow of his canoe, he would literally feel the motion of his vessel.

It was not until 1862 that this unique navigational system was revealed in a public notice prepared by a resident missionary. It was an additional 30 years before it was comprehensively described by Captain Johann Winkler of the German Navy. He became so intrigued by the stick charts that he made a major effort to determine the navigational principles behind them and convinced the navigators to share how the stick charts were used. He recognized that the stick charts represented a significant contribution to the history of both navigation and cartography because they symbolized something that had never before been accomplished—a system of mapping and navigating by ocean swells. They are an indication that ancient maps may have looked far different, encoding different aspects from the natural world, than the maps commonly used today. The use of stick charts and navigation by swells apparently came to an end shortly after World War II. The venerable stick chart and ocean-going canoe were no match for large motorized vessels with modern navigational devices. They do, however, continue to be made in the Marshall Islands, though very few people are able to use them as navigation aids. They are primarily made and sold instead as tourist souvenirs.

Further Reading

Clark, Megan. "Cultural Cross-Purposes and Expectation as Barriers to Success in Mathematics." *Proceedings of the Ninth International Congress on Mathematical Education* 3 (2004).

Tee, Garry. "Mathematics in the Pacific Basin." *British Journal for the History of Science* 21 (1988).

Thomas W. Hair

Payroll

Category: Business, Economics, and Marketing.
Fields of Study: Algebra; Number and Operations.
Summary: Various payroll systems employ different mathematical calculations.

A variety of pay practices date back to ancient times, including compensation for services in the form of food, commodities, land, or livestock. Payroll systems are connected with the history of bookkeeping, which can be traced back to 4000 B.C.E. Paymasters were responsible for paying workers. Governments kept financial records called "pipe rolls" at least as early as the eleventh century. In 1494, Franciscan friar and mathematician Luca Pacioli published the book *Summa de Arithmetica, Geometria, Proportioni et Proportionalita*, which contained double-entry bookkeeping. The term *payroll* dates back to the seventeenth century, and compensation gradually changed from goods to money. In the mid-twentieth century, mathematician Grace Murray Hopper developed a compiler, later known as the FLOW-MATIC, which could be used for payroll calculations. When the U.S. Navy could not develop a working payroll plan, they called Hopper back to active duty. In the early twenty-first century, a payroll specialist is listed by some schools as a career option for mathematics majors. Accountants and actuaries calculate quantitative measures and predictions based on historic payroll information and salary increases. For example, the pensionable payroll is calculated as an integral that takes salary increases into account. In payroll analysis, the impact of changing salary expenses is compared to other factors, such as sales or profit.

Frequency

Some employees are paid each day they work; however, in many cases, an employer will withhold daily earnings and pay the cumulative amount earned at a later time as a lump sum. Common payroll frequencies include weekly, bi-weekly (every other week), semi-monthly (twice a month), and monthly. Each of these frequencies would correspond to receiving 52, 26, 24, and 12 paychecks each year, respectively, assuming a full year of work. Some seasonal jobs pay only for part of the year, but still use the standard payroll frequencies. For example, teachers often receive pay for only nine months. Some schools offer for that pay to be spread over a full year to guarantee consistent income during the summer months when teachers are not actually working.

On payday, the employee will receive earned wages for the previous pay period. Rather than receiving cash, sometimes an employee will receive a check that can be exchanged for an equivalent amount of cash. Other times, an employee will receive income as a "direct deposit" where the income is automatically deposited into the employee's checking or savings account.

Earning Money

Some employees work for an hourly wage—for every hour of work they perform, they get paid a specified amount of money. Suppose that a worker had an hourly wage of $10 and worked for 20 hours. To find the total amount of the paycheck, the worker would multiply the hourly wage by the number of hours worked. For example, $10 × 20 = $200.

Sometimes, contracts or laws dictate the number of hours a person can work per week and—should they work more than that amount—his or her income increases. For example, in the United States, 40 hours is a common workweek. A person working over 40 hours often gets paid "time and a half" or "wage and a half" for the number of hours over 40 that he or she works (called "overtime"). Again, assuming an hourly wage of $10, an employee who worked 48 hours in one week would earn $10 × 40 = $400 for the first 40 hours they worked. The eight hours he or she worked beyond 40 hours would earn him or her extra money. If the employee earns "time and a half," the time would be multiplied by 1.5 before being multiplied by his or her hourly wage. If he or she earns "wage and a half," the wage would be multiplied by 1.5 before being multiplied by the number of hours worked beyond 40. In reality, the method of calculating overtime earnings is irrelevant since multiplication is associative. Time and a half would be calculated as $\$10 \times (1.5 \times 8) = \$10 \times 12 = \$120$, and wage and a half would be calculated as $(\$10 \times 1.5) \times 8 = \$15 \times 8 = \$120$. The total earnings for that week would be found by taking the sum of these wages: $400 + $120 = $520.

Another method for earning money is a salary. Unlike the hourly wage, a salary is a predetermined amount of money that the worker earns regardless of how long (or how short) it takes the worker to accomplish those tasks. Often, salary is determined based on how much a person will make over a year's time. However, rarely does a person only receive one paycheck a year. The amount of money earned on each paycheck is calculated by taking the salary and dividing it by the number of pay periods in a year. That number will vary depending on how often a person gets paid. Suppose an employee agreed to work for a salary of $31,200 each year. Looking at the common pay periods, weekly, bi-weekly, semi-monthly, and monthly, this employee would earn $600, $1,200, $1,300, or $2,600, respectively, for each paycheck during the year.

A worker earning commission does not actually get paid based on how long it takes to do the job, but by how productive the worker is (oftentimes based on the amount of items the worker sells). Sometimes, commission is a flat fee per item sold, other times, it is a percentage of sales. For example, if an employee earned 7% commission on sales and sold $1,250 worth of merchandise on a given day, then pay would be calculated $1,250 × 7% = $1,250 × 0.07 = $87.50. Some jobs combine an hourly rate and commission—the employee earns a certain amount of money for every hour they are at the job, but then also earns commission on top of that wage to determine the total money earned.

Payroll Withholdings

Upon receipt of a paycheck or notice of direct deposit, usually the amount paid to the employee (the net pay) is less than what is calculated as his or her earnings for the pay period (the gross pay). Before being issued money, an employee may have his or her income reduced by certain amounts—some voluntary, others involuntary. In order to pay for various levels of government (and the benefits they offer), income and payroll taxes are frequently withheld from earnings. Some employees pay premiums for different insurances (such as medical, life, or disability) from their pay. Sometimes, money is withheld as a long-term savings for eventual retirement of the employee. Job-related expenses can also be withheld, such as for dues or charges for employee uniforms.

Further Reading

Booth, Phillip, Robert Chadburn, Steven Haberman, Dewi James, Zaki Khorasanee, Robert Plumb, and Ben Rickayzen. *Modern Actuarial Theory and Practice*. 2nd ed. Boca Raton, FL: CRC Press, 2004.

Bragg, Steven M. *Essentials of Payroll: Management and Accounting*. Hoboken, NJ: Wiley, 2003.

Haug, Leonard. *The History of Payroll in the U.S.* San Antonio, TX: American Payroll Association, 2000.

Chad T. Lower

Pearl Harbor, Attack on

Category: Government, Politics, and History.
Fields of Study: Geometry; Measurement; Number and Operations; Problem Solving.
Summary: Mathematicians were involved in both the planning of and the response to Pearl Harbor.

The attack on Pearl Harbor, a major engagement of World War II and the impetus for the United States' entry into the war, took place early Sunday morning, December 7, 1941, on the island of Oahu, Hawaii. The Japanese Navy, commanded by Admiral Isoroku Yamamoto, planned and executed the surprise attack against the U.S. naval base and nearby army air fields. As a result, the United States declared war on Japan. In his address to Congress, President Franklin D. Roosevelt famously proclaimed December 7 "a date which will live in infamy."

Both leading up to and as a result of the attack on Pearl Harbor, mathematicians in Japan and the United States mobilized for the war. For instance, after Pearl Harbor, the American Mathematical Society and the Mathematical Association of America converted their War Preparedness Committee to a War Policy Committee to increase research on "mathematical problems for military or naval science, or rearmament" and to strengthen mathematics education in order to prepare undergraduate students for military service. The attack has also been surrounded by speculation as to how the United States could have been caught off guard so easily. The naval base had been designed as nearly impenetrable to surprise attack because of the geography and geometry of the island. However, new technologies made the attack possible: the aircraft carrier could bring low-flying aircraft within attack range, and the Japanese development of shallow-running torpedoes could skim the surface of the harbor's relatively shallow water. One of the largest controversies involves U.S. efforts to decode Japanese communications that may have given forewarning of the attack.

Japanese Mathematicians

Leading up to Pearl Harbor, the number of Japanese graduate students increased and several studied in Germany. Mathematicians applied lattice theory and logic to the design of circuits. In the 1930s, both the United States and Japan successfully built a cyclotron, an early particle accelerator. Mathematics was also important in electrical engineering and airplane design. With a focus on aerodynamics and science and technology policy, the Japanese Technology Board was founded in 1941. A statistical institute contributed to war production. Japanese cryptologists also created many variations of military codes that were in use prior to Pearl Harbor, such as Kaigun Ango—Sho D, later referred to as "JN-25B" by cryptanalysts in the United States. Before committing to the attack on Pearl Harbor, the Japanese Navy conducted feasibility studies that included calculations and considerations of their current military resources; the need for a longer, circuitous route outside the customary naval traffic lanes to avoid detection by both military and civilian ships,; the probability of encountering severe winter storms and critical data obtained from spies in Hawaii, such as the patterns of military activity at Pearl Harbor. They concluded that the attack was possible, if dangerous, and they originally intended to specifically target U.S. aircraft carriers to optimize the long-term effects of the attack. Experimentation and simulated training attacks yielded a satisfactory plan only a few weeks before the event.

U.S. Mathematicians

In the United States, mathematicians conducted ballistics research at Aberdeen Proving Ground. Max Munk used the calculus of variations in airfoil design at the National Advisory Committee for Aeronautics, a precursor to the National Aeronautics and Space Administration (NASA). Technology such as radar, developed by scientists and mathematicians including Christian Doppler and Luis Alvarez, served military uses, though it was still in its infancy at the time of Pearl Harbor. Responsibility for compiling codes for military use and using cryptology to decipher codes shifted from Military Intelligence to the Army Signal Corps in 1929. William Friedman was the chief civilian cryptologist at the Signal Intelligence Service. The U.S. Army at that time realized the importance of mathematics in deciphering, and the first three civilian cryptanalysts hired by the U.S. Army were mathematics teachers.

Forewarning

Many wonder how the United States could not have known of the impending Japanese attack, which had been planned and practiced months in advance. The new radar installation on Opana Point did, in fact, detect the incoming Japanese attack planes, but they were ultimately mistaken for a group of U.S. planes that were due to arrive from the mainland that morning. A U.S. destroyer also spotted a Japanese submarine attempting to enter the harbor, which it reported, but the information was not acted upon immediately. Both would have given at least short-term warnings to the ships and personnel. However, much of the accountability is assigned to the U.S. and Japanese intelligence and counter-intelligence efforts. Correspondence declassified many years after the war suggests that the United States could at least partially understand the codes needed to monitor Japanese naval movements on the eve of Pearl Harbor. While U.S. and British cryptanalysts had successfully broken some Japanese codes, such as the MAGIC code, the United States was not able to determine from those messages that the attack was about to happen. The broken codes were the ones used primarily for diplomatic messages sent by the Japanese Foreign Office and military strategy was rarely shared with the Japanese Foreign Office. The U.S. Navy had three cryptanalysis centers devoted to breaking Japanese naval codes. Prior to the attack, American cryptanalysts had been using traffic analysis to follow Japanese naval movements. Traffic analysis is the process of looking for patterns in communications to infer if an attack is about to occur. According to the National Security Agency, the Japanese, aware that their communications were being monitored, issued "dummy traffic to mislead the eavesdroppers into thinking that some of the ships sailing through the North Pacific were still in home waters." Additionally, as Japanese forces were preparing for the attack, radio traffic was limited, greatly reducing the ability of American intelligence to determine a pattern. These efforts to stymie cryptologists were effective in keeping the impending attack a secret from the United States. Mathematicians and historians continue to analyze whether signal intelligence techniques could have revealed Japan's intentions.

Further Reading

Booß-Bavnbek, Bernhelm, and Jens Høyrup. *Mathematics and War*. Basel, Switzerland: Birkhäuser, 2003.

National Security Agency. "Pearl Harbor Review." http://www.nsa.gov/about/cryptologic_heritage/center_crypt_history/pearl_harbor_review.

PearlHarbor.org. "Why Did Japan Attack Pearl Harbor?" http://www.pearlharbor.org.

Wilford, Timothy. "Decoding Pearl Harbor: USN Cryptanalysis and the Challenge of JN-25B in 1941." *The Northern Mariner* XII, no. 1 (2002).

Calli A. Holaway
Michael G. Lovorn

Pensions, IRAs, and Social Security

Category: Business, Economics, and Marketing.
Fields of Study: Algebra; Data Analysis and Probability; Measurement; Number and Operations.
Summary: The development and allocation of retirement income can involve significant mathematical analysis.

Planning for retirement is one of the most important financial responsibilities a person faces. Ideally, after working for several decades, a person will be in a financial position to sustain a desired lifestyle during retirement. In the United States, the source of retirement income can be from a combination of one or more of the following: Social Security; an employer-sponsored pension plan; individual savings, including individual retirement accounts (IRAs); or other mechanisms. The U.S. government has provided military pensions to disabled veterans and widows since the Revolutionary War. This benefit expanded after the U.S. Civil War to include nearly any veteran who had served honorably for some minimum time. Southern states also paid Confederate veterans.

By the early twentieth century, state, municipal, and city governments were paying pensions to their employees, especially firemen and policemen. Teachers were the next large group to receive benefits. Private pensions started in the late nineteenth century with the American Express Company and several railroads. When the 1926 Revenue Act exempted pension trust income from taxes, companies had a new incentive to provide employee pensions, which became commonplace by the 1930s. Social Security was designed in 1935 to extend pension benefits to those not covered by a private pension plan. In the early twenty-first century, Social Security benefits are the main source of retirement income for most retirees, though this varies greatly depending on income from earnings, assets, and private pensions.

Each of these potential sources of retirement income can involve significant mathematical and financial analysis to estimate an individual's retirement needs, determine necessary pre-retirement financial planning, and evaluate the potential uncertainty associated with personal and economic factors. Specialized mathematicians known as "actuaries" work for governments and industries to design financially sound insurance and pension programs that help meet people's retirement needs.

At the same time, as the professionals at the American Pension Corporation (a major pension administrator) assert, good actuaries "are more than just mathematicians—[they] take great pride in [their] ability to dissect and communicate the intricacies of pension administration in layman's language."

Pensions

A pension provides a stream of income during retirement. It is typically sponsored by a person's employer—either a corporation or governmental entity. The amount and timing of the retirement income stream provided by a pension are a function of several factors, such as the worker's salary, the proportion of that salary invested into the pension plan, any matching funds or contributions to the pension fund provided by the employer, the length of the worker's tenure with the employer, and the investment performance of the pension fund. There are two types of pension plans: defined-benefit (DB), and defined-contribution (DC). DB plans, which have to some extent been phased out in the private sector but are still common in the public sector, define the benefits that will be paid to the worker during retirement. Assuming the solvency of the pension plan—a significant issue in itself—a worker covered under a DB plan is guaranteed to receive the benefits defined by the plan.

Because of potential difficulties in adequately funding DB plans, many (particularly private sector) employers converted to DC plans during the last several decades of the twentieth century. With DC plans, the retirement benefits are not specified; rather, the plan defines the periodic contributions to be invested during the worker's life, and then the retiree receives an income stream based on the actual accumulated amount of the investment fund. Relative to DB plans, this means that the employer's risk of inadequate retirement benefit funding is reduced, and that some risk has been transferred to the employee, who faces an uncertain pension income stream.

Mathematics of Pensions

The mathematics associated with pensions involves both "future (or accumulated) value" and "present value" concepts. The general idea is that a worker (or the sponsoring employer) accumulates a retirement fund by setting aside and investing periodic amounts during the working years. Then, upon retirement, this accumulated amount ideally represents sufficient funds with which to provide the retiree an adequate stream of income until death. This retirement income stream may be obtained by leaving the funds invested and withdrawing a certain amount per year, or through the purchase of an annuity, which provides the payment stream. In most cases, these two approaches are mathematically equivalent.

While somewhat straightforward conceptually, achieving an adequately funded and effective retirement plan (especially DB plans, which generally involve more sophisticated and extensive mathematical and financial considerations than DC plans) is a challenging mathematical and actuarial problem. Some of the parameters involved in a pension analysis, and for which assumptions must be made, include the following:

1. Periodic contributions to the pension investment fund—usually expressed as a percentage of worker salary during employment.
2. Size of the retirement income stream needed or desired—generally estimated as a percentage of projected salary immediately prior to retirement.
3. Rate of return on the invested retirement funds, both before and after retirement.
4. Changes in worker salary throughout employment.

5. Impact of inflation on the worker's buying power.
6. Taxation rules and regulations, both during employment and in retirement.
7. Longevity and mortality.

Along with these assumptions, actuaries use mathematics and computer modeling to determine potential answers to questions such as how much must a worker (or employer) invest every month (or year) into a retirement plan in order to successfully achieve that worker's financial goals in retirement?

IRAs and Social Security

In addition to having an employer-sponsored pension plan, a worker can supplement retiree income with personal savings. One such mechanism is one or more types of IRA. While the rules surrounding IRAs are extensive, they can have potential advantages for some people, including certain tax-advantaged properties.

Social Security is a particularly contentious issue in the twenty-first century. Some have compared Social Security to a type of scam called a "Ponzi scheme" in which a growing pool of new investors' money is used to pay the promised returns to previous investors. Despite superficial resemblances (for example, current taxpayer money is used to pay variable benefits to others), Social Security is not a savings plan or investment account, but rather a tax, which nullifies the comparison. However, there have been proposals to replace Social Security with an investment program, using a variety of calculations and probabilistic mathematical models to try to demonstrate its cost-effectiveness and the likelihood of the system's impending failure.

Another major financial issue related to social security is the potential misuse of Social Security numbers. Initially issued to track workers for taxation and benefits, these nine-digit numbers are now assigned routinely at birth and have grown over time into the role of a unique identifier for creditors, schools, employers, and others who want to assign codes to individuals. Modern identity theft, which usually involves a person using a fake or stolen social security number to obtain credit or other benefits, has been on the rise as a result of Internet growth and the widespread collection of personal data. Mathematicians have calculated that a person making up a false social security number in 2010 has about a 50% chance of matching a real

> ## Stochastic Variables
>
> What makes such a quantitative analysis particularly challenging is that many of these parameters are stochastic rather than deterministic—their future values are uncertain, and they can (and do) change value over time. Data analysis and probability concepts are used to account for this uncertainty. For example, inflation and investment rate of return are both stochastic variables, with considerable uncertainty regarding their values in both the short- and long-term.
>
> Estimates of possible future values and the relative probabilities or likelihoods of those values can be made by analyzing historical data. These estimates can then be used to project future scenarios and quantify the potential impact of possible future values on the retirement funding process.
>
> Another critical stochastic variable in retirement planning is the age at death of the retiree. The number of years that a retiree lives beyond the date of retirement is an essential factor in determining the total amount of income needed during the retirement years. Actuaries research and analyze historical mortality data for people with various identifiable attributes. From these analyses, a probability distribution of possible ages at death, with their relative likelihoods, can be developed.

number. Faking multiple numbers results in an almost-guaranteed match very quickly.

These calculations have been used to counter thieves' assertions that they did not know numbers they were using were real. Social Security numbers themselves are not random (for example, the first three digits are a numerical code for geographic location), and mathematical and computer methods have used publicly available data, like date and place of birth, to successfully predict most or all of a person's social security number. There are also concerns that the government will run out of Social Security numbers, which are not reused after a person dies. Some calculations suggest that the supply will be exhausted early in first half of the twenty-second century. Alternative proposals include using alphanumeric or hexadecimal strings, which offer more permutations for a series of nine "digits." Others suggest including a security checksum in the number to decrease fraudulent use.

Further Reading

Anderson, Arthur. *Pension Mathematics for Actuaries*. Winsted, CT: ACTEX Publications, 2006.

Harding, Ann, and Anil Gupta. *Modeling our Future: Population Ageing, Social Security and Taxation*. Amsterdam, The Netherlands: Elsevier Science, 2007.

Muksian, Robert. *Mathematics of Interest Rates, Insurance, Social Security, and Pensions*. Upper Saddle River, NJ: Prentice Hall, 2002.

Rick Gorvett

Predicting Attacks

Category: Government, Politics, and History.
Fields of Study: Algebra; Data Analysis and Probability.
Summary: Predictive mathematical models can be used to attempt to foresee and counter various types of attacks.

An increasing area of interest in mathematics is the use of algorithms and computer models to predict attacks—military attacks, terrorist attacks, and even attacks on Web servers. As with meteorology, a model is a probabilistic statement; the future cannot be predicted with absolute certainty but probable causes, patterns, and outcomes can be quantified and mathematically modeled to extrapolate the likelihood of new events. Humankind has been trying to predict attacks ever since one group first fought another using some combination of observation and subjective judgment. However, formal prediction of attacks using mathematical methods appears to have originated only within the last two centuries and has escalated with advances in technology and data gathering.

Mathematician Lewis Richardson made contributions to many areas within and outside mathematics, such as numerical weather prediction, in the first half of the twentieth century. The Richardson iteration is one method for solving systems of linear equations, while the Richardson effect refers to the apparently infinite limit of coastline lengths as the unit of measure decreases, a precursor to the modern study of fractals. Richardson spent many years analyzing data on wars from the early nineteenth century onward, using mathematical methods such as probability theory and differential equations, often quantifying psychological variables, such as mood. He identified several patterns in war and identified some variables likely to prevent conflict. He is often credited with first introducing the notion of power laws to relate conflict size, frequency, and death toll. At the start of the twenty-first century, models had grown in complexity. In 2009, a University of Maryland team developed a model that uses 150 variables and data accumulated from the activity of 100 insurgent groups in the Middle East in order to model their reactions to Israeli activities. Other models have been developed to attempt to predict violence and attacks in Iraq and continue to be refined. Statistical methods like data mining and power law functions are prevalent in modern predictive modeling.

Data Mining

Data mining is the process of extracting patterns from large to enormous bodies of data. Isaac Asimov's *Foundation* stories, the first of which was published in 1942, depicted a future where "psychohistory" was the study of the future using the body of history as data from which to extrapolate the future. Modern data mining is quite similar to Asimov's predictions and may be accomplished by many mathematical methods. For example, many use artificial neural networks, which are computational models that mimic neuron behavior. Genetic algorithms, credited to scientist John Holland, are search heuristics inspired by the processes of gene recombination and evolution. Decision trees may be used to determine conditional probabilities. In the 1980s, support vector machines (SVMs) were developed to analyze data to find patterns for statistical classification. All of these developments greatly advanced the state and potential of machine learning and facilitated rapid processing of increasingly larger and frequently interlinked databases from sources such as credit card companies, telecommunications businesses, and government intelligence agencies. Within the U.S. government, the Department of Defense began using data mining in the late 1990s in its Able Danger program, which gathered counterterrorism data, including data about the Al Qaeda terrorist group. Some asserted that the program uncovered the names of four of the alleged September 11, 2001, hijackers a year before the attacks. In February 2002, the U.S. Office of Science and Technology Policy convened a panel of government and industry leaders to discuss data mining as a counterterrorism tool. While it is now widely used, some criticize it because the sparsity of some information and the relative infrequency of terrorist attacks make identifying statistically significant patterns, which are critical to finding the anomalies that signal an attack, prone to unacceptable levels of error.

Cyber Security

Mathematicians, computer scientists, and others are continually working on new methods to predict and counter attacks on Web servers, e-mail, and digital records of all kinds. The Internet is filled with malicious activity, from phishing and identity theft to distributed denial of service attacks. Electronic attacks are facilitated by the same computer technology that is used to predict attacks. The traditional guard has been to block a source of malice after the attack, by e-mail as spam or blocking an IP address after harmful activity originates from it. These methods are commonly known as *blacklists* and are now widely compiled and shared. However, they are by definition reactive measures to attacks. Just as e-mail spam filters have become preemptive, marking mail as "spam" automatically based on a number of factors, IP-blocking can also be conducted preemptively.

The method of predictive blacklisting uses shared attack logs as the basis for a predictive system, like the customer recommendation systems employed by Amazon or Netflix. Computer scientists Fabio Soldo, Anh Le, and Athina Markopoulou developed what is known as an "implicit recommendation system"—implicit because ratings are inferred rather than given directly by the subjects of the model. Their multilevel prediction model uses mathematical methods, such as time series analysis and neighborhood models, adjusted specifically for attack forecasting. Inputs to the model include factors such as attacker-victim history and interactions between pairs or groups of attackers and victims. Similar models—using different types of data—can be built to predict terrorist attacks and the behavior of enemy forces, and such models are included in the standard order of battle intelligence reports used by the U.S. Army.

The data needed to predict attacks are not restricted to private databases. Information is widely available from the Internet or the scrolling news banners of 24-hour news networks. Neil Johnson used a variety of sources to investigate insurgent wars, employing some of the same mathematical techniques as Richardson in his analyses and modeling. After gathering and analyzing data for almost 60,000 insurgent attacks occurring in multiple conflicts around the world, he and his collaborators discovered similarities between the frequency and intensity of attacks in all conflicts. Further, they found that the statistical distribution for insurgency attacks was significantly different from the distribution of attacks in traditional war. The model quantifies connection between insurgency, global terrorism, and ecology, and counters the common theory of rigid hierarchies and networks in insurgencies. Johnson notes:

> Despite the many different discussions of various wars, different historical features, tribes, geography and cause, we find that the way humans fight modern (present and probably future) wars is the same, just like traffic patterns in Tokyo, London, and Miami are pretty much the same.

Further Reading

Jakobsson, Markus, and Zulfikar Ramzan. *Crimeware: Understanding New Attacks and Defenses*. Boston: Addison-Wesley, 2008.

Memon, Nasrullah, Jonathan Farley, David Hicks, and Torben Rosenorn. *Mathematical Methods in Counterterrorism*. New York: Springer, 2009.

Bill Kte'pi

Predicting Preferences

Category: Business, Economics, and Marketing.
Fields of Study: Data Analysis and Probability; Geometry; Measurement.
Summary: Psychology of choice and predictive models of preferences are exciting areas of mathematics blending social science, economics, and commerce.

Mathematically, preference is an ordering of alternative possibilities. It can refer to conscious choices based on ideas and beliefs, positive emotional responses or liking, or biologically mandated behaviors. Preferences are usually determined statistically: for individuals, based on multiple instances of decisions over time; and for groups, based on aggregated data of members. In 2009, the Netflix Prize contest awarded a team called BellKor's Pragmatic Chaos $1 million for their preference-predicting algorithm.

Theoretical and Behavioral Economics

Among all sciences that deal with predicting preferences, such as social psychology and education theory, the most developed mathematical apparatus can be found in economics. As any branch of mathematics, theories of economic preferences start with axiomatic assumptions. These abstract axioms do not always apply to all real situations. Economic theories that take into account psychological factors, such as cognitive limitations and emotions, are developed within an interdisciplinary area called "behavioral economics."

Most abstract theories of preference prediction assume most parts of the so-called total order, which is a group of mathematical axioms and properties from set theory. Let A, B, and C be different choices. Total order assumes that either $A \leq B$ or $B \leq A$. In real life, this assumption is a statistical statement at best: today a person can prefer apples, but might prefer bananas tomorrow. The property of transitivity says that if $A \leq B$ and $B \leq C$, then $A \leq C$. This property works in some situations; for example, if one prefers $20 over $10, and $100 over $20, it is likely the person will prefer $100 over $10. However, in complex situations with multiple choices, such as elections, transitivity fails to describe real human behavior. Experiments show that, given a choice between one pair of candidates at a time, people may prefer Beth over Alice, Carol over Beth, and Alice over Carol. One axiom of total order, called "antisymmetry," that almost never makes sense in preference theories is that if $D \leq E$ and $E \leq D$, then $E = D$. For example, when group data shows that people think diesels are worse or the same than electric cars, and electric cars are worse or the same as diesels, it does not mean that diesel cars and electric cars are the same entity. It means that people prefer them about the same. Economic theories call this situation "indifference" and use a separate symbol for it: $E \sim D$.

Another assumption frequently made in economic preference predictions comes from topology and is called "continuity." It is the assumption that if A is preferred over B, then an option that is very similar (close) to A will also be preferred over at option that is very similar to B. Many complex phenomena, including preferences, are discontinuous. They exhibit various "tipping points," near which minute differences cause radical changes in preferences. These non-continuous phenomena are studied using models from calculus or chaos theory, a branch of differential equations. One frequent example of noncontinuous preference is price near powers of 10: many people choose to buy an object that costs $999 over a similar object that costs $1,001 even though the difference in prices is minuscule compared to the total. Behavioral economics explains this by cognitive limitations: people see 1001 as thousands and 999 as hundreds, which is technically correct but makes less of a difference in this case than intuition leads one to believe.

Paradoxical Preferences

A paradox is a false or contradictory statement that logically follows from a set of true statements. Preference prediction leads to several types of paradoxes. A very frequent type is the situation when an initial model describes the reality well, but its mathematical corollaries do not. Another type, a true logical paradox, occurs when mathematical corollaries contradict one another.

For example, the expected value is the sum of products of probabilities and payoffs. Suppose a fair coin is flipped in a hypothetical game and the player is paid $10 if the coin lands on heads and $20 if it lands on tails. The expected value of winning is $15 because $0.5(10) + 0.5(20) = 15$. When the same game is played many times, it is rational to prefer options with higher expected values. Under this assumption, it is better to play the game where the player is paid nothing for heads and $40 for tails than the first game, because the expected value of winning is higher: $0.5(0) + 0.5(40) = 20$. However, in real life, risk aversion will make many people choose the first game.

To resolve this and other related paradoxes, many preference models account for risk aversion as a separate variable. A utility function is the measure of relative satisfaction of a range of choices. An assumption that people will only want to maximize utility is not realistic, because it does not account for risk aversion. Because marginal choices usually come with higher risks, the utility function that accounts for risk aversion will look like a hump, being concave.

Bounded rationality principle is commonly used to explain paradoxical preferences by taking into account limited information, time, and cognitive abilities of people. Models based on bounded rationality include human limitations, such as computational capacity, and are based on computer science, statistics, and psychology.

Information Theory and Aesthetic Preferences

Information theory is a mathematical science that studies storing, compressing, and processing of data. In the 1990s, its branch called "algorithmic information theory," which deals with the complexity of algorithms, was applied to explain some aspects of the human sense of beauty and of aesthetic preferences. According to this theory, objects that have shorter algorithmic descriptions in terms of observer's knowledge will seem more beautiful, compared to objects with longer algorithmic descriptions. For example, it is easier to remember an object with mirror symmetry because only half of the information is original—symmetry provides information compressibility. Therefore, symmetric objects, as well as objects with patterns or fractal self-similarity, are seen as more beautiful.

Algorithmic information theory also models preferences by interest, which are separate from preferences based on beauty. Within these models, interest can be compared to the first derivative of beauty, showing the observer's perception of change in understanding. People prefer an experience on the basis of interest when it involves better compressibility or predictability of information than before. For example, noticing a new pattern (and therefore better organizing an image) is preferred because it is interesting.

Preferences, Desires, and Motivation

Many preferences and choices are based on needs, wants, and desires, which are explained in theories of motivation. Researching motivation is challenging because of individual differences among people, as well as language ambiguity. There are disagreements among researchers even over relatively straightforward terminology, such as intrinsic and extrinsic motivation.

(Photos.com)

Psychology of Choices

Statistical analysis of real situations, such as elections, as well as results of experiments and questionnaires, allow scientists to aggregate increasingly sophisticated knowledge of human mechanisms of choice and preference. For example, from the purely mathematical viewpoint, gaining an amount and avoiding loss of the same amount are equivalent. However, most people regret loss more strongly than they regret missed opportunity—a fact extensively used in advertisements of savings and discounts.

Preferences are very strongly influenced by power over the situation. Most people accept much higher risks for given gains if they enter the situation of their free will, compared to risks of mandated behaviors. This phenomenon comes up, for example, when mandatory immunizations are proposed—the fact that people would not have a choice makes very small risks unacceptable.

Many motivation theories include taxonomies of needs and desires. For example, in Maslow's hierarchy, named after Abraham Maslow, unsatisfied physiological needs, such as hunger or thirst, have higher priority than unsatisfied self-esteem needs, such as recognition. Some theories identify long lists of motivators, such as curiosity, tranquility, order, and independence. Other theories only define a few broad classes of needs.

Each category of need can be considered a variable. Graphs of values of these variables versus levels of motivation often demonstrate the characteristic "mirrored C" shape called a "backward bending curve." For example, as activities provide more order, they first become more motivating (and preferred), but beyond a certain point, more order becomes less motivating. This curve is famously described in the baseball manager Lawrence "Yogi" Berra's joke about a restaurant: "Nobody goes there anymore. It's too crowded." People usually prefer restaurants that are not too empty or too full.

Preferences and Demographics

A number of statistical studies find significant differences in preferences of different demographics within populations, such as males and females, socioeconomic classes, ages, and political affiliations. Because statistical packages make many types of mathematical and statistical analyses of databases very easy, there are many results that demonstrate significant differences in preferences among different demographics. However, determining meanings of these differences is a significantly more difficult research problem. Demographic differences in preferences can also vary from culture to culture. In some cultures, for example, more females than males prefer bright colors in clothes, and in other cultures, it is reversed.

Further Reading

Anthony, Martin, and Norman Biggs. *Mathematics for Economics and Finance: Methods and Modeling.* New York: Cambridge University Press, 1996.

Berry, M. A. J., and G. Linoff. *Data Mining Techniques For Marketing, Sales and Customer Support.* Hoboken, NJ: Wiley, 1997.

Netflix Prize. http://www.netflixprize.com.

Maria Droujkova

Prehistory

Category: Government, Politics and History.
Fields of Study: Measurement; Number and Operations.
Summary: Historians believe that even the earliest people used mathematics.

Many books on the history of mathematics begin with the ancient Egyptians and Babylonians, but those civilizations did not begin until about 5000 years ago. Although historians do not know many details, human life had been progressing for several millennia prior to that time. Even archeology offers little detail on the earliest mathematics, so most knowledge comes from speculation. However, from what is known about human beings in general, and especially about prehistoric life, even the earliest people must have known and used some mathematics.

The use of "mathematics" probably even precedes the development of modern human beings. Studies of animal behavior have shown that animals, and especially birds, seem to possess limited number sense, recognizing the difference between groups of two and three and even larger sets. Bees can recognize and even communicate information about the location of orchards and fields for pollination, displaying a sense of space that could be called "geometry." Even more spectacular are the long migratory trips of herd animals, flocks of birds, and groups of butterflies, often traveling thousands of miles to return to the same fields every year. These examples certainly do not represent a sophisticated concept of mathematics and are instinctual, but they show a mathematical organization in the brain.

Language, Counting, and Quantities

The earliest humans (wherever the line is drawn between pre-human and human) continued the mathematical thinking shown in animals. As their brains developed, their mathematics also grew stronger and more sophisticated. This progression continued as early grunts become proto-languages, for a key part of mathematics is not only having the concepts in one's head, but also representing and communicating the concepts to others. Hence, language was a key ingredient in prehistoric mathematics (as it remains today).

A concept of counting must have come early, as people began to distinguish quantity. Even if they did not have linguistic terms for numbers beyond three or four, they would at least be able to make rough comparisons of large quantities and much larger quantities—consider that even modern humans often need notations, pictures, or concrete examples to handle specific large quantities, but certainly can tell the difference between a dozen and a hundred and a million. Many aspects of life require at least limited counting—to make sure all one's goats (or children) are present, to share items fairly in a group or to calculate the size of a load to be carried, and many other applications.

It is only a small jump of abstraction to begin to record quantities with tally marks. It is likely that people first collected stones or other small objects to represent quantities and later began to "write" them as tallies. Tally marks have been found in many parts of the world scratched on cave walls or carved onto wooden sticks and were also likely written in sand or clay, which shifted to destroy the writing. Probably the most famous prehistoric mathematical object is the Ishango bone, found in south-central Africa, and thought to be at least 15,000 years old. The bone has several sets of tallies scratched onto it—some have pointed out that they are mostly prime numbers, but that is probably a coincidence. Using tallies quickly leads to a problem: a long line of marks is hard to deal with, even if one had some limited counting words. Probably, many people around the world recognized that some structure helped handle large quantities of tally marks—especially collecting them into groups of the same size. Not only does this make counting more efficient but it also leads to the concept of multiplication. In nearly all modern languages—most derived from ancient or even prehistoric languages—the higher counting words use a system of groups and groups of groups, now called "place-value," but they reach back to the prehistoric convenience of putting tally marks together.

Measurement and Geometry

Closely tied to counting was the use of comparative relationships—especially large and small, tall (or long) and short, and even old and young. These may have come when exact counts were difficult, but the comparisons were obvious and usually visual. A tall stack of blocks would easily be seen to have more items than a short stack; a long line of tally marks (grouped or ungrouped) was a greater quantity than a short line. As actual counting developed and numbers were applied to comparisons, the beginnings of measurement occurred—measurement is really just comparisons of quantities where one side of the comparison is a defined unit. To make comparisons easier, certain items of specific size or quantity became units, and as people reached farther to wider audiences, units became at least roughly standardized. Often, body parts were used both for counting tabulations and as "standard" units. For example, the distance from the elbow to the fingertips was approximately the same for most adults, so in the Middle East, this length became the "cubit."

Geometry also has deep roots in the human story. Circles must have been recognized in the shape of the sun and full moon and the apparent edge of the horizon. Efficiency caused people to arrange objects to fit together well in patterns—often circular but sometimes rectangular. The first tools used sharp angles, heavy weights, and tall, thin cylinders. The beginnings of farms led to more organized geometrical arrangements in the shapes of fields and structures. Often, the "invention" of the wheel is considered one of the big milestones of the start of civilization, and this represents a practical understanding of the geometry of circles. As objects became more sophisticated—woven mats, farming tools, larger structures, and even bridges—many more geometrical relationships and properties were discovered. These might be considered the beginnings of engineering—using mathematical properties in practical applications.

Pure Mathematics

Archeologists have also noted some prehistoric mathematics that may have been closer to pure mathematics. Cave paintings, carved sculptures, and textile patterns show contemporary mathematical objects such as circles, triangles, parallel lines, quadrangles, symmetric patterns, and the crosshatch. However, no one has yet deciphered what the geometric signs meant to prehistoric peoples. Some symbols appeared repeatedly in various parts of the world. They may have served practical or religious values, but they also were art—perhaps art for its own sake, for beauty. Certain numbers may have had mystical meanings that were seemingly less useful for day-to-day activity but important for esthetics and spirituality.

The overlap between this pure mathematics and the practical needs of early farmers was the use of mathematics in astronomy and calendars. Could the gods show the times for planting and harvesting? Could humans discern the plans of these gods and use them in practice? Most of the spectacular prehistoric structures, from Stonehenge in England to the huge geometrical patterns of Nazca in Peru, have been linked to measures of the sun's movement and the seasons. Mathematics led prehistoric peoples in solving their daily problems and to thinking of the universe and infinity. Mathematics still serves modern humans in the same ways.

Further Reading

Boyer, Carl. *A History of Mathematics*. Hoboken, NJ: Wiley, 1991.

Burton, David M. *The History of Mathematics: An Introduction*. New York: McGraw-Hill, 2007.

Eves, Howard. *Introduction to the History of Mathematics*. New York: Saunders College Publishing, 1990.

Ifrah, Georges. *The Universal History of Numbers*. Hoboken, NJ: Wiley, 1994.

Von Petzinger, Genevieve. "Geometric Signs in Rock Art & Cave Paintings." http://www.bradshawfoundation.com/geometric_signs/geometric_signs.php.

Lawrence H. Shirley

Quality Control

Category: Business, Economics, and Marketing.
Fields of Study: Data Analysis and Probability; Measurement.
Summary: Industrial productions and processes can be mathematically studied to help ensure quality.

Statistical quality control, or more broadly, quality assurance, seeks to improve and stabilize the production and delivery of goods and services. A central concern of quality control is the testing and reporting of measurements of quality—typically as part of a monitoring process—to ensure that the quality of the item being studied meets certain standards.

Quality standards are determined by those who produce the goods or services. Some standards are specification limits imposed by engineering or design concerns that define conformance to a standard. For example, in making airplane engines, a certain part may need to have a diameter between 12 and 14 millimeters or it will not fit into a housing. However, for many processes, there are no specification limits and quality standards may be defined internally from data on past behavior of a process that is judged to be "in control" or "stable." For example, in examining the safety of a large production line, it may be that in each week of the last five years, the average number of person hours lost to accidents has been 1.3. There is no specification limit for this quantity, but control limits can be based on this historical average.

In order to analyze a process for statistical quality control effectively, a process must first be declared to be "in control." To be in statistical control, the vast majority of the products or services must be of sufficient quality for the producers to be satisfied. Moreover, the process must be stable (the mean and variance of the quality measurements must be roughly constant). If a process is in control, then statistical analysis can provide meaningful control limits to the process for monitoring. Graphical methods play a significant role in statistical quality control.

History

Some measure of quality control was in evidence during the building of the Great Pyramids of Egypt. Archeologists have long been impressed not only with the complexity of the construction process, but also by its precision. In the Middle Ages, medieval guilds were formed, in part, to ensure some level of quality of goods and services. The use of statistical methods in quality control—also called "statistical process control" or (SPC)—is more recent, with most of the development in the twentieth century. Graphical methods for quality control were introduced in a series of memos and papers in the 1920s by Walter E. Shewhart of Bell Telephone Laboratories. The charts he developed and promoted are known today as "Shewhart control charts." H. F. Dodge and H. G. Romig, also of Bell Laboratories, applied statistical theory to sampling inspection, defining rules for the acceptance of many products. Joseph M. Juran, whose focus was more on quality management, rather than SPC, was another early quality pioneer at Bell Laboratories and later Western Electric.

W. Edwards Deming applied SPC to manufacturing during World War II and was instrumental in introducing these methods to Japanese industry after the war ended. He and Juran are generally credited with helping Japanese manufacturing shed the negative image that "made in Japan" had in the 1950s and transforming the country into a source of high quality goods consumed all over the world. In the early twenty-first century, quality control issues continue to appear in the media as concerns proliferate over the quality of goods produced in China.

Common-Cause and Special-Cause Variation
Shewhart and Deming defined two types of variation that occur in all manufacturing and service processes in their 1939 book *Statistical Methods from the Viewpoint of Quality Control*. A certain amount of variation is a part of all processes and can be tolerated even when the goal is to produce goods and services of high quality. This variation is called "common-cause variation," and it comprises all the natural variation in the process. The second variation, called "special-cause variation," is unusual and is not part of the natural variation. Special-cause variation needs to be detected as soon as possible. Quality control charts are designed to detect special-cause variation and distinguish it from common-cause variation.

Quality Control Charts
A quality control chart plots a summary of the quality measurements from each item (or a sample) in sequence against the sample number (or time). A center line is drawn at the mean, or at the desired center of this statistic. Upper and lower control limits are drawn indicating thresholds above or below which will signal an "out of control" measurement. Sometimes, various warning lines are drawn as well, and a variety of rules for deciding if the measurement is really out of control are available. The simplest chart, called an "individual" (or "runs") "chart," plots a single measurement for each item. The control limits are based on the Normal probability model, which implies that for a process in control, only 0.27% of the observations will lie more than three standard deviations (σ) from the center. Therefore, if the process stays in control, a false alarm will occur only once in about 1/0.0027 or once every 370.4 observations. The central idea of a control chart is that a special cause will cause the mean to shift (or the standard deviation to increase), and so the measurement will fall outside the 3σ limits with higher probability. If the shift is great enough, the time to detection will be very short. However, if the special cause results in a subtle shift, it may take many observations before such a signal is detected. Various other types of charts are available that have generally better performance in terms of both false alarm rates and failure to detect shifts.

Total Quality Management and Philosophy
The ideas of Deming, Juran, Shewhart, and others have inspired numerous other people and quality movements. One such movement is total quality management (TQM) also known as "total quality" and "continuous quality improvement." As the name implies, this approach to quality involves more than the monitoring of manufacturing or service processes. It includes all parts of the organization and, specifically, the role of management to help ensure that in providing goods or services, that "all things are done right the first time." Implementing these ideas throughout a large organization gave rise to an abundance of books, experts, and quality "gurus" in the latter part of the twentieth century. One approach to total quality focuses on reducing variation (decreasing σ). If the common-cause variation can be reduced enough, while the process is in control, essentially no measurements will fall outside the 3σ limits. This notion is the essential idea behind the 6σ approach, first popularized by the Motorola company and later the General Electric Company in the 1980s. By the late 1990s, a majority of the Fortune 500 companies were using some form of the 6σ approach.

Further Reading
Deming, W. Edwards. "Walter A. Shewhart, 1891–1967." *American Statistician,* 21 (1967).
———. *Out of the Crisis*. Cambridge, MA: MIT Press, 2000.
Juran, Joseph M. *Quality Control Handbook*. New York: McGraw-Hill, 1999.
———. *Management of Quality Control*. New York: Joseph M. Juran, 1967
Snee, Ronald D., and Roger W. Hoerl. *Leading Six Sigma: A Step-by-Step Guide Based on Experience With GE and Other Six Sigma Companies*. Upper Saddle River, NJ: FT Press, 2002.

RICHARD DE VEAUX

Renaissance

Category: Government, Politics, and History.
Fields of Study: Geometry; Representations.
Summary: The Renaissance's resurgence in humanism also benefited mathematics and engineering.

The Renaissance or Rinascimento (both words mean "rebirth") was a flourishing of philosophy, art, architecture, science, and high culture more generally beginning in fourteenth-century Europe. Renaissance thinkers thought of themselves as restoring the civilization of Greece and Rome after what they called "the Middle Ages." The Renaissance saw the rise of humanism, hermeticism, Neoplatonism, and realist art involving optical perspective; the decline of feudalism; increased circulation of ideas due to printing; the Protestant Reformation; a strong interest in classical literature and history; a strengthened interest in science and mathematics and their applications; and world exploration.

Early Renaissance (c. 1300–1450)

The Renaissance can be traced back to the thirteenth-century writings of Dante Alighieri, Francesco Petrarca, and Brunetto Latini and the paintings of Giotto di Bodone. Such work was sponsored by bankers, merchants, and industrialists who rose to great wealth and influence, displacing the Church and landed nobility as primary sponsors of high culture.

Starting in the mid-fourteenth century, humanist scholars searched libraries to recover the lost texts of classical Rome. Many edited texts went to print, increasing their accessibility at (relatively) low cost. After approximately 50 years, attention turned to recovering the Greek heritage, which—though mostly lost in the West—had continued on in Byzantium. Many Greek scholars migrated west at this time, bringing their expertise and manuscripts to Venice, in particular. The recovery and translation of Plato's works, along with several tracts in neoplatonism and hermeticism, fueled an interest in applying simple numerical ratios and geometric regularity in fields as diverse as art and architecture, cosmology, alchemy, and musical tuning. The intentions included occult efforts to replicate cosmic structures, invoking astral influences at the human scale. More visceral results were achieved by composers, such as Josquin des Prez, who brought polyphonic techniques to Italy from the Low Countries, laying foundations for important Italian composers (such as Giovanni Pierluigi di Palestrina) toward the end of the sixteenth century.

Renaissance (c. 1450–1500)

The Renaissance spread north from Tuscany and across the Alps during the second half of the fifteenth century. Political philosophy, exemplified by Niccolò Machiavelli's *Prince* and *Discourses on Livy*, attempted a rational analysis of political structures contextualized by cultural difference and the practicalities of everyday life. Vernacular languages came to be used for scholarly writing, making texts more widely readable as did printing, which advanced rapidly with the establishment of fine publishing houses in the Veneto. Examples include the Aldine Press, where italic typefaces were invented and Erhard Ratdolt's press, which pioneered the printing of mathematical diagrams when producing the first edition of Euclid's *Elements* in 1482.

The mid-Renaissance was centered on the Republic of Florence, largely sponsored by a powerful banking family, the Medici. The ideals of this period are expressed in Florentine architecture, such as Filippo Brunelleschi's Church of San Lorenzo, which has a legible geometric regularity, bright and even light, openness, and a delicately balanced stillness. Ideals in painting included realism based on optical theory. Artists could occupy the leading edge of mathematical research; Piero della Francesca, for example, produced treatises on perspective theory in addition to painting with perspective techniques. Sculpture also developed a scholarly foundation through both historical study of the classical texts that had survived and hands-on dissection of fresh cadavers.

High Renaissance (c. 1500)

The High Renaissance lasted only briefly before transforming into Mannerism. It was focused on Rome, owing to the patronage of Pope Julius II. Art gained a level of dynamism best known through the works of Rafaello Sanzio (Raphael) and Michelangelo Buonarotti in Rome, and Tiziano Vecelli (Titian) and Giorgione in Venice. Leonardo da Vinci's *Last Supper*, Raphael's *School of Athens*, and Michelangelo's ceiling in the Sistine Chapel were painted during the High Renaissance.

Further north, the Renaissance adapted to local cultures and circumstances. In Germany, for example, goldsmiths crafted clocks, automata, and mathematical and astronomical instruments for their patrons. Reformation printers published a wide range of medieval texts alongside Lutheran tracts, largely shedding the refined typography of Venice in favor of speed and quantity. Gothic elements remained strong in the art and architecture of England, the Netherlands, and Scandinavia and Renaissance influences reached those countries only after they had become Mannerist. Because of Protestantism, secular authorities replaced the Catholic Church as the primary sponsor of cultural works.

Renaissance Science and Mathematics

Renaissance scholars initially reacted against Scholastic natural philosophy by turning to Neoplatonism, taking an often mystical and magical approach to nature, often with practical goals. This shift can be seen in the intertwining of alchemy and astrology, for example, and in the wide range of applications described in Giambattista della Porta's 1558 book *Natural Magic*. The title reflects a distinction drawn between natural magic, which invoked empirical knowledge of nature to achieve results; in contrast to spiritual magic, which regulated astral influence using amulets and talismans; and demonic magic, which invoked supernatural beings.

The Church's need for calendrical reform led Nicolaus Copernicus to develop heliocentric astronomy as an improvement upon the Hellenistic methods maintained and developed throughout the Middle Ages. Astronomy was favored also in Protestant territories owing to the educational reformer Philip Melanchthon arguing that it was an ideal way to learn about divine creation.

Artillery motivated studies in ballistics, leading to stellated polygonal designs for fortresses, such as Naarden in the Netherlands and the Kronborg in Denmark. Aristotelianism, however, still provided qualitative theory for ballistics and other practical endeavors, such as hydraulic engineering.

The development of machines and engineering techniques inspired efforts to classify and theorize about them, as shown by the published "theaters of machines" by Jacques Besson and Agostino Ramelli.

The influences of exploration can be dated at least as far back as 1488, when Bartholomeo Dias found a connection between the Atlantic and the Indian Ocean that led to trade routes established beginning in 1498 with Vasco da Gama's arrival in Calicut, six years before Christopher Columbus found the West Indies. Such journeys motivated developments in navigation and shipbuilding as well as an outward-looking attitude. Trade expanded, especially in Spain, Portugal, and—as the new knowledge spread north—the Netherlands. Descriptions and specimens brought back from foreign regions caused disputes and reforms in biological taxonomy that were eventually settled in the eighteenth century by Charles Linnaeus.

Progressive rational problem-solving, combined with the growth of theoretical method and a growing preference for naturalistic rather than occult explanations, provided many elements needed for the eventual emergence of modern empirical science.

Mathematics was boosted early by the ascendance of merchants and bankers who needed computational methods to manage money and later to solve problems in navigation and cartography. Some advanced material was assimilated from Arabic sources, such as geometric methods and high-precision trigonometric tables. Solving polynomial equations became a display of virtuosity; the quadratic had been solved in antiquity, now Girolamo Cardano and other mathematicians developed solutions for cubics and higher order problems. As algebra developed, many algebraic symbols were invented and evolved into the forms used today. Hindu-Arabic numerals replaced Roman numerals but the calculation of the products, ratios, and square roots of large numbers in astronomy and navigation was still onerous and error-prone. These operations were facilitated by conversion into addition and subtraction problems using prosthaphaeresis (based on trigonometric transforms), and later through the invention of logarithms.

Further Reading

Field, J. V. *The Invention of Infinity: Mathematics and Art in the Renaissance*. Oxford, England: Oxford University Press, 1997.

Goulding, Robert. *Defending Hypatia: Ramus, Savile, and the Renaissance Rediscovery of Mathematical History*. New York: Springer, 2010.

Hall, Marie Boas. *The Scientific Renaissance, 1450–1630*. New York: Dover Publications, 1994.

Hay, Cynthia. *Mathematics from Manuscript to Print 1300–1600*. Oxford, England: Oxford University Press, 1988.

ALISTAIR KWAN

Revolutionary War, U.S.

Category: Government, Politics, and History.
Fields of Study: All.
Summary: The American Revolutionary War saw advances in mathematics cryptography and education.

The American Revolutionary War was a political and armed conflict between Great Britain and the British colonies on the North American continent between 1775 and 1783. Colonists who sought to end British rule and declare their political and economic independence supported the establishment of 13 colonial governments, each of which in turn sent representatives to Philadelphia to set up the Second Continental Congress.

This congress debated the state of political and economic ties to Britain, plied for support from other European powers, and discussed the possibilities and potential of a collective effort to make the separation official. Shortly after its inception, the Second Continental Congress formed a Continental Army and issued the Declaration of Independence. These actions announced the birth of a new nation: the United States of America. The "War of American Independence," as the American Revolutionary War is also called, saw fierce fighting in a wide variety of locations throughout the new nation and on the soil of virtually every new state. Some key battles were fought in Lexington, Concord, and Boston, Massachusetts; Saratoga and Ticonderoga, New York; Trenton, New Jersey; King's Mountain and Cowpens, South Carolina; and Yorktown, Virginia; among many other places.

The war lasted almost a decade and ended with the Treaty of Paris, which was signed at the Palace of Versailles in 1783 and recognized the sovereignty of the United States of America. There are many statistics available that relate to aspects of the war, including casualties and cost. For instance, some report that the British spent about £80 million while incurring a national debt of 250 million pounds, while the United States spent approximately $135 million, of which $37 million became the national debt. Mathematics was used in a wide variety of ways, including in the design and implementation of artillery and in planning strategy and tactics. Mathematicians fought in the war, conducted surveys, and created and decoded ciphers. The mathematics educational system also changed significantly as a result of the war.

Louis-Antoine de Bougainville

Many historians agree that the Americans would have been unable to win the war without the political and

---

Cryptography

Early U.S. military intelligence began during the Revolutionary War. Paul Revere, William Dawes, and others used light signals to warn of invading forces before the battles at Lexington and Concord, which are generally considered to be the first military engagements of the war. James Lovell, who has been called the "father of American cryptanalysis," broke the British ciphers, which were rearrangements of letters. He used a method known as frequency analysis, which involves determining letters based on the frequency of symbols in the coded message.

Lovell discovered that the British often changed ciphers by shifting them instead of creating a new rearrangement and this made them easier to decode. Lovell also created his own cipher forms but these were deemed too confusing for those wanting to send and receive messages.

This belief was even true for Benjamin Franklin, who was well versed in mathematics and enjoyed magic squares recreationally. Franklin commented, "If you can find the key & decypher it, I shall be glad, having myself try'd in vain." American diplomats began to rely increasingly on replacements of words and other techniques instead of alphabet substitutions, and spies for both sides conveyed information about supplies and troop movements using codes. For instance, U.S. spy Benedict Arnold used book ciphering, in which a word is represented by a number that corresponds to a location in a book, in his communication with British intelligence officer John Andre.

military support of France and other allies. Louis-Antoine de Bougainville was a French mathematician who became the first Frenchman to sail around the world. In 1752, he wrote a calculus book, *Traité du calcul–intégral*, which brought him recognition within the mathematical community for his clear exposition and updates to differential and integral calculus. After a second edition and election to the Royal Society of London in 1756, he turned to a career in which he participated in numerous wars, including the Revolutionary War.

His astronomical observations became important to later explorers. He stated, "geography is a science of facts: one cannot speculate from an armchair without the risk of making mistakes which are often corrected only at the expense of the sailors." During the Revolutionary War, he was a commodore who supported the U.S. side.

Simeon DeWitt

U.S. Army geographer Simeon DeWitt subscribed to *The Mathematical Correspondent*, generally regarded as the first U.S. special-interest scientific publication. DeWitt was a student at Rutgers University when British troops burned the college buildings. He continued his study of mathematics and surveying on his own and was appointed the geographer of the army by General George Washington. After the war, he became surveyor-general of New York State.

Education

Mathematics education changed dramatically in the United States during and after the war. Before the war, students usually learned mathematics from British works, although Americans like Isaac Greenwood had written arithmetic texts. Advanced mathematics included algebra, geometry, trigonometry, calculus, and surveying techniques. Many colleges were shut down during the war because students and professors served as soldiers, and buildings were used for other purposes. However, some members of the army were trained in mathematics during the war. After the war, new primary schools and colleges were established. Between 1776 and 1815, numerous mathematics texts were published in the United States. Some of these were reprints of English works, and others were compilations or new works by American writers. In 1788, American Nicholas Pike published his text, *The New and Complete System of Arithmetick: Composed for the Use of the Citizens of the United States*, which contained both arithmetic and geometry. It was popularized by patriotic recommendations. There was also a change in the education of women. Prior to the war, it was thought that mathematics beyond simple arithmetic was unnecessary for women. After the war, mathematics educational opportunities began slowly to increase, as women were educated in mathematics to help in family businesses.

Further Reading

Weber, Ralph. "James Lovell and Secret Ciphers During the American Revolution." *Cryptologia* 2, no. 1 (1978).

Tarwater, Dalton. *The Bicentennial Tribute to American Mathematics*. Washington, DC: The Mathematical Association of America, 1977.

Tolley, Kim. *The Science Education of American Girls*. New York: Routledge, 2003.

Zitarelli, David. "The Bicentennial of American Mathematics Journals." *The College Mathematics Journal* 36, no. 1 (2005).

Calli A. Holaway
Michael G. Lovorn

Roman Mathematics

Category: Government, Politics, and History.
Fields of Study: Connections; Number and Operations; Representations.
Summary: The ancient Romans, who are often remembered for their applied mathematics, made important contributions to surveying, time-keeping, and astronomy.

The Roman period for mathematics could be said to have started when a Roman soldier was sent to seize Archimedes during the capture of Syracuse. Told by Archimedes to wait as he finished his diagrams, the soldier lost patience with the old man and slew him. The popular stereotype of the Romans is that they did little to advance Greek discoveries in mathematics, instead merely applying Greek methods to practical problems. This conception is not entirely fair. The Roman Empire was not one homogenous zone, but was rather a collection of culturally diverse provinces. For this reason, many works produced during the time of Roman rule, like the books of Ptolemy, writing in Alexandria, Egypt, are written in ancient Greek rather than Latin. Therefore, these books could be considered Greek, Roman, or Greco-Roman depending on the context. However, despite this diversity, the Roman period led to the dominance of some mathematical practices that still have an influence in the twenty-first century.

Roman Numerals

One of the most distinctive remnants of Roman mathematics is the use of Roman numerals, which are letters that stand for specific values and usually work as additive values. The numerals are

I = 1   V = 5   X = 10
L = 50   C = 100   D = 500
M = 1000.

So: LXXVII $= 50 + 2(10) + 5 + 2(1) = 77$.

The numerals are written with the largest values at the left, proceeding to the smaller values. They can also have subtractive constructions. I preceding subtracts one from a 10 to make nine. X before an L or C produces 40 or 90, and C before D or M produces 400 or 900. So:

MCMXLVIII =

$1000 + (1000 - 100) + (50 - 10) + 5 + 3(1) = 1948$.

The origins of the system are unknown. It has been proposed that they were based on tally marks, with I being a notch, V being a double notch to mark five, and 10 as crossed-notches (though it could also be that X was formed from two V symbols). The number IV to represent 4 is a later addition based on medieval Latin and does not seem to have been used by the Romans, who instead used IIII.

This system is not very helpful for arithmetic, and so it is little surprise to find that the Romans developed the portable abacus to ease mathematical operations. This device was a tray with a number of columns etched into it that could hold pebbles. A pebble (in Latin, the word "calculus") had a value depending on the column that held it. Moving a pebble a column to the left increased its value by a factor of 10. Such an abacus could be used by merchants in the city or by surveyors working for the military.

Survey

Roman surveyors employed geometry to divide the landscape and lay out cities with effects that can still be seen in the twenty-first century. The key to Roman survey was a tool called a *groma*, which was a tall staff with a beam, known as a *rostro*, at right-angles to the staff at the top. The rostro supported a wooden cross, and at each end of the cross-beams was hung a plumb line. Sighting across these lines allowed Roman surveyors to lay out grids of perpendicular lines in the landscape. Surveyors could then divide land for agricultural purposes, and some field systems in Europe are based on these ancient surveys. The *groma* also left an impression on modern cities. The Romans frequently built new cities in conquered territories, for either native inhabitants or new settlements of veteran soldiers. At the heart of a Roman settlement lay the forum, the central civic space, which usually lay at the intersection of the Cardo maximus (the main north-south street) and the Decumanus maximus (the main east-west street). This system created new cities with grid-plans in which the main intersection was laid out by a *groma*. These perpendicular grids were the origins of many European settlements and was adopted in the planning of many U.S. cities in the nineteenth century.

The Roman Calendar

The Roman calendar instituted by Julius Caesar made a radical change to time-reckoning in Europe. Before this development, European calendars outside Rome were usually luni-solar calendars. As such, each month was related to the lunar cycle, which is not commensurate with the solar year, and so periodically whole months, known as "inter-calary months" would be inserted into the year to keep the months in step with the seasons. Insertions would usually have to be done every two or three years. Even ancient authors recognized that this system was inefficient, including Herodotus, who wrote in the late fifth century B.C.E. that the Egyptians had a much more accurate solar calendar. In 45 B.C.E., Julius Caesar adapted the Egyptian method of timekeeping for Roman use.

Each month was counted as a period of days, usually 30 or 31 but with 28 or 29 in February. In addition, Julius Caesar laid down rules for when an inter-calary day would be added to February. The Egyptians corrected the calendar by adding a day every fourth year. Unfortunately, the Romans counted inclusively, meaning that the leap year was in the fourth year, rather than after the fourth year. For example, 2020 is a leap year. For the ancient Romans, the second year in the cycle is 2021 and the third is 2022. Therefore, 2023 is the fourth and the Romans of Julius Caesar's time would have made this a leap year, rather than 2024. Augustus Caesar corrected this error in the early years of the first century C.E.

This method of keeping the years remained until the reforms of Pope Gregory XIII in 1582, though Britain and the American colonies did not implement the Gregorian calendar until 1752. The difference between the two calendars is that years divisible by 100 are not leap years, unless the year is divisible by 400. Otherwise, years are marked by the same cycle of months as the ancient Romans did.

Mathematics and the Cosmos

Even though ancient mathematicians had a relatively small set of tools based in geometry and arithmetic, these could be used to create incredibly intricate models. Ptolemy proposed a model of the universe that contained circles rotating upon circles to reproduce the movement of the planets. The connections between mathematics and cosmology made mathematics attractive to philosophers of the Roman period. The assertion that mathematics could reveal truth became increasingly contentious in late antiquity. Pagan philosophers came into conflict with a new religious sect, Christianity, which was increasingly powerful. One notorious incident was the killing of Hypatia, a female mathematician philosopher, in the city of Alexandria by a Christian mob. For some ancient historians, her death marks the end of the period known as classical antiquity.

Further Reading

Cuomo, Serafina. *Ancient Mathematics*. London: Routledge, 2001.

Dilke, Oswald. *Roman Land Surveyors: Introduction to the Agrimensores*. Newton Abbot, England: David and Charles, 1971.

Hannah, R. *Time in Antiquity*. London: Routledge, 2009.

Jaeger, Mary. *Archimedes and the Roman Imagination*. Ann Arbor: University of Michigan Press, 2008.

ALUN SALT

Sales Tax and Shipping Fees

Category: Business, Economics, and Marketing.
Field of Study: Number and Operations; Measurement.
Summary: Different types of sales taxes and shipping fees affect the final price of a purchase.

Benjamin Franklin famously noted, "Our Constitution is in actual operation; everything appears to promise that it will last; but in this world nothing is certain but death and taxes." When someone makes a purchase, often times there are extra charges added to the customer's bill. These costs may include a tax, shipping charges, or fees. These extra amounts, however, have a special purpose and they are each computed differently. For example, a sales tax is based on a percentage of the total amount of the sale and that percent is regulated by local and state governments. On the other hand, shipping is charged to cover the delivery of merchandise from the retailer to the customer's location. These fees are based on the policies of the company selling the goods as well as how quickly the customer would like their purchase delivered. Lastly, fees can be special charges; for example, insurance might be added to a purchase to cover the cost of the merchandise in the event it is lost or damaged during delivery. Albert Einstein commented that preparing a tax return "is too difficult for a mathematician. It takes a philosopher." The calculations to determine sales tax and shipping fees utilize percentages, multiplication, and addition, but Einstein may have been referring to the ever-changing instructions.

Both mathematicians and philosophers have long been involved in issues related to taxation. The *Jiuzhang suanshu* (*Nine Chapters on the Mathematical Art*) contains related problems. In the tenth century, astronomer and mathematician Abu'l-Wafa wrote a text on mathematics for scribes and businessmen, with part four of the book containing seven chapters devoted to various kinds of taxes and related calculations. In the seventeenth century, lawyer and amateur mathematician Étienne Pascal worked as a tax assessor and was appointed as the chief tax officer. In order to help his father in his tax work, mathematician and philosopher Blaise Pascal invented the Pascaline, which is reported to be the first digital calculator. In the twenty-first century, financial planners, mathematicians, and actuaries create mathematical models and investigate a variety of mathematical concepts related to taxes and fees, including the impact of flat rate, progressive, symmetric, or asymmetric taxation; and game theory applied to the interaction between taxpayers and tax collectors. They also investigate equilibrium states and how increasing or decreasing sales taxes or shipping and handling fees or using a nonlinear structure impacts consumer decisions about purchases and business sales.

Sales Tax

Many states, counties, and municipalities levy a sales tax as a way to increase revenues for their government or to balance their budget; however, not every state or local government charges a sales tax. The rate of the tax varies depending on the laws of the governmental unit. In other words, a purchaser will encounter different sales tax rates throughout the United States. The charges in 2010 varied from 0% in states like Alaska or Delaware to a high of 8.25% in California. This means that a person in Alaska who pays $100 for an mp3 player would not be required to pay any tax on the sale. However, a person buying that same mp3 player in California would be required to pay this tax. In other words, that $100.00 purchase would have an 8.25% tax added to the cost, meaning the new purchase price would be the original cost ($100.00) plus the sales tax ($8.25) for a total of $108.25.

Many localities exempt certain classifications of goods from their sales tax. Some common exceptions include groceries and prescriptions. On the other hand, special items such as gasoline, cigarettes, and alcohol have a significantly higher sales tax, as they have the potential to add sizeable revenue to a state's budget. A federal law called the Internet Tax Freedom Act (ITFA) specifically addresses sales over the Internet. The law provides that no governmental unit is allowed to add any special or additional tax on Internet purchases. This means that a sales tax may be charged on Internet purchases at the same rate as items purchased in person or by phone but no extra tax charge can be added.

Shipping and Handling Fees

Shipping and handling fees vary dramatically by seller as well as by the type of shipping the buyer requests. Common factors used to compute delivery costs include (1) how many items are being purchased, (2) how much the order weighs, and (3) how quickly the customer would like to receive their merchandise. However, common shipping types include free shipping, overnight delivery, two day or expedited delivery, and standard shipping, which may vary from three to seven days. In addition, the cost may change based on the number of items purchased or the weight of the merchandise. The following three examples illustrate different types of shipping options:

- *Flat fee*: The seller charges a flat shipping fee for all purchases regardless of price, weight, or number of items.
- *Progressive*: The seller charges a progressively larger shipping charge based on the cost of the purchase. Shipping for a $50 purchase might cost $5, while shipping for a $100 purchase might cost $10.
- *Flat fee and item charge*: The seller charges a flat shipping rate plus an item charge (shipping + charge × number of items). Assume that the base shipping is $3.99, and there is a charge of $0.99 for each item. A one item purchase would have a charge of $3.99 + $0.99 = $4.98. However, suppose the purchaser buys three items. In that case, the charge would be $3.99 + 3($0.99) = $6.96.

Shipping and fees are often grouped together as one charge; however, some vendors are known to charge each of these as separate and distinct charges. Vendors often add an additional charge to deliver a purchase. One example would be a package that requires special handling based on size or weight, such as a piece of furniture. Higher cost items such as jewelry might have an insurance charge added to the customer's total.

Further Reading

Anderson, Patrick. *Business Economics and Finance With MATLAB, GIS and Simulation Models*. Boca Raton, FL: CRC Press, 2000.

Consortium for Mathematics and Its Applications. *Mathematical Models with Applications*. New York: W. H. Freeman & Company, 2002.

Marks, Gene. "Don't Forget the Handling!" *Accounting Today* 23 (2009).

Scanlan, M. "Use Tax History and Its Implications for Electronic Commerce." *The Information Society* 25 (2009).

Konnie G. Kustron

Scheduling

Category: Business, Economics, and Marketing.
Fields of Study: Data Analysis and Probability; Number and Operations.
Summary: Scheduling can be a complex mathematical exercise and is necessary to keep businesses and supply chains running efficiently.

Intense competitiveness forces companies to optimize performance in terms of cost, time, and resources. Scheduling is the process of developing and implementing optimal operational plans. Formal concepts of scheduling date to the Industrial Revolution and innovations like Henry Ford's assembly line, although the basic ideas probably existed from antiquity in any society where people manufactured goods.

In manufacturing, multiple tasks are carried out in sequence to produce a final output from raw materials. Further, steps in a manufacturing process may be performed on different machines that require variable time to deliver outputs and it is possible that materials will be transported between facilities. A mathematically determined schedule that takes into account all relevant variables in the process serves to optimally allocate resources with respect to demand of the tasks, including shortening time intervals to reduce unproductive time and minimizing costs from wasted time and materials. Operations research is a field of applied mathematics and science that uses mathematical tools, such as simulation and modeling, linear programming, numerical analysis, graph theory, and statistical analysis, to arrive at optimal or near-optimal solutions to complex problems like scheduling. It may also tackle problems in which the resources are not materials but people. The scheduling of airplane crews is a highly constrained and difficult problem because of legal limits on work and rest times as well as the need for crews to return to a home base. Allocation of police, fire, and ambulance services is also a widely used and very important application of scheduling theory.

Production Management

As a part of production management, scheduling interferes with many different aspects of business such as the supply chain, inventory maintenance, and accounting. For example, consider a paint company that makes provisions of sales for the next month by analyzing previous data. In light of these provisions, schedulers determine the expected arrival time and amount of different types of chemicals, which have different delivery times.

The supply chain should be able to deliver the correct amounts of chemicals in time. In a similar way, accounting of the cost of supply and inventory should be accessible for the schedulers. Because of the number of operational parts of business that scheduling is related with, it is apparent that scheduling is a very complex process. It gets more complex with larger variation in types of products and larger numbers of machines varying in processing times. Thus, schedulers demand thorough knowledge of factors such as the processing time of each machine, delivery time, the amount of resources to allocate among machines, and the size and flow of operations for each product.

Manufacturing

In many manufacturing processes, different machines might share the same input, or inputs of a machine might consist of outputs from multiple machines. Scheduling operations in these type of cases requires extensive mathematical modeling. Two basic types of modeling for production scheduling are distinguished by the presence of randomness within. Deterministic models do not include the probability of faults in processes or critical changes in capacity or resource availability. They are based on previous averages of production figures and output rates, so they do not easily adapt to changes in demand or capacity constraints. In these cases, rescheduling is needed, which causes time and resource loss if repeated too many times. They are best suited to manufacturing productions that involve less risk of defects. Stochastic models, on the other hand, involve the probability of unexpected malfunctions or critical changes by distributing probability analytically to individual steps of the schedule. Usually, they are appropriate for processes consisting of many individual operations. For example, these models examine machine failure rates and aim to provide options for when a breakdown occurs. Also, these models maintain an inventory of materials, which may prove critical in maintaining production. Simulations of models provide schedulers an environment to test possibilities that can obstruct the flow of production.

Further Reading

Conway, Richard W., William L. Maxwell, and Louis W. Miller. *Theory of Scheduling*. New York: Dover Publications, 2003.

Pinedo, Michael. *Scheduling: Theory, Algorithms, and Systems*. New York: Springer, 2008.

Blazewicz, J., K. H. Ecker, E. Pesch, G. Schmidt, and J. Weglarz. *Scheduling Computer and Manufacturing Processes*. New York: Springer, 2001.

Kogan, K., and E. Khmelnitsky. *Scheduling: Control-Based Theory and Polynomial-Time Algorithms*. New York: Springer, 2000.

Ugur Kaplan

Stock Market Indices

Category: Business, Economics, and Marketing.
Fields of Study: Algebra; Data Analysis and Probability; Measurement.
Summary: Stock market indices use sophisticated mathematical formulas to track the performance of the stock market and to help inform investors.

Mathematical stock market indices are used for a variety of purposes: as indicators of overall market health and activity, as measures of specific corporate profitability and activity, as performance metrics against which institutional investors (such as mutual fund managers) are measured, and for individual portfolio optimization and risk assessment. Some mathematicians and economists were developing price-based indices as early as the nineteenth century as well as analyzing pricing trends for explanation and prediction of market behavior. The Dow Jones Industrial Average (DJIA), named for journalist Charles Dow and statistician Edward Jones, appeared in 1896. Initially, it was a simple sum or average of the stock prices from 12 large companies. Since then, stock market indices have increased in their variety and mathematical complexity. For example, technical analysts use Fibonacci retracement levels, named after mathematician Leonardo Pisano Fibonacci, in order to model support and resistance levels in the currency market. Mathematicians and statisticians are instrumental in producing these indices. They also conduct theoretical and applied studies of market performance using these indices as data. In 1999, French-American mathematician Benoit Mandelbrot showed that market volatility can be modeled by fractal geometry, which contradicted some aspects of modern portfolio theory. Author and mathematician John Allen Paulos addressed many mathematical stock market issues in his popular book *A Mathematician Plays the Stock Market*.

Definition and Examples

When describing the performance of the stock market as a whole (or a segment of the market, such as selected large-company stocks, or all small-company stocks, or stocks of all companies belonging to a particular industry), one is usually referring to a stock market index. Such an index is a representation of a hypothetical portfolio that contains a certain quantity of each of the stocks in the market (or market segment). The quantity of each stock in the fictitious portfolio depends upon the "weighting" technique employed.

Some of the more commonly encountered stock indexes include the following:

- S&P 500, comprised of 500 large-company U.S. stocks that cover about 75% of U.S. equities
- DJIA, comprised of 30 large-company U.S. stocks
- Wilshire 5000, comprised of the most common stocks in the United States (although not necessarily exactly 5000 of them)
- Nikkei 225, an index of Japanese equities
- FTSE, a collection of indices of British stocks

Building a Stock Market Index

The wide variety of stock market indices fall into several weighting categories, each involving a different mathematical approach to combining stocks within a hypothetical portfolio. One can imagine a potentially unlimited number of ways of creating a portfolio that includes numerous company stocks: for example, a portfolio comprised of one share of each stock, a portfolio comprised of the same dollar amount of each stock, and so on. The most common methods of weighting stocks within an index are price-weighting and market-value-weighting. (To simplify, stock performance is treated as only a function of changes in the stock price over time—as capital gains and losses. In reality, dividends, stock splits, and a variety of other issues must be taken into account, which makes the specific mathematical applications more complex than represented in this entry.)

Price-Weighted Indices

A price-weighted stock index represents a theoretical portfolio that includes one share of each stock comprising the index. The price or value of the index is then equal to the average of individual stock prices. Therefore, the relative impact of a given company stock on the index is a function of the company's stock price per share: larger prices per share imply greater influence on the index.

Suppose that $S_i(t)$ represents the per-share price of stock i at time t, and let $S_I(t)$ be the value of the index at time t. Then, the price of a price-weighted index could be defined as simply the arithmetic average of the stock prices in the index:

$$S_I(t) = \frac{\sum_{i=1}^{n} S_i(t)}{n}(t)$$

where n is the number of stocks comprising the index.

While the value of a price-weighted index is simple to calculate, typically the measure of most interest to an investor is not the actual price of the index, but rather the percentage change (the rate of return) in the index over a period of time. Let $r_i(t, t+1)$ be the rate of return on stock i during the period from time t to time $t+1$, and let $r_I(t, t+1)$ be the return on the index between times t and $t+1$ (assume an annual return period for purposes of this discussion, but returns can also be calculated daily, monthly, quarterly, or over any other period of time).

Then, the return on a price-weighted index is

$$r_I(t, t+1) = \frac{S_I(t+1)}{S_I(t)} - 1 = \frac{\sum_{i=1}^{n}\left[r_i(t, t+1) \times S_i(t)\right]}{\sum_{i=1}^{n} S_i(t)}.$$

Multiplying this value by 100 yields the return expressed as a percentage change. The DJIA and other Dow Jones averages are examples of price-weighted indices.

Market-Value-Weighted Indices

A market-value-weighted (also called "value-weighted") stock index is one that weights the individual per-share stock prices according to the relative market values, or market capitalizations (called "market cap" for short), of the component stocks. A company's market cap is simply the totally value of its outstanding equity and is calculated as the per share stock price multiplied by the number of stock shares outstanding. Thus, an individual company's influence on a value-weighted index is a function of the overall equity value, or size, of the company—larger companies have greater influence on the movement of the index.

Using the notation introduced above, and letting N_i represent the number of shares of stock i outstanding, the rate of return on a value-weighted stock index would be

$$r_I(t, t+1) = \frac{\sum_{i=1}^{n}\left[r_i(t, t+1) \times S_i(t) \times N_i\right]}{\sum_{i=1}^{n}\left[S_i(t) \times N_i\right]}.$$

The S&P 500 and other Standard & Poor's indices are examples of market-value-weighted indices.

Other Types of Index Weightings

While price-weighted and value-weighted indices are common, there are other weighting techniques that can be used. For example, it is possible to create an index that gives equal weight to the return of each stock comprising the index. In such a case, the return on the index would be calculated as

$$r_I(t, t+1) = \frac{\sum_{i=1}^{n} r_i(t, t+1)}{n}.$$

With such an index, the performance of each stock has the same impact on the overall index return as every other stock.

Another possibility in creating an index would be to use geometric, as opposed to arithmetic, averaging. A geometric average is calculated by multiplying n numbers together and taking the n-th root of the product (as opposed to summing the numbers and dividing by *n*, as with an arithmetic average).

The key in interpreting the various types of stock market indices is to know their underlying construction and to understand and interpret them appropriately. Price-weighting and equal-weighting, for example, can result in very different index performance indications than value-weighting, even relative to the same underlying stock return data. The appropriate index to use in a given situation depends upon the specific purpose in mind. If one wants a measure of market performance that is more influenced by the price movements in the stocks of larger companies, for example, a value-weighted index may be most appropriate. If the sizes of companies are not relevant for analytical purposes, or if the companies that comprise an index are very similar in size and other attributes, a price-weighted or value-weighted index may be appropriate.

Further Reading
Bodie, Zvi, Alex Kane, and Alan Marcus. *Investments*. New York: McGraw-Hill/Irwin, 2008.
Paulos, John Allen. *A Mathematician Plays the Stock Market*. New York: Basic Books, 2003.

Rick Gorvett

Strategy and Tactics

Category: Government, Politics, and History.
Fields of Study: Geometry; Measurement; Problem Solving; Representations.
Summary: Mathematical concepts and processes can be used to analyze optimal strategies in a variety of situations.

In a competitive situation, such as businesses selling similar products, armies engaged in battle, opponents playing games, oil companies deciding where to drill, and employees bargaining for better salaries, successful outcomes depend on choosing the best plan of action from among a set of strategies to achieve a specific outcome. In many cases, mathematics can be used to analyze the situation and help to choose the best strategy. Mathematical techniques have been—and will continue to be—developed to address a wide range of problems in areas such as military logistics, intelligence, and counterintelligence.

The first step in the process is to determine the objective. That goal may be to maximize profit, beat the opposing army, or win the game. Next, the possible strategies to choose from and the limitations or constraints that may affect the choice of strategy need to be identified.

In competitive situations, the opponent's choice of strategy must be taken into consideration as well. While there are many examples of systematically analyzing and selecting the "best" strategies throughout history, the twentieth century—especially the World War II era—saw the emergence of operations research as the discipline that explores and develops systematic techniques for making decisions that are the "best" in some sense, usually maximizing profits/benefits or minimizing costs/liabilities.

Decision making can be approached mathematically in a number of ways depending upon the situation involved and the information available.

Linear Programming: Choosing the Best Option When Resources are Limited

Many decision problems arose out of troop supply needs during World War II. With a war on several fronts, deciding how to ship the limited troops and supplies to maximize their effectiveness was daunting. Many of the situations had the following characteristics:

- There were resources needed in specific combinations by a number of end users, and the amount of each resource was limited.
- The resources were used proportionally for each combination (in other words, to assemble whole units from raw materials, the number of raw materials needed was the same for each unit produced).
- The goal was to maximize the benefit or minimize the cost, and the cost or benefit was proportionally related to the number of units produced (in other words, the more produced, the higher the benefit or cost).

These characteristics yield a mathematical structure that is linear. Each resource corresponds to an equation or inequality that is a linear combination of unknown quantities representing the units to be combined or produced The objective function is also a linear combination of the number of units. See Example 1 for a very simple, classic example that involves deciding how to prepare a "balanced" meal.

Example 1. A linear programming problem.

A dietician wishes to prepare a salad meal that has a minimum amount of calories but still satisfies nutritional requirements. In particular, it must have at least 30 grams of protein and at most 9 grams of fat. The foods available are an ounce of lettuce with 4 calories, no fat, and 1 gram of protein; and slices of roast beef with 90 calories, 3 grams of fat, and 16 grams of protein. What amounts of lettuce and beef should the dietician serve with a diet salad dressing? Minimize calories = $4L + 90B$, where $1L + 16B \geq 30$ and $3B \leq 9$.

These problems are easy to solve when they are small, like the problem in Example 1. The problems that arise in practice—such as those under consideration during World War II—are usually much larger and can involve hundreds of unknowns. During World War II, British and U.S. mathematicians looked for an approach that could make use of computers, which were being developed at that time and offered the possibility of performing many simple calculations quickly. In 1947, too late for the war effort, U.S. mathematician George Danzig (1914–2005) developed the simplex algorithm for solving linear programming problems. The simplex algorithm is an efficient recipe for solving linear programming problems of any size and is very easy to program on a computer. In the decades since the development of the simplex algorithm, many industries have used this procedure to solve problems in fields as diverse as banking, natural resources, manufacturing, and farming.

Linear programming problems are usually used to model static situations in that the final solution is essentially the result of one decision made under a clear set of assumptions. Many decision problems are more complicated, with a number of intermediate decisions to be made. These more dynamic problems often involve a probabilistic component as well, with uncertainty playing a complicating role in each decision.

Game Theory

Often, people are faced with a decision in which the resulting payoff will depend on external forces that are hard to predict (like natural forces). One option may always be best, but it is more likely that the best choice will simply "depend" on other factors. For example, when deciding which crop to plant, a farmer can list seed costs and profits based upon yield, but the yield will depend on the weather. A table can be made for each crop choice based upon several different weather scenarios, with past experience used to assign a probability to each possible weather scenario. Example 2 provides a standard format, usually called the "payoff matrix."

Example 2. A payoff matrix.

	List the possible states of the external forces
List the possible actions to choose from in making the decision	List the gain (profit, benefit, etc.) for each combination of actions and states.

Many decisions can be similarly structured, including determining what stocks to buy, what products to market, and what wars to wage. Different people will make different decisions depending upon their comfort level with risk.

Strategies for systematic decision making can be placed in four categories:

1. *Optimist strategy*: "MaxiMax" (Maximize the maximum gain). Find the best gain for each possible action and choose the largest of these maximums. Of course, that action may have the most risk associated with it, since the maximum gain may also coincide with the least likely state for the external force. In this case, the farmer may plant something that would have huge profits but only in the most unlikely weather conditions.

2. *Pessimist strategy*: "MaxiMin" (Maximize the minimum gain). Find the smallest gain for each action, and choose the largest of these minimums. This is a safe choice because it yields the minimum guaranteed gain regardless of external forces. In this case, the

farmer may choose a "safe" crop to plant. If weather is really good, another crop would have been a better choice.

3. *Balanced strategy*: "MiniMax Regret." Calculate the "regret" for each possible action by determining the cost of choosing that action compared to benefits of the best state of the external forces. Find the worst (largest) regret for each action and pick the action with the smallest worst-case regret.

4. *Averaging strategy*: "Expected Value." Use the probabilities governing the external forces to determine the expected gain for each action and choose the highest one. Expected gain or payoff is calculated as a weighted average of the gain for each state of the external force where the weight for each state is the probability of that state occurring. This strategy can be thought of as determining the action that, when chosen repeatedly, provides the best average benefit over the long term. For the farmer, this may not seem reasonable, since the decision under consideration is what to plant in a single, given year.

When the external force is an opponent with choices to make rather than a natural phenomenon with a random component, these decision situations can be examined as mathematical games. Two-person games can be represented with a payoff matrix as in Figure 2. The "row" player lists strategies on the left and the "column" player lists strategies across the top. The entries of the matrix are pairs of numbers, the row player's payoff, and the column player's payoff, respectively. In situations where the row player's winnings are equal to the column player's losses, and vice versa, the payoff matrix entries can be completely defined with one number, conventionally the row player's payoff. These games are called "zero-sum games" because for a particular pair of strategies, the row player's payoff and the column player's payoff, being negatives of each other, sum to zero.

Example 3. The prisoner's dilemma.

Two suspects are arrested by the police. They are each offered the same deal: Confess and receive a reduced sentence. If one confesses and the other does not, the confesser goes free and the other gets a 10-year sentence. If both confess, each gets a five-year sentence. If neither confesses, both get a one-year sentence on reduced charges. Neither prisoner knows what the other will say. What should they do?

	Confess	Refuse
Confess	$(-5,-5)$	$(0,-10)$
Refuse	$(-10,0)$	$(-1,-1)$

While mathematicians have been studying decision making and games of strategy systematically for several centuries, game theory emerged as a recognized mathematical approach to analyzing these decision processes in the 1930s and 1940s through research published by John von Neumann (1903–1957). The "prisoner's dilemma" (Example 3) was investigated in the 1950s and led to additional interest in the field.

The prisoner's dilemma captures many interesting features of competitive situations. Analysis shows that the intelligent prisoner should always confess, since the "best" outcome will occur no matter what the other prisoner decides to do: –5 is better than –10 if the other prisoner confesses; 0 is better than –1 if the other prisoner refuses to talk. However, this individual "best choice" results in each prisoner confessing and getting a five-year sentence, whereas if neither confesses, they only get one-year sentences. This feature of competitive behavior and strategies can be thought of as the friction between basing strategic decisions on individual goals or on the common good.

With appropriate choices for the values in the table, these games could model a number of competitive situations, such as two companies trying to determine what price to set for competing products or two armies determining how to wage war.

Decision Trees

In situations where the ultimate decision depends on an intermediate choice, a decision tree can help to organize the information and facilitate a systematic analysis. A company may be ready to bring a product to market and needs to decide whether or not to invest funds up front in a test market exercise. The test market may bring in better information about how to market

Figure 1. A "decision tree."

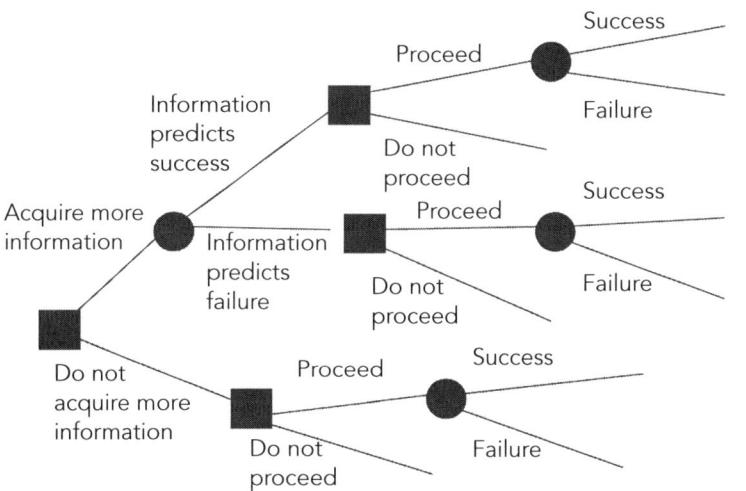

the product on a larger scale, thus increasing profit, but the cost of the test market exercise would also take away from the profit. An oil drilling company could choose to invest funding in test wells before determining the final drilling location. A university may be trying to hire a senior administrator and could choose to invest funds in a head-hunter search firm.

In all of these situations, the outcomes can be organized into a tree diagram like the one in Figure 1. Each "decision fork" is represented by a square, and each event fork—governed by external, possibly random forces—is represented by a circle. The branches leading from the event forks have probabilities assigned based upon the likelihood that an outcome will occur. Typically, acquiring additional information will result in an increased probability of success (or failure), and so the probabilities of success and failure will be different for different event forks.

Each terminal branch represents a final outcome. If current assets, the cost of the information acquisition, and the gains or losses under success and failure are known, then each terminal branch can be labeled with the net gain (or loss) for that option. Once those values are determined, the tree can be "folded back" through calculating the expected outcomes from the probabilities to determine which decisions to make to maximize the gain.

The decision points and events may include more than two options or outcomes, and there may be more than two decisions to be made before the final outcome, so the tree may have more forks and branches than the one in Figure 1 but the analysis process is the same.

From these trees, the value of the additional information acquired can be calculated. This calculation can assist companies in determining how much they should be willing to pay for that information. Also, the amount of risk a company is willing to assume can be incorporated into the process, allowing companies that are willing to shoulder a larger risk for the (slimmer) chance of a larger gain to include that information into the analysis.

Further Reading

Mesterton-Gibbons, M. *An Introduction to Game-Theoretic Modeling.* Redwood City, CA: Addison-Wesley, 1992.

Raifa, H. *Decision Analysis: Introductory Lectures on Choices Under Uncertainty.* Reading MA: Addison-Wesley 1968.

Winston, W. L. *Operations Research: Applications and Algorithms.* 4th ed. Belmont, CA: Brooks Cole-Thompson Learning, 2004.

Holly Hirst

Unemployment, Estimating

Category: Government, Politics, and History.
Fields of study: Algebra; Data Analysis and Probability.
Summary: Unemployment rates are calculated using intricate statistical models and sampling methods.

An unemployed person is generally defined as an individual who is available for work but who currently does not have a job. Overall unemployment is typically quantified using the unemployment rate, which represents the number unemployed people as a percent of the labor force. The Bureau of Labor Statistics is an independent statistical agency of the U.S. federal government primarily responsible for measuring labor market activity. Many mathematicians and statisticians are involved in data collection, modeling, and estimation of employment activity, including the highest levels of direction and management. For example, Janet Norwood was the first woman commissioner of the U.S. Bureau of Labor Statistics and frequently spoke to the Joint Economic Committee and other congressional Committees. She was also president of the American Statistical Association and chair of the Advisory Council on Unemployment Compensation. Regarding her work, she noted, "These data figure very prominently in most of the political debates, so it is extremely important that they be accurate and of high quality, and that they be released in a manner that is totally objective."

Economist John Maynard Keynes's revolutionary work, *The General Theory of Employment, Interest and Money*, was published in 1935–1936. The Industrial Revolution and shift away from an agrarian economy had significantly changed the way in which researchers in many fields looked at economic measures, including employment, and the Great Depression brought even greater attention and emphasis to these concepts. Because of labor-market volatility in the late 1920s, the 1930 U.S. census attempted the first comprehensive federal measure of unemployment, but data from the decennial census were not timely enough to be useful in assessing the effectiveness of Depression legislation to aid unemployed workers. Statisticians used newly emerging polling methods to develop better measures and mathematical models. Better methods also changed, at times, the definition of unemployment. Overall, it is commonly accepted that unemployment induces negative effects on the financial and economic status of societies and individuals with respect to many variables. As workers become unemployed, the goods and services that they could have produced are lost along with the purchasing power of these workers, thus leading to the unemployment of more workers. In addition, a large unemployment rate can induce significant social changes and has been the foundation of civil unrest and revolutions. Mathematicians and statisticians continue to create explanatory and forecasting models that are used to guide policies and decisions intended to stabilize economies and aid unemployed workers at local, state, and national levels. These models draw from mathematical ideas and techniques in a wide range of areas, including time series analyses, equilibrium modeling, structural component modeling, neural networks, and simulation.

Sample Design and Collection of Unemployment Data

In most countries, the task of collecting and analyzing unemployment-related information is assigned to certain governmental agencies. In the United States, the Current Population Survey (CPS), conducted by the Census Bureau for the Bureau of Labor Statistics since the mid-twentieth century, provides most of the necessary data. Counting every unemployed person each month is impractical in terms of both cost and time, so the Census Bureau conducts a monthly survey of the population using a sample of households that is designed to represent the civilian population of the United States. At the start of the twenty-first century, the (CPS) surveyed about 50,000 households per month. The selection is generally a multistage stratified sample selected from many different sample areas. The sample provides estimates for the nation and serves as part of model-based estimates for individual states and other geographic areas.

In the first stage of sampling, the United States is divided into primary sampling units (PSUs) that usually consist of a metropolitan area, a large county, or a group of smaller counties. PSUs are then grouped into strata based on some factor that divides the population into mutually exclusive homogeneous groups. The homogeneity of the stratum ensures that the within-strata variability is very small compared to the variability between strata. One PSU is then randomly selected from each stratum with a probability of selection proportional to the PSU's population size. The second stage of sampling consists of randomly selecting small groups of housing units from the sample PSUs. Elements from this sample of housing units are called "secondary sampling units" (SSUs). These households are usually selected from the lists of addresses obtained from the last decennial census of the population. Housing units from blocks with similar demographic composition and geographic proximity are grouped together in the list. The final sample is usually described as a two-stage sample but occasionally, a third stage of sampling is necessary when actual SSU size is extremely large. In this situation, a third stage, called "field subsampling," is needed in order to keep the surveyor's workload manageable. This involves selecting a systematic subsample of the SSU to reduce the number of sample housing units to a more convenient number. Once a survey is designed and the sample is drawn, field representatives and computer-assisted telephone interviewers contact and interview a responsible person living in each of the sample units selected to complete the interview.

Seasonal Adjustment of Unemployment Data

The collected data by the CPS are subjected to a series of transformations and adjustments before the analytical tools are applied to fit adequate models to the unemployment rate and explain its behavior in terms of relevant factors. Because some types of employment are seasonal or cyclical over time, such as December holiday retail sales or fall farm harvesting, adjustments must often be made to account for such cycles. In fact, throughout a one-year period, the level of unemployment experiences continuous variations because of such seasonal events as changes in weather, major holidays, agricultural harvesting, and school openings and closings. Since seasonal events follow an almost regular periodic pattern each year, their influence on the overall pattern can be easily estimated and eliminated. There are two popular methods for removing seasonality. The first estimates the seasonal component using a regression model with time series errors. The explanatory variables in the regression equation are 12-period harmonic terms. Once the regression coefficients are estimated, the fitted values are evaluated for each month subtracted from the corresponding actual values leading to seasonally adjusted series. The second method consists of simply taking seasonal differences of the unemployment series. The removal of the anticipated seasonal component makes it easier for data analysts to observe fundamental variations in the unemployment level, such as trends, gains, non-seasonal intrinsic cycles, and effects of external events, especially those related to economic factors.

Rate Estimation and Prediction

Since the unemployment survey is conducted in the same manner on a monthly basis, the type of data collected is called "time series data." Dependence or autocorrelation among the observations in such data is common, which means that most classical mean-variance types of statistical models are not applicable for estimation and prediction with most unemployment data. Mathematical and statistical models that take into account the particularity of time-dependent data are called "time series models." Among the most popular and useful are autoregressive integrated moving average (ARIMA) models and their seasonal extension (SARIMA). Such models can be used to describe the relationship between a current unemployment rate and past ones using differencing operations and linear equations. As a consequence, the model can also be used to predict future realizations of the unemployment rate. The ARIMA models are very flexible in the sense that they allow for the inclusion of external factors, which can help explain the movement of the unemployment rate and lead to estimators and predictors with smaller variability errors.

Further Reading

Downey, Kirstin. *The Woman Behind the New Deal*. New York: Anchor Books, 2010.

Flenberg, Stephen. "A Conversation With Janet L. Norwood." *Statistical Science* 9, no. 4 (1994).

Pissarides, Christopher. *Equilibrium Unemployment Theory*. Cambridge, MA: The MIT Press, 2000.

Zbikowski, Andrew, et al. *The Current Population Survey: Design and Methodology*. Technical Paper 40. Washington, DC: Government Printing Office, 2006.

Mohamed Amezziane

Vedic Mathematics

Category: Government, Politics, and History.
Fields of Study: Number and Operations; Problem Solving.
Summary: Vedic mathematics involves challenging mental calculations and was transmitted orally.

Vedic mathematics is a system of mathematics associated with India's Upper Indus Valley prior to 1000 b.c.e. Originally transmitted orally, the Vedic mathematics known in the twenty-first century was abstracted from ancient Sanskrit texts, known as "Vedas." Sri Bharati Krsna Tirthaji rediscovered the Vedas in the early 1900s, but his scholarly results were not published until 1965.

The Vedas covered all areas of knowledge, with the mathematics created to support this knowledge. Since recording mechanisms were not available, Vedic mathematics involves creative mental calculations, often at very challenging levels. Through Arab and Islamic writers in the 770s C.E., some Vedic mathematics was transmitted and became part of European mathematics, including elements such as the Arabic numerals, the multiplication sign, and a symbol for zero. However, the mental aspects of Vedic mathematics were not known until 1965, and these "secrets" have provided scholars, mathematicians, and students interesting explorations into multiple areas, including basic arithmetic computations, factoring, exponents, algebra in the form of linear through cubic equations, elementary number theory, analytic geometry involving the conic sections, the Pythagorean theorem, and differential calculus.

The Sutras

Sixteen formulas (or *Sutras*, which means "thread") form the foundation of Vedic mathematics, along with fourteen "sub-Sutra" corollaries. Expressed as word phrases, each formula acts as a "thread" woven throughout the Vedic mathematics system, assuming the role of a unifying element.

For example, Sutra #2 states: "All from 9 and the Last from 10." Sutra #3 states: "Vertically and Crosswise." The combined importance of both Sutras is best explained within the context of mental multiplication, such as finding the "sum" 88 × 98. Both numbers are close to the "base" 100, involving "deficiencies of 12 and 2," respectively. The desired product is obtained using these deficiencies (Sutra #2), then represented either mentally or symbolically (by Sutra #3):

$$88 -- 12$$
$$98 -- 2$$
$$86/24$$

In these operations, the deficiencies 12 and 2 are placed to the right of the original numbers, 88 and 98. The 86 is found by subtracting a deficiency from the other number in the product (98 − 12 = 86 = 88 − 2), while the 24 is the product of the deficiencies. Finally, the desired result is found: 88 × 98 = 8624, as the 86 actually represented 8600. Though this process involves a sense of magic, it is much easier than the modern computational algorithm commonly used in the twenty-first century.

It is not only important to investigate why this Sutra-based technique works, but also determine possible constraints or exceptions. For example, applying the Sutra to the product 25 × 57, the process becomes:

$$25 -- 75$$
$$57 -- 43$$
$$-18/3225$$

Because the desired product can be obtained via −1800 + 3225 = 1425, the power and the limitations of the Sutra become more evident, especially the emphasis on the numbers 9 and 10. The technique is not useful in this example because the large internal products and need for a negative quantity become obtrusive. However, the method does work, and it can be proven true algebraically. Suppose the desired product is $a \times b$, where a and b are whole numbers less than 100. Using the respective deficiencies ($100 - a$ and $100 - b$), the Sutra's process leads to the algebraic identity $ab = 100[b - (100 - a)] + (100 - a)(100 - b)$. Thus, the numbers a and b could be any numerical values—positive, negative, fractions, irrational, or even complex numbers.

Finding Decimals

As another example, Sutra #1 states: "By One More than the One Before." This Sutra is used in the construction of the number system, as each whole number is one greater than its predecessor (akin to the Peano postulates formulated in the nineteenth century). However, the Sutra's power is its application in other situations as well. Suppose the problem was to find the repeating decimal equivalent to the "vulgar" fraction 1/19, usually obtained by laboriously dividing 19 into 1. The Sutra suggests a focus on "one more than the number before" the 9, or the number 2, which is one more than the 1 which appears before the 9. The 2 (called *Ekadhika* for "one more") becomes the new divisor in lieu of the troublesome 19. The "strange" decimal resulting from this division of 2 into 1 is

$$0._1 0 5_1 2 6 3_1 1_1 5_1 7_1 8 9_1 4 7_1 3_1 6 8 4 2 1 \ldots.$$

To explain this strange expression, start with a 0 and a decimal point. Then, 1 divided by 2 is 0 remainder 1, represented by placing a 0 in the decimal expression, preceded by a subscripted 1 as the remainder. The process is repeated, where 10 (or the visual of the subscripted 1 and adjacent 0) is divided by 2, resulting in 5 with remainder 0. Thus, the 5 in the decimal expression now is not preceded by a subscripted number. Next, 5 divided by 2 results in 2 remainder 1, which are represented as before with the remainder becoming the preceding subscript. And, in subsequent divisions, 12 divided by 2 is 6 remainder 0, 6 divided by 2 is 3 remainder 0, 3 divided by 2 is 1 remainder 1, and so on. Finally, to get the final value of the decimal expression for 1/19, the subscripted values are removed: 1/19 = 0.052631578947368421, as they are needed only as "mental" reminders of the division process by 2. The mathematical explanation underlying this process is quite complex, but can be found in Chapter 26 of Tirthaji's *Vedic Mathematics*.

These two examples illustrate the enjoyment of investigating Vedic mathematics. On one level, the 16 Sutra and their corollaries provide efficient mental algorithms that become very powerful and efficient in special instances. On a second level, the careful examination of the Sutra and its application provides a rich opportunity to understand the role of generalization and algebraic identities.

Further Reading

Bathia, Dhaval. *Vedic Mathematics Made Easy*. Mumbai, India: Jaico Publishing, 2006.

Howse, Joseph. *Maths or Magic? Simple Vedic Arithmetic Methods*. London: Watkins Publishing, 1976.

Tirthaji, B. K. *Vedic Mathematics*. Delhi: Motilal Banarsidass, 1965.

Williams, Kenneth, and Mark Gaskell. *The Cosmic Calculator: A Vedic Mathematics Course for Schools*. (Books 1, 2, and 3). Delhi: Motilal Banarsidass, 2002.

Jerry Johnson

Vietnam War

Category: Government, Politics, and History.
Fields of Study: All.
Summary: Because of the importance of cryptography in World War II and the emergence of game theory in the 1950s, mathematics was heavily involved in the Vietnam War.

The Vietnam War, a conflict transpiring in Vietnam, Cambodia, and Laos from 1955 to 1975, involved the Communist forces of North Vietnam, the Viet Cong, the Khmer Rouge, the Pathet Lao, the People's Republic of China, the Soviet Union, and North Korea, against the anti-Communist forces of South Vietnam, the United States, South Korea, Australia, the Philippines, New Zealand, Thailand, the Khmer Republic, Laos, and the Republic of China. Most American involvement was concentrated from 1963 to 1973, with the last U.S. troops leaving with the fall of Saigon in 1975. It eventually resulted in a Communist victory, with U.S. forces and their allies withdrawing, Communist parties taking control of Laos and Cambodia, and South Vietnam unified with the North under Communist rule.

Mathematicians and the War

Mathematicians fell on both sides of the disagreement regarding the Vietnam War. Some served in the war effort, such as William Corson, an economist with an undergraduate degree in mathematics who later wrote the book *The Betrayal*. Grace Murray Hopper returned to active duty in 1967 because of an increased demand for naval computer systems. Others engaged in war-related research. Warren Henry helped develop the hovercraft for nighttime fighting during the 1960s while working at Lockheed Space and Missile Company and this was used in the war.

In 1966 and 1970, mathematicians at the International Congress of Mathematicians appealed to their colleagues to avoid war-related work. Mathematicians around the world organized or participated in protests, including Alexander Grothendieck in France and Steven Smale in the United States. Mathematicians in Japan at the University of Kyushu in South Japan organized "demonstrations of the 10" against the war on the 10th, 20th, and 30th of the month. Funding originally designated for teacher development during the New Math movement was instead directed to the war. Some have asserted that this diversion of funds was one of the main reasons that the educational movement failed. Mathematics played a role in the war in a number of ways, including war strategy, precision weapons, airplane computers, cryptography, and a statistically flawed 1969 draft drawing. Statisticians and others have used statistical techniques to study the long-term effects of Agent Orange on soldiers. Decision theory has been used to model the war. Systems analysis and game theory may have contributed to U.S. involvement and defeat, such as in the decisions of Secretary of Defense Robert McNamara.

Game Theory

One of the key political leaders of the American forces during the Vietnam War was Robert McNamara, a student of game theory, who served as the secretary of defense from 1961 to 1968—the period corresponding with the nation's first serious engagement with the war and its major expansions and escalations. McNamara was also responsible for the policy of Mutually Assured Destruction (MAD), a nuclear policy grounded in game theory. It said that the best deterrent to full-scale use of nuclear weapons was for opposing sides to each possess sufficient firepower to completely destroy the other so that neither side dares attack, knowing it cannot survive the counterattack. A chilling take on foreign policy, history may be on McNamara's side with the Cold War. The escalating war in Vietnam is another story. From a game theory perspective, those escalations make perfect sense. Consider that fact that North Vietnam had the options to escalate or to negotiate a peace. The United States also had those options as well as the option to pull out. The only way for the United States to gain a military advantage—and potential victory—was to escalate, with the worst possible outcome of such escalation being a stalemate. Despite increased desertion and plummeting morale, as well as growing anti-war sentiment at home, McNamara continued to escalate the engagement because it was the most promising option he was trained to see.

This was later used as an example of "escalation of commitment," a phenomenon identified in Barry Straw's 1976 paper "Knee Deep in the Big Muddy: A Study of Escalating Commitment to a Chosen Course of Action," wherein cumulative prior investment becomes the motive to continue to escalate one's investment even when rational thought says it is the wrong choice. That initial error of judgment becomes the motive to continue, to stay committed to the course of action, in order to justify it. The more one continues, the greater error one must admit to if one disengages, which is why psychologists sometimes refer to this phenomenon as the "commitment bias," a natural tendency to want to believe that one has been making the right choices and to ignore evidence to the contrary.

Further Reading

Batterson, Steve. *Stephen Smale: The Mathematician Who Broke the Dimension Barrier*. Providence, RI: American Mathematical Society, 2000.

Bosse, Michael. "The NCTM Standards in Light of the New Math Movement: A Warning!" *The Journal of Mathematical Behavior* 14, no. 2 (1995).

Bunge, Mario. "A Decision Theoretic Model of the American War in Vietnam." *Theory and Decision* 3, no. 4 (1973).

Starr, Norton. "Nonrandom Risk: The 1970 Draft Lottery." *Journal of Statistics Education* 5, no. 2 (1997).

Bill Kte'pi

Voting Methods

Category: Government, Politics, and History.
Fields of Study: Algebra; Number and Operations; Problem Solving.
Summary: Social choice theory concerns itself with the mechanics of group decisions such as elections and the impact methodology can have.

Voting theory (also known as "social choice theory") is concerned with how group decisions are made when there are a number of alternatives from which to choose (for example, finding the winner of an election). When there are only two options, voting is straightforward—the winning alternative (also called the "social choice") should be the one that receives the most votes. However, when the choice is among three or more alternatives, determining the social choice is significantly more complex. There are many reasonable methods for selecting a winner and the methods can produce different winners even when given the same sets of votes. All voting methods have inherent flaws and, regardless of the method used, strange and paradoxical situations can occur. For example, in the 2000 U.S. presidential election, George W. Bush and Al Gore were major party candidates, while Ralph Nader, representing the Green Party, had much less support. Although Bush won the election, exit polls at the time indicate that had Nader not been on the ballot in some states, Gore almost surely would have won the election.

In other words, in the U.S. electoral system, the presence (or lack thereof) of an "also-ran" candidate can have a profound outcome on the winner. This disturbing property is one of many that interests mathematicians, economists, and political scientists who study voting theory.

Preference Ballots

Preference ballots, where voters rank the alternatives in order of preference, are among the most useful ways of gathering information from voters. A voting method aggregates these preferences in some way and determines a social choice (or choices, in the case of ties). In this way, a voting method can be thought of as a function whose typical input is a set of individual ballots and whose output is the winning alternative, or—in the case of a social welfare function—a ranking of the alternatives, perhaps with ties. Many such functions are possible:

- *Plurality method*: A procedure that returns as the social choice the alternative that is the top preference on the most ballots (the candidate with the most first place votes).
- *Weighted voting method*: Also called the "positional method," this process assigns points to an alternative based on its position on a ballot, with higher placings on a ballot earning more points. The winning alternative is the one having the most points.
- *Borda count*: A special positional method whereupon a voter's lowest-ranked alternative earns zero points, the voter's second lowest-ranked alternative earns one point, and so on, with the voter's top choice earning $n-1$ points, assuming n candidates.
- *Hare system*: Also called "instant runoff voting" or "plurality with elimination," this method arrives at the social choice by successively eliminating less desirable outcomes. In this procedure, ballot-counting proceeds in rounds, with the candidate having the fewest first-place votes eliminated at the end of each round. A ballot on which an eliminated candidate was the top choice has its vote transferred to the highest ranking remaining candidate on the ballot. The process of elimination continues until one candidate has more than half the first place votes (a "majority"), in which case that candidate is declared the winner.
- *Dictatorship*: In a dictatorship, one voter is specially designated so that the social choice is always the alternative that this voter has at the top of his or her ballot.

For example, suppose that there are 100 voters in an election, and three candidates (A, B, and C). Suppose that the voters express their votes as shown in the following table:

Number of Voters	40	35	25
1st Choice	A	B	C
2nd Choice	C	C	B
3rd Choice	B	A	A

Note that 40 of the voters prefer A as their top choice, 35 prefer B as their top choice, and 25 prefer C as their top choice. If this election were decided using the plurality method, then candidate A would win with 40 first place votes (with B and C earning 35 and 25 first place votes, respectively). Using the Borda count, A would tally $40 \times 2 = 80$ points, B would earn $(35 \times 2) + 25 = 95$ points, and C would win with $(25 \times 2) + 75 = 125$ points. Using the Hare system, candidate C would be eliminated in Round 1, and C's votes would transfer to candidate B, because B is second on all 25 ballots. In Round 2, B has 60 first place votes to A's 40, so B is the winner.

This example demonstrates that different methods can yield different results. As Donald Saari writes, "Rather than reflecting the voters' preferences, the outcome may more accurately reflect which election procedure was used."

It should be noted that there are other methods of voting that do not require preference ballots. In a system called "approval voting," a voter may vote for as many candidates as desired. The winner is the candidate receiving the most votes. No distinction is made among the candidates of which the voter "approves," and the voter can vote for any combination of the candidates.

Fairness

By aggregating voters' preferences and producing a social choice, an election method should reflect, in some way, the will of the people. Given the vast library of possible election methods, it is natural to ask whether there is a method that captures this will in an ideal way. Social choice experts have developed different ways of assessing the quality of voting methods, and the notion of "fairness" has emerged as a prime consideration. When there are two alternatives, it can be expected that any reasonable voting method will be anonymous (all voters are treated equally), neutral (the two candidates are treated equally), and monotonic (if a voter changes his vote from candidate A to candidate B, then that should not hurt candidate B). Mathematician Kenneth May proved in 1952 that if the number of voters is odd and ties are not allowed, then only one voting method is anonymous, neutral, and monotonic: "majority rule," the procedure where the candidate with more than half the first place votes is declared the winner.

When there are three or more alternatives, there are many desirable properties for voting methods. The following list of criteria is far from exhaustive:

Majority criterion: This method requires that when some alternative is the first choice on more than half the ballots, that alternative should be the social choice. The plurality method satisfies this criterion, for if a candidate has a majority, no other candidate can have as many first place votes. On the other hand, the Borda count violates the majority criterion, for there are elections where a candidate can have a majority but still lose.

Condorcet winner criterion: This is a slightly weaker condition: if an alternative is preferred head-to-head over every other alternative in a one-on-one matchup that ignores the other alternatives, then that candidate should win the election. The example above shows that the plurality method violates this criterion. While candidate C is preferred over A on 60 of the ballots, and C is preferred over B on 65 of the ballots, C loses the plurality election to A. The Hare system and the Borda count fail the Condorcet winner criterion as well.

Pareto condition: This method asserts that for every pair x and y of candidates, if all voters prefer x to y, then y should not be a social choice. This is a relatively weak criterion, and all of the methods described above satisfy it.

Monotonicity criterion: According to this method, if x is a social choice and someone changes a ballot in such a way that x is moved up one spot (in other words, x exchanged with the alternative immediately above x on the ballot), then x should still be a social choice. In other words, making a change to a ballot that is favorable only to a winning candidate should not hurt the candidate. The plurality method and the positional voting methods satisfy monotonicity, but the Hare system does not.

Independence of Irrelevant Alternatives: Also called "binary independence," this method states that if x is a social choice while y is not, and if a voter changes a ballot in a way that does not change the relative positions of x and y on the ballot, then y should still not be a social choice. In other words, changing the positions of other "irrelevant" candidates on a ballot should not affect the relative position of x over y or y over x in the outcome. This is precisely the difficulty that occurred in the 2000 U.S. presidential election, where Nader's presence in the election affected the relative rankings of Bush and Gore.

Although each of these criteria is, in turn, a reasonable expectation of a voting method, Kenneth Arrow, in 1952, proved the mutual exclusivity of them. In his "impossibility theorem," Arrow showed that if there are at least three alternatives and a finite number of voters, then the only social welfare function that satisfies both the Pareto condition and "independence of irrelevant alternatives" is a dictatorship. This profound result, which earned Arrow the Nobel Prize in Economics in 1972, argues against the possibility of a theoretically perfect democracy. Nevertheless, Arrow himself encourages continuing to search for voting methods that work well most of the time. He writes:

> My theorem is not a completely destructive or negative feature any more than the second law of thermodynamics means that people don't work on improving the efficiency of engines. We're told that you'll never get 100% efficient engines . . . It doesn't mean you wouldn't like to go from 40% to 50%.

Sincere and Strategic Voting

Strategic voting is the practice of voting against one's true preferences in order to achieve a better outcome in an election. This contrasts with sincere voting, where one votes according to one's true preferences. Strategic voting most often occurs in situations where a voter's preferred candidate has little chance of winning, or where the voter's top candidate is most threatened by his second or third candidate. While strategic voting can affect the outcome of an election, its effects can be disastrous. Election results should reflect the aggregate will of the people, and if voters do not express their individual preferences truthfully, then the voting method has little hope of determining the socially desired outcome. Therefore, voting methods that tend to encourage strategic voting are unattractive. It should be noted that for strategic voting to be at all effective, there must be at least three candidates in the election, and the voters need a thorough understanding of both the voting method being used and the preferences of other voters.

For example, in the 2000 election, exit polls in Florida indicated that Nader voters widely supported Gore as their second choice, far beyond both the margin of error of the polls and Bush's margin of victory. Had these voters instead voted strategically for Gore, Gore would likely have carried Florida and its 25 electoral votes, thereby winning the presidency. The U.S. electoral college notwithstanding, this shows how powerfully the plurality method encourages strategic voting. Had the 2000 election been decided by the Borda count, one could imagine that a conservative voter might have Bush as the top choice, Gore as the second choice, and Nader last, but might insincerely rank the candidates in the sequence Bush, Nader, Gore in an attempt to maximize the point differential between Bush and Gore.

Some voting methods prove resistant to strategic voting. One of the major advantages of the Hare system is that it tends to encourage sincere voting. In the 2000 election, for example, a Nader supporter would have less reason to vote strategically for Gore if it is known that the vote will transfer to Gore should Nader be eliminated. Nevertheless, there are situations where even with the Hare system, strategic voting can prove beneficial to a voter.

In the 1970s, Allan Gibbard and Mark Satterthwaite proved that no voting method is completely immune to strategic voting. Any non-dictatorial system that uses preference ballots and allows at least the possibility of any candidate winning will necessarily lead to situations, however hypothetical, where strategic voting can be beneficial. This proof serves as a result analogous to Arrow's, but in the realm of strategic voting. As with Arrow's result in fairness, it is important to note that the degree to which a voting method encourages sincerity still serves as an important criterion for selection.

Further Reading

Brams, Steven. *Mathematics and Democracy: Designing Better Voting and Fair-Division Procedures*. Princeton, NJ: Princeton University Press, 2007.

Saari, Donald. *Basic Geometry of Voting*. New York: Springer, 2003.

Taylor, Alan, and Allison Pacelli. *Mathematics and Politics: Strategy, Voting, Power and Proof*. New York: Springer, 2008.

STEPHEN SZYDLIK

World War I

Category: Government, Politics, and History.
Fields of Study: All.
Summary: World War I saw an increased emphasis on applied mathematics but ultimately disrupted mathematics research.

Although mathematicians were not as heavily involved with the conduct of World War I as they would be with World War II, the four years of conflict impacted the field of mathematics in two main ways: they severed international ties among researchers, thus slowing collaborative research efforts; and the war provided the circumstances for applied mathematics to develop more fully through military research. Many mathematicians contributed their knowledge and abilities to the war effort. At the same time, others published papers unrelated to the military, worked to encourage reconciliation among mathematicians of warring nations, or strove to end the war outright. World War I, which was fought from 1914 to 1918, was precipitated by the assassination of Archduke Franz Ferdinand of Austria. After the initial declaration of war on Serbia by Austria, countries with various political alliances joined the fighting, with the result that more than 30 countries on five continents were ultimately named as combatants. The massive scope of this first truly global war led U.S. President Woodrow Wilson to refer to it as the "war to end all wars."

Mathematics Applied to Military Research

Some mathematicians turned their attention to more practical and applied uses of the field. World War I saw extensive use of both trench warfare, which the United States had already experienced somewhat during the U.S. Civil War; and potent chemical weapons, like mustard gas. In the United States and in Europe, mathematicians researched ballistics and aeronautics as the warring countries sought advantages in firepower on land and began to realize the potential of air power. Mathematician John Littlewood performed research on ballistics and improved tables for the British Royal Garrison Artillery. In the United States, important figures such as Gilbert Bliss, Oswald Veblen, Norbert Wiener, and Forest Ray Moulton worked at the U.S. Army's Aberdeen Proving Ground, Maryland, in ordnance and improvement in ballistics calculations. The American Mathematical Society published, in 1919, a list of over 175 mathematicians working in some capacity to support the war effort. The National Advisory Committee on Aeronautics also began construction of the Langley Laboratory in 1917, although research did not fully get underway until a few years later.

Similarly, Europeans conducted research with the aim of improving military operations. The British mathematician Frederick William Lanchester devised a formula to calculate the likely outcome of a battle between opponents of different strengths. He also published a series of articles on the military potential of aeronautics, which in 1916 were collected into a book. At Göttingen, Germany, Felix Klein and others instituted the Aerodynamic Proving Ground in 1917. In Italy, Mauro Picone investigated new methods for calculating ballistics tables, and Vito Volterra proposed using helium in airships.

As was the case in many wars dating back into antiquity, codes and cyphers played an important role. For example, "trench codes" consisting of three number or letter groups were used for rapid communications of tactical situations but they were fairly easily cracked and were quickly supplanted by more complex structures. The Germans widely employed the ADFGVX cypher, so named because only those six letters were used in coded messages. They had been chosen to minimize operator error because when those letters are sent by Morse code, they sound very different from one other. The code was a fractionating transposition cipher using a modified Polybius square, named for second-century B.C.E. historian Polybius of Megalopolis, with a single columnar transposition.

The cypher keys were typically changed every few days and the code was broken in only a few isolated cases during the war. A general solution was found in the 1930s by William Friedman, who is often referred to as the "father of modern U.S. cryptography." The Germans also used some double transposition cyphers, which applied the same transposition key horizontally and vertically to the same matrix. In addition, they proved to be skillful in deciphering the codes of others, and the U.S. Army began to experiment with using Native American languages as military code. Several Choctaw soldiers served in the U.S. Army in Europe during World War I and are credited with helping to win some major battles.

The goal of war-related mathematicians was to improve the efficiency of military action. In the United States, this goal also applied to the home front. Allyn A. Young, the president of the American Statistical Association, proposed in a December 1917 address that a central statistical office or commission be established to aid the coordination of various boards and agencies then gathering statistics related to the war.

A greater division between mathematics research and teaching concerns also occurred around the time of World War I, as evidenced by the founding and branching off of the Mathematician Association of America in 1915 and the National Council of Teachers of Mathematics in 1920 from the more research-focused American Mathematical Society.

Non-Military Research During the War Years

Although much mathematical work from 1914 to 1918 related to improving military capability, there were many other notable advances that did not have immediate effects on war power. For instance, Albert Einstein published his general theory of relativity in 1915. David Hilbert also published field equations about that time. While a prisoner of war in Russia, the Polish mathematician Waclaw Sierpinski published a paper on his fractal triangle. Together with Godfrey Harold Hardy, after arriving at Cambridge University on Hardy's invitation, Srinivasa Iyengar Ramanujan published a series of papers on number theory during the war.

Efforts for Peace and Reconciliation

At the same time, some mathematicians focused not on improving the conduct of war or other research, but instead on ending the conflict and reconciling with their colleagues in the peace that would follow. Perhaps the most famous case is that of the British mathematician Bertrand Russell, who soon after the turn of the century had identified a paradox that challenged assumptions of set theory and in the years immediately before the war had co-authored *Principia Mathematica* with Alfred North Whitehead. Repulsed by the battlefield slaughters and the general support of his countrymen for the war, Russell became an increasingly active pacifist, eventually taking part in public demonstrations and spending six months in prison for his antiwar writings.

Less dramatically, but still forcefully, the German David Hilbert made a point of recognizing the accomplishment of colleagues in enemy countries. The Dutch geometer Luitzen Egbertus Jan "L. E. J." Brouwer worked after the war to bring German mathematicians back into recognition. Gosta Mittag-Leffler, a Swedish mathematician, deliberately published English, French, and German papers in his journal *Acta Mathematica*. After the war, he and Godfrey Harold Hardy worked to encourage reconciliation with German researchers.

Approaching the cause of peace from another angle, the Quaker mathematician Lewis Fry Richardson, who had served in an ambulance unit in France during the war, worked to understand the causes of wars so as to better prevent them. A limited printing of his first paper on the subject, "The Mathematical Psychology of War," appeared in 1919. In later decades, as World War II loomed, Richardson would return to the subject.

Conclusion

The death of possible future contributors to the field of mathematics during World War I as a whole was, of course, an incalculable loss. By disrupting the continuity of research and discovery, the war also delayed advances in areas of mathematics such as topology and set theory. At the same time, however, the possible applied uses of mathematics began to receive more attention and appreciation. In addition, national governments became more aware of the military value of mathematicians—a value that they would exploit much more thoroughly and effectively in World War II.

Suspension of International Cooperation

The war ended or made much more difficult the international relations among mathematicians that had developed in previous decades. National organizations of mathematicians publicly condemned their colleagues in enemy countries. International meetings were abandoned. Even after the war, an international congress did not fully accept German members again until 1928. A mathematician of one nation working in or visiting a hostile country might run the risk of being stranded, or worse, face arrest and imprisonment. As a whole, there were few mathematicians who made efforts during the war to maintain relations with their counterparts and such efforts were sometimes limited to individual statements of protest against a severing of ties among nations. The division of researchers slowed the development of some fields, like topology and set theory.

Further Reading

Dauben, Joseph W. "Mathematicians and World War I: The International Diplomacy of G. H. Hardy and Gosta Mittag-Leffler as Reflected in Their Personal Correspondence." *Historia Mathematica* 7 (1980).

Newman, James R. "Commentary on a Distinguished Quaker and War." In *The World of Mathematics*. Vol. 2. Edited by James R. Newman. New York: Simon & Schuster, 1956.

Price, G. Baley. "American Mathematicians in World War I." In *AMS History of Mathematics, Volume I: A Century of Mathematics in America, Part I*. Providence, RI: American Mathematical Society, 1988.

Siegmund-Schultze, Reinhard. "Military Work in Mathematics 1914–1945: An Attempt at an International Perspective." In *Mathematics and War*. Edited by Bernhelm Booss-Bavnbek and Jens Hoyrup. Basel, Switzerland: Birkhauser Verlag, 2003.

United States Cryptologic Museum. *The Friedman Legacy: A Tribute to William and Elizabeth Friedman*. 3rd ed. Fort Meade, MD: National Security Agency,

2006. http://www.nsa.gov/about/_files/cryptologic
_heritage/publications/prewii/friedman_legacy.pdf.

CHRISTOPHER J. WEINMANN

World War II

Category: Government, Politics, and History.
Fields of Study: All.
Summary: World War II saw significant mathematical advances in cryptography, operations research, and navigation.

World War II was fought between two major alliances of countries, the "Axis" and the "Allies." The beginning might be traced to pacts signed in 1936 and 1937 by the three primary Axis powers: Germany, which came to control much of the European continent; Italy, which influenced the Mediterranean; and Japan, which governed much of East Asia and the Pacific. The ultimately victorious Allies coalition, led by Great Britain, the United States, and the Soviet Union, gained the surrender of Italy in 1943 and Germany and Japan in 1945. Well over 50 countries participated in the war, and there were millions of military and civilian deaths, some of the most controversial being those that resulted from the United States' use of the atomic bomb in Japan. Mathematics played a critical role in many aspects of the war effort, notably in coding and encryption, which achieved levels unseen in previous wars and led to additional developments in the subsequent cold war era, such as mathematician Claude Shannon's ideas on information theory. New areas of applied mathematics, such as operations research, also emerged from technologies and problems created during or inspired by the war. Many mathematicians served in the military or worked for military agencies, such as the U.S. Aberdeen Proving Grounds. An Applied Mathematics Panel was formed in 1942 to solve war-related mathematical problems. Mathematicians were involved in the Manhattan Project to develop the atomic bomb, a matter that is widely discussed even in the twenty-first century with regard to the ethics of mathematics research and social obligations of mathematicians as citizens of the world. The immediate prewar era and wartime would also result in a flood of mathematicians and scientists emigrating to the United States and many other Allied countries, fleeing religious or political persecution, particularly in Nazi-controlled Europe. It also likely accelerated the growth of participation of women in mathematical and scientific careers. These individuals would shape both research and teaching for decades to come.

Codes and Cyphers

Through World War I, most encrypted messages either used a paper-and-pencil cipher or a "book code" in which the enciphered version of each word was looked up in a codebook. Between the world wars, two new types of cryptography emerged: superencypherment and rotor machines.

With superencypherment, the text to be enciphered was converted into a string of digits. Then, a string of random digits (known as "additives") was added with non-carrying addition. If the additives were never used again, the result was the "one-time pad" cipher. However, if the string of additives digits is reused, it is possible for code-breakers to break the cipher. In the 1930s, American cryptographer William Friedman developed the "kappa test," a statistical test to determine when a superencypherment string was being reused.

The Japanese Navy used a codebook to convert plain text into numeric code groups, which were then superencyphered using a book of 50,000 random digits. During wartime, the number of encrypted messages sent was such that any string of these digits was reused, and the U.S. Navy was able to break the Japanese code.

The main technique was to search for so-called double hits. Suppose two encrypted messages read:

… 77899 45616 27249 31464 68461 …
… 77899 81957 27249 81279 59138 …

The double hit is underlined. It could be because of chance but the cryptographer assumes that it is because of the same code words being enciphered by the same stretch of additive. With enough double hits, the cryptographer can recover portions of the additive and start decoding the underlying code words, as well as locating the so-called indicator (numbers hidden in the message to tell the recipient where in the book of additives the sender started). It took months of traffic for enough double hits to appear to break the Japanese naval code, which was changed several times a year.

The kappa test could also be used to locate re-used stretches of additive. In 1943, in a project later code-named VENONA, the U.S. Army spotted seven double hits in 10,000 Soviet diplomatic messages. The Soviets, who used the unbreakable one-time pad system, had blundered by re-issuing some 30,000 pages of random additive, and VENONA succeeded in breaking some 2900 Soviet messages.

The Germans and Italians used the "Engima" cipher machine, which consisted of three rotors plus a steckerboard (a plugboard), which added a monalphabetic substitution to the polyalphabetic generated by the rotors. A rotor was a disc with 26 electrical contacts (for the Roman alphabet) on each side. Wiring inside the rotor connected the contacts. Such a rotor creates a monalphabetic cipher—each letter would always be replaced with the same letter. If the rotor is allowed to rotate one contract between letters, it generates a polyalphabetic cipher with a period of 26. If two rotors are connected together, so that the second one advances one space after the first one completes a rotation (in the same way as the rotating numbers in a mechanical car odometer), then the two rotors generate a polyalphabetic cipher with a period of 26×26 (sometimes 26×25, depending on how the two rotors were geared together). Three rotors generate a period of $26 \times 26 \times 26$, and so on. The operator had up to eight rotors available, giving up to

$$\frac{8!}{5!} = 336$$

possibilities for the rotors. For each day, there was a prearranged rotor selection and steckerboard setup and the operator would choose at random an initial rotation for each of the three rotors of the day. An "indicator" giving this random initial position had to be inserted into the message.

In the 1930s, three mathematicians, Marian Rejewski, Zerzy Rozycki, and Henyrk Zygalski of the Polish *Biuro Szyfrow* (Cypher Bureau) had figured out the wiring of the rotors in the Enigma, had worked out techniques for deciphering this indicator; which had been enciphered using the same Enigma, and had invented a machine called a "bomby," which automated much of the work. With these tools and techniques, they were able to read German Enigma messages until the Germans introduced changes in 1938 that defeated the Polish techniques.

The Poles then turned over their work to the British and French. The British took over an estate north of London called Bletchley Park and brought in mathematicians to work on the Enigma and other ciphers. The first four mathematicians were Alan Turing (whose Turing Machine, of 1936 formed the theoretical basis of later computers), Gordon Welchman, John Jeffreys, and Peter Twinn. Bletchley Park's main method for breaking Enigma was to find a crib (a word or words that were highly likely to be in a particular place in the message). Despite the features of Enigma that were supposed to hide any evidence of the plain text, there were certain relationships among the letters of the cyphertext that had to occur when the crib was enciphered. A machine called a "Bombe" then ran through all 26^3 positions of the three rotors, finding the very few that would produce these relationships. Multiple runs would be required for different choices of rotors but Bletchley also developed a statistical technique that— with luck—would eliminate numerous rotor choices.

Searching for a code that would be difficult to break using mathematically based cryptography methods, the U.S. government recruited native Navajo speakers. The Navajo language is very complex with unique phonetics, grammar, and syntax and no written or symbolic alphabet, making it nearly impossible for someone without substantial exposure to understand (no Axis linguists had such exposure) and providing no written cypher that could be analyzed. Several hundred Navajo code talkers served with the U.S. Marines, most in the Pacific theater.

Computers

While general-purpose electronic computers did not exist until after World War II, work during the war helped lead to their development. By 1940, analog computers of considerable sophistication existed. However, there were only a handful of digital computers, all of them electromechanical and not differing much in concept from Babbage's analytical machine of the nineteenth century. At that time, the only design for an electronic computer was from John V. Atanasoff of Iowa State College (now Iowa State University), who with Clifford Berry designed the Atanasoff–Berry Computer (ABC). It was not a general-purpose computer, limited to the solution of sets of linear equations.

In Germany, Konrad Zuse began working on computers in 1936. In 1941, he constructed the electromechanical Z3, which was the first general-purpose programmable computer. It was used for calculations for aircraft design and was destroyed by Allied bombing in 1943. After the war, Zuse built computers commercially and also developed the first programming language, Plankalkül.

In 1941, the Germans invented a new type of cypher for high-level communications. Instead of replacing or scrambling letters, a machine was developed that worked on the bits of the five-bit teletype (Baudot–Murray) code. In principle, this process was a superencypherment in which the bits of the teletype code were superenciphered by a string of binary additives. The additives were not random but were produced by a set of 10 wheels that rotated with different periods.

To solve this cipher, Bletchley Park constructed an electronic device called the "Colossus." Ten were built, each having from 1500 to 2500 vacuum tubes apiece. It was not a general-purpose computer since it could solve only one particular problem but the experience with electronic circuits and the knowledge that a device with thousands of vacuum tubes would work inspired, after the war, three successful British efforts (Turing's ACE, Cambridge University's EDSAC, and Manchester University's Mark I) to build general-purpose electronic computers. This kept the United Kingdom competitive in computer design with the United States through the beginning of the 1960s.

The Ordnance Department of the U.S. Army had the task of computing large numbers of range tables for artillery. Its Ballistic Research Laboratory, in cooperation with the Moore School of Engineering at the University of Pennsylvania, had the foresight—and ambition—to contract for an electronic computer, to be known as Electrical Numerical Integrator and Computer (ENIAC). The principal designers of the ENIAC were John Mauchly and John Presper Eckert (later developers of the UNIVAC line of computers), although many of the ideas of the design came from Atanasoff's ABC. The ENIAC did not become operational until 1945. One of its first uses was in designing the hydrogen bomb.

By 1944, the shortcomings of this pioneering design had been realized. It could not handle the workload required for numerical solution of partial differential equations and plans were started for a more advanced computer to be known as EDVAC. In 1945, John von Neumann combined his own ideas, those of Alan Turing, and those of the ENIAC developers into the paper, "First Draft of a Report on the EDVAC," which laid out the principles of the modern computer. This paper led to the "Von Neumann machine" model, still used in the twenty-first century, although most of the ideas came from Turing.

Operations Research

In June 1941, Coastal Command (that portion of the Royal Air Force that operated over the seas from land bases) brought in physicist Patrick M. S. Blackett as an advisor. Blackett decided that instead of designing new weapons, his duty was to analyze how Coastal Command performed its operations and see what he could recommend to improve them. Hence, his work became known to the British as "operational research" (also called "operations research").

Blackett and his colleagues investigated a wide variety of submarine and anti-submarine operations. In one such project, the group figured out that a submarine attacked by an aircraft would not have time to dive very deep (indeed, it might still be on the surface), and that a setting of 25 feet for the depth charges the aircraft dropped had the best chance of lethality to the submarine. Another project was to figure out the optimum size of a convoy. It turned out that the larger the convoy was, the better. A convoy, even a large one, had almost the same chance of avoiding being seen by a submarine as a single ship did. What mattered was not the area of sea the convoy covered but its perimeter, where the escorts were stationed. The perimeter increased much slower than did the number of ships, so if both the number of ships and the number of escorts were doubled, each escort had a smaller length of the perimeter to cover, which gave it a better chance to catch enemy submarines trying to penetrate its portion of the perimeter.

The success of Blackett's original group led to operational research's extension to many other parts of the British forces. In April 1942, the U.S. Navy founded its own Anti-Submarine Warfare Operations Research Group, originally for antisubmarine warfare and later for work throughout the Navy. As Admiral King reported:

> The knowledge . . . made it possible to work out improvements in tactics which sometimes increased the effectiveness of weapons by factor

or three or five, to detect changes in the enemy's tactics in time to counter them before they became dangerous, and to calculate force requirements for future operations.

Navigation

World War II presented navigation problems not seen in prewar flying, such as how to find a target at night from the air. In the Battle of Britain, the Germans first used the "Knickebein" system for target location at night. Knickebein and it successor "X-gerät" used narrow radio beams that crossed over the target. Later, the Germans introduced "Y-gerät," which used a single ground station, with the aircraft transmitting a return signal from which the distance from the aircraft to the transmitter could be determined by the ground station.

The Allies also developed targeting systems. One was the British "OBOE" in which two stations broadcast signals to which the aircraft responded, allowing each station to determine the distance to the aircraft. The aircraft flew a fixed distance in a circular arc from the first station until it was at a specified distance from the second station. The intersection of these two arcs was the target location. This Y-gerät/OBOE technique, except with the aircraft transmitting and the ground station responding, is still used in the twenty-first century in the Distance Measuring Equipment (DME) system widely used by both military and civilian aircraft for navigating over land.

The British also developed the "GEE" system, which used a different mathematical technique. There was no transmitter on the aircraft. Instead, there was a "primary" or "master" transmitter and at least two "secondary" or "slave" transmitters on the ground. The primary would broadcast a signal, and each secondary would broadcast its own signal as soon as it received the signal from the primary. Any given difference between the arrival times of the signal from a primary and secondary defined one branch of a hyperbola (since a hyperbola is the locus of all points the difference of whose distance from two foci is constant and whether the primary or secondary signal arrived first tells which branch of the hyperbola). The second primary-secondary pair defined one branch of a second hyperbola, and these two branches intersect in exactly two points. Either dead reckoning or a third pair could then be used to determine which of these two intersection points was the aircraft's position.

GEE was soon developed into the Long Range Navigation (LORAN) system, which is still used worldwide for navigation at sea within approximately 1000 kilometers of the LORAN stations. Beyond that distance, the ionospheric bounce of the signals interferes with the ground wave.

The Mathematics Community in World War II

Mathematicians participated in both military service and multiple civilian roles during World War II. Some enlisted voluntarily or were drafted, such as Herman Goldstine, who worked as the army liaison to the ENIAC project. Many stayed in their academic positions, continuing to prepare students and working on war-related training programs in mathematics. Others left their colleges and universities to work for government programs related to the war effort, including the growing area of operations research, such as G. Baley Price, who worked on applications like bomber accuracy and Philip Morse, who is sometimes referred to as the "father of U.S. operations research" and is credited with organizing the U.S. Anti-Submarine Warfare Operations Research Group. Companies like the Radio Corporation of America (RCA), Westinghouse Electric Corporation, Bell Laboratories, Bell Aircraft Corporation, Grumman Aircraft Engineering Corporation, and Lockheed Corporation recruited mathematicians to help fulfill war contracts. The government also widely recruited nonmilitary mathematicians for groups like the Office of Scientific Research and Development, which had branches conducting medical research, fuse research, and a multi-application area looking at problems like submarine warfare, radar, and rocketry. This body came to include the Applied Mathematics Panel in 1942.

Mathematician and scientist Warren Weaver, a pioneer in the field of machine translation, headed the panel. Some of the problems investigated included gas dynamics and compressible fluids, underwater ballistics and explosions, shock waves in air and water, mechanics and damage in air-to-air combat and anti-aircraft fire, ballistics and firing tables, torpedo spread angles, land mine clearance techniques, and statistical methods. In this time period, women also experienced increasing opportunities to pursue and contribute to a diverse range of careers, including science and mathematics. Hunter College professor Mina Rees took a leave of absence during World War II to contribute to the war effort, working with the Applied Mathematics Panel. Following the war, she became head of the mathematics branch of the Office of Naval Research. The American Mathematical Society said

> ... the whole postwar development of mathematical research in the United States owes an immeasurable debt to the pioneer work of the Office of Naval Research and to the alert, vigorous and far-sighted policy conducted by Miss Rees.

Further Reading

Budiansky, Stephen. *Battle of Wits: The Complete Story of Codebreaking in World War II*. New York: The Free Press, 2000.

Goldstine, H. *The Computer From Pascal to von Neumann*. Princeton, NJ: Princeton Univeristy Press, 1972.

Haufler, Hervie. *Codebreakers' Victory: How the Allied Cryptogaphers Won World War II*. New York: New American Library, 2003.

Hodges, Andrew. *Alan Turing: The Enigma*. New York: Simon & Schuster, 1983.

Rees, Mina. "Mathematical Sciences and World War II." *The American Mathematical Monthly* 87, no. 8 (1980).

JAMES A. LANDAU

Resource Guide

Books

Aaboe, Asger. *Episodes From the Early History of Mathematics*. Washington, DC: Mathematical Association of America, 1975.

Adrian, Yeo. *The Pleasures of Pi and Other Interesting Numbers*. Singapore: World Scientific Publishing, 2006.

Agresti, A. *Categorical Data Analysis*. Hoboken, NJ: Wiley, 2002.

Aho, A. V., J. E. Hopcrotf, and J. D. Ullman. *The Design and Analysis of Computer Algorithms*. Reading, MA: Addison-Wesley, 1976.

Albert, Jim, and Jay Bennett. *Curve Ball: Baseball, Statistics, and the Role of Chance in the Game*. New York: Springer-Verlag, 2001.

Ascher, Marcia. *Mathematics Is Everywhere: An Exploration of Ideas Across Cultures*. Princeton, NJ: Princeton University Press, 2002.

Ball, W. W. Rouse. *A Short Account of the History of Mathematics*. New York: Sterling Publishing Company, 2001.

Barnett, Raymond, Michael Ziegler, and Karl Byleen. *Calculus for Business, Economics, Life Science, and Social Science*. Upper Saddle River, NJ: Prentice-Hall, 2005.

Baumohl, Bernard. *The Secrets of Economic Indicators: Hidden Clues to Future Economic Trends and Investment Opportunities*. 2nd ed. Upper Saddle River, NJ: Pearson Education, 2008.

Beckmann, Petr. *A History of π (Pi)*. New York: Barnes & Noble, 1971.

Behrends, Ehrhard. *Five-Minute Mathematics*. Providence, RI: American Mathematical Society, 2008.

Bell, Eric Temple. *Men of Mathematics*. New York: Simon & Schuster, 1937.

Bennett, Jay, and James Cochran. *Anthology of Statistics in Sports*. Philadelphia, PA: Society for Industrial and Applied Mathematics, 2005.

Berggren, Lennart, Jon Borwein, and Peter Borwein. *Pi: A Source Book*. New York: Springer-Verlag, 1997.

Berlekamp, Elwyn R., John H. Conway, and Richard K. Guy. *Winning Ways for Your Mathematical Plays*. Natick, MA: AK Peters, 2001.

Blackwell, William. *Geometry in Architecture*. Hoboken, NJ: Wiley, 1984.

Blatner, David. *The Joy of π*. New York: Walker & Co., 1997.

Blue, Ron, and Jeremy White. *The New Master Your Money: A Step-by-Step Plan for Gaining and Enjoying Financial Freedom*. Chicago: Moody, 2004.

Blum, Raymond. *Mathemagic*. New York: Sterling Publishing, 1992.

Bodie, Zvi, Alex Kane, and Alan Marcus. *Investments*. Chicago, IL: McGraw-Hill/Irwin, 2008.

Borwein, Jonathan, and Peter Borwein. *A Dictionary of Real Numbers*. Pacific Grove, CA: Brooks/Cole Publishing Co., 1990.

Boyer, C. B. *A History of Mathematics*. Hoboken, NJ: Wiley, 1968.

Boyer, C. B. *The History of the Calculus and Its Conceptual Development*. New York: Dover Publications, 1949.

Brealey, Richard A., Stewart C. Myers, and Franklin Allen. *Principles of Corporate Finance*. 9th ed. New York: McGraw-Hill, 2008.

Bressoud, David. *The Queen of the Sciences: A History of Mathematics*. Chantilly, VA: The

Teaching Company, 2008.

Broverman, Samuel A. *Mathematics of Investment and Credit.* Winsted, CT: ACTEX Publications, 2008.

Burkett, Larry, and Brenda Armstrong. *Making Ends Meet: Budgeting Made Easy.* Gainesville, GA: Crown Financial Ministries, 2004.

Burton, David M. *The History of Mathematics: An Introduction.* New York: McGraw-Hill, 2005.

Calinger, Ronald. *A Contextual History of Mathematics.* Upper Saddle River, NJ: Prentice-Hall, 1999.

Clagett, Marshall. *Archimedes in the Middle Ages.* Madison: University of Wisconsin Press, 1964.

Closs, Michael. *A Survey of Mathematics Development in the New World.* Ottawa: University of Ottawa, 1977.

Closs, Michael, ed. *Native American Mathematics.* Austin: University of Texas Press, 1986.

Coe, Michael D. *Breaking the Maya Code.* New York: Thames and Hudson, 1992.

Copeland, Thomas E., J. Fred Weston, and Kuldeep Shastri. *Financial Theory and Corporate Policy.* 4th ed. Upper Saddle River, NJ: Pearson Education, 2005.

Cullen, Christopher. *Astronomy and Mathematics in Ancient China: The Zhou Bi Suan Jing.* Cambridge, England: Cambridge University Press, 1996.

Cuomo, Serafina. *Ancient Mathematics.* London: Routledge, 2001.

Davenport, Harold. *The Higher Arithmetic: An Introduction to the Theory of Numbers.* Cambridge, England: Cambridge University Press, 1999.

Davis, Morton D. *The Math of Money: Making Mathematical Sense of Your Personal Finances.* New York: Copernicus, 2001.

De Mestre, Neville. *The Mathematics of Projectiles in Sport.* Cambridge, England: Cambridge University Press, 1990.

Devlin, Keith. *The Math Gene: How Mathematical Thinking Evolved and Why Numbers Are Like Gossip.* New York: Basic Books, 2001.

———. *The Unfinished Game: Pascal, Fermat, and the Seventeenth-Century Letter That Made the World Modern.* New York: Basic Books, 2008.

Drobat, Stefan. *Real Numbers.* Upper Saddle River, NJ: Prentice-Hall, 1964.

Dudley, Underwood. *Numerology or What Pythagoras Wrought.* Washington, DC: Mathematical Association of America, 1997.

Eastway, Rob, and John Haigh. *Beating the Odds: The Hidden Mathematics of Sport.* London: Robson Books, 2007.

Eglash, Ron. *African Fractals: Modern Computing and Indigenous Design.* New Brunswick, NJ: Rutgers University Press, 1999.

Eves, Howard. *An Introduction to the History of Mathematics.* New York: Saunders College Publishing, 1990.

Flegg, G. *Numbers: Their History and Meaning.* New York: Schocken Books, 1983.

Friberg, Jöran. *Unexpected Links Between Egyptian and Babylonian Mathematics.* Singapore: World Scientific Publishing Co., 2005.

Friedman, Arthur. *World of Sports Statistics: How the Fans and Professionals Record, Compile, and Use Information.* New York: Athenaeum, 1978.

Fries, Christian. *Mathematical Finance: Theory, Modeling, Implementation.* Hoboken, NJ: Wiley, 2007.

Frumkin, Norman. *Guide to Economic Indicators.* Armonk, NY: M. E Sharpe, 2000.

Gamow, George. *One, Two, Three... Infinity.* New York: Viking Press, 1947.

Gardner, David, and Tom Gardner. *The Motley Fool Personal Finance Workbook: A Foolproof Guide to Organizing Your Cash and Building Wealth.* New York: Fireside Books, 2003.

Gardner, Martin. *Mathematics, Magic and Mystery.* New York: Dover, 1956.

Gay, Timothy. *The Physics of Football.* New York: HarperCollins, 2005.

Gerdes, Paulus. *Geometry From Africa: Mathematical and Educational Explorations.* Washington, DC: Mathematical Association of America, 1999.

Gillings, R. J. *Mathematics in the Time of the Pharaohs.* New York: Dover Publications, 1982.

Gutstein, Eric, and Bob Peterson, eds. *Rethinking Mathematics: Teaching Social Justice by the Numbers.* Milwaukee, WI: Rethinking Schools, 2005.

Hadamard, Jacques. *A Mathematician's Mind.* Princeton, NJ: Princeton University Press, 1996.

Hardy, G. H. *A Mathematician's Apology.* Cambridge, England: Cambridge University Press, 1941.

Henry, Granville C. *Logos: Mathematics and Christian Theology.* Lewisburg, PA: Bucknell University Press, 1976.

Hersh, Rueben. *What Is Mathematics, Really?* New York: Oxford University Press, 1997.

Hoyle, Joe Ben, Thomas F. Schaefer, and Timothy S. Doupnik. *Fundamentals of Advanced Accounting*. New York: McGraw-Hill, 2010.

Kalbfleisch, John D., and Ross L. Prentice. *The Statistical Analysis of Failure Time Data*. Hoboken, NJ: Wiley, 2002.

Katz, Victor J., ed. *Mathematics of Egypt, Mesopotamia, China, India, and Islam: A Sourcebook*. Princeton, NJ: Princeton University Press, 2007.

Kellison, Stephen G. *Theory of Interest*. New York: McGraw-Hill, 2009.

Kimmel, Paul D., Jerry J. Weygandt, and Donald E. Keiso. *Financial Accounting: Tools for Business Decision Making*. Hoboken, NJ: Wiley, 2009.

King, Jerry. *The Art of Mathematics*. New York: Plenum Press, 1992.

Klein, John P., and Melvin L. Moeschberger. *Survival Analysis: Techniques for Censored and Truncated Data*. New York: Springer-Verlag, 1997.

Kline, M., *Mathematical Thought From Ancient to Modern Times*. New York: Oxford University Press, 1972.

Koetsier, T., and L. Bergmans, eds. *Mathematics and the Divine: A Historical Study*. Amsterdam: Elsevier, 2005.

Longe, Bob. *The Magical Math Book*. New York: Sterling Publishing, 1997.

Martzloff, Jean-Claude. *A History of Chinese Mathematics*. New York: Springer-Verlag, 1987.

Moses, Robert P., and Charles E. Cobb, Jr. *Radical Equations: Civil Rights From Mississippi to the Algebra Project*. Boston: Beacon Press, 2001.

Mullis, Darrell, and Judith Handler Orloff. *The Accounting Game: Basic Accounting Fresh From the Lemonade Stand*. Naperville, IL: Sourcebooks, 2008.

Nahin, Paul J. *Dr. Euler's Fabulous Formula*. Princeton, NJ: Princeton University Press, 2006.

Nasar, Sylvia. *A Beautiful Mind: The Life of Mathematical Genius and Nobel Laureate John Nash*. New York: Simon & Schuster, 2001.

Oliver, Dean. *Basketball on Paper: Rules and Tools for Performance Analysis*. Washington, DC: Brassey's, 2004.

Pullan, J. M. *The History of the Abacus*. New York: F. A. Praeger, 1969.

Rafiquzzaman, M. *Fundamentals of Digital Logic and Microcomputer Design*. Hoboken, NJ: Wiley, 2005.

Rudin, W. *Principles of Mathematical Analysis*. New York: McGraw-Hill, 1953.

Salem, Lionel, Frédéric Testard, and Coralie Salem. *The Most Beautiful Mathematical Formulas*. Hoboken, NJ: Wiley, 1992.

Schwarz, Alan. *The Numbers Game: Baseball's Lifelong Fascination with Statistics*. New York: St. Martin's Press, 2004.

Smith, D. E. *History of Mathematics*. Vol. 2. New York: Dover Publications, 1958.

Solow, Daniel. *How to Read and Do Proofs: An Introduction to Mathematical Thought Process*. Hoboken, NJ: Wiley, 1982.

Steen, Lynn A. *On the Shoulders of Giants: New Approaches to Numeracy*. Washington, DC: National Academy Press, 1990.

Sterrett, Andrew. *101 Careers in Mathematics*. Washington, DC: The Mathematical Association of America, 1996.

Suzuki, Jeff. *A History of Mathematics*. Upper Saddle River, NJ: Prentice Hall, 2002.

Taylor, Alan D. *Mathematics and Politics: Strategy, Voting Power, and Proof*. New York: Springer-Verlag, 1995.

van der Waerden, B. L. *Geometry and Algebra in Ancient Civilizations*. Berlin: Springer, 1983.

Venema, G.A. *The Foundations of Geometry*. Upper Saddle River, NJ: Pearson Prentice Hall, 2006.

Weygandt, Jerry J., Paul D. Kimmel, and Donald E. Keiso. *Managerial Accounting: Tools for Business Decision Making*. Hoboken, NJ: Wiley, 2008.

Winkler, Peter. *Mathematical Puzzles: A Connoisseur's Collection*. Natick, MA: AK Peters, 2004.

Wright, Tommy, and Joyce Farmer. *A Bibliography of Selected Statistical Methods and Development Related to Census 2000*. Washington, DC: U.S. Bureau of the Census, 2000.

Yeldham, F. A. *The Teaching of Arithmetic Through Four Hundred Years (1535–1935)*. London: G. G. Harrap & Company, 1935.

Yong, L. L., and A. T. Se. *Fleeting Footsteps*. Singapore: Word Scientific Publications, 2004.

Zaslavsky, Claudia. *Africa Counts: Number and Pattern in African Culture*. Chicago: Lawrence Hill Books, 1999.

Zill, D. G. *Calculus with Analytic Geometry*. Boston: Prindle, Weber & Schmidt, 1985.

Journals and Magazines

The AMATYC Review
The American Mathematical Monthly
Association for Women in Mathematics Newsletter

Biometrics
Chance
The College Mathematics Journal
Experimental Mathematics
The Fibonacci Quarterly
Historia Mathematica
IMU-Net
Involve
Journal of Humanistic Mathematics
Journal of Integer Sequences
Journal of Recreational Mathematics
Journal of Statistics Education
Loci
MAA FOCUS
Math Horizons
Mathematics Magazine
Mathematics Teacher
NAM Newsletter
Notices of the American Mathematics Society
The Pentagon
Pi Mu Epsilon Journal
Plus Magazine
PRIMUS
Rose-Hulman Undergraduate Mathematics Journal
SIAM Review
Scholastic Math
Significance
Teaching Children Mathematics
Undergraduate Mathematics and Its Applications

Internet
American Institute of Mathematics
    www.aimath.org
The Algebra Project
    www.algebra.org
AMATYC
    www.amatyc.org
American Mathematical Society
    www.ams.org
American Statistical Association
    www.amstat.org
Association for Women in Mathematics
    www.awm-math.org
CryptoKids
    www.nsa.gov/kids
Datamath Calculator Museum
    www.datamath.org
Illuminations
    illuminations.nctm.org
MacTutor History of Mathematics
    www-history.mcs.st-and.ac.uk
Mathematical Fiction
    http://kasmana.people.cofc.edu/MATHFICT
Math for America
    www.mathforamerica.org
Math Forum
    www.mathforum.com
Math Fun Facts!
    www.math.hmc.edu/funfacts
MathDL
    mathdl.maa.org/mathDL
Mathematical Association of America
    www.maa.org
Mathematical Science Research Institute
    www.msri.org
The Museum of Mathematics
    www.momath.org
National Association of Mathematicians
    www.nam-math.org
National Council of Teachers of Mathematics
    www.nctm.org
RadicalMath
    www.radicalmath.org
Society for Industrial and Applied Mathematics
    www.siam.org
We Use Math
    www.weusemath.org
Wolfram MathWorld
    www.mathworld.wolfram.com

Index

Text and page numbers in **boldface** refer to main topics.

10-10-80 principle, 44, 45

abacus, 26, 27, 29, 51
Abd al-Latif (Muhammad Taragay Ulughbek), 25
Abel, Niels Henrik, 81, 85
Abel Prize, 85
Aberdeen Proving Ground, 134
Able Danger program, 138
Aboriginal kinship system, 129
Aboriginal paintings, 130
Abu Abdallah Book of Addition and Subtraction by the Indian Method (al-Khwarizmi), 25
Abu al-Wafa Buzjani, 23
Abu Kamil Shuja ibn Aslam, 12, 23
accounting, 1–4
    assets/liabilities in, 1, 3
    Benford's Law, 2
    as record keeping, 2
ACNielsen Corporation, 114
Acta Mathematica (journal), 169
actuarial science, 114
Adelstein, Abraham Manie, 13
ADFGVX code, 168
Adibi, Jafar, 109
adjustable rate mortgages, 95
advertising, 4–7
Aerodynamic Proving Ground, 168
Africa, Central, 7–9
Africa, Eastern, 9–11
Africa Counts (Zaslavsky), 7, 10
African Fractals (Eglash), 8

African Institute for Mathematical Sciences, 13
African Mathematical Union (AMU), 15
African Mathematics Olympiads, 15
Africa, Northern, 11–12
Africa, Southern, 12–14
Africa, Western, 14–15
Afrika Matematica (journal), 15
Afrikaners, 13
Agent Orange, 163
aha calculations, 70
Ahlfors, Lars, 82
Ahmed ibn Yusuf, Abu Ja'far, 12
aircraft design
    helium airships, 168
airplanes/flight
    World War II and, 173
al-Banna, al-Marrakushi ibn, 12
Albategnius, 119
al-Biruni, Abu Arrayhan, 24, 25
algebra and algebra education
    al-Khwarizmi and, 23, 25
    Arabic/Islamic mathematics, 22
    Babylonian mathematics, 38
algorithms
    genetic, 138
    information theory and, 141
    iterative, 43
Alhazen, 24
Alighieri, Dante, 146
al-Kashi, Jamshid, 23
    Miftah al-Hisab (*Calculators' Key*), 24

    The Key to Arithmetic, 31
    Treatise on the Circumference, 31
al-Khujandi, Abu Mahmud, 25
al-Khwarizmi, 119
    Al-kitab al-muhtasar fi hisab al-jabr wa-l-muqabala (*Compendium on Calculation by Completion and Reduction*), 23
    Book of Addition and Subtraction by the Indian Method, 25
Al-kitab al-muhtasar fi hisab al-jabr wa-l-muqabala (*Compendium on Calculation by Completion and Reduction*) (al-Khwarizmi), 23
Almagest, The (Ptolemy), 11
al-Maghribi, Samu'il (al-Samaw'al), 23, 24
Al Qaeda, 109, 139
al-Samaw'al, 23
al-Tusi, Nasir, 31
    Treatise on the Quadrilateral, 24
Alvarez, Luis, 104, 105
al-Zarqali (Arzarchel), 119
A Mathematician Plays the Stock Market (Paulos), 154
America, Caribbean, 15–17
America, Central, 17–18
American Express Corporation, 63
American Mathematical Society (AMS), 19
America, North, 18–20
American Pension Corporation, 136
America, South, 20–22

179

Amitsur, Shimshon Avraham, 34
amortization, 95, 96, 112
"A Multivariate EWMA Approach to Monitor Process Dispersion" (Bernard), 17
André, Désiré, 73
Andre, John, 148
Angkor Wat, 28
Anh Le, 139
An Introduction to Algebra (Day), 54
annuity tables, 107
Anti-Submarine Warfare Operations Research Group, 172, 173
Appendix (Bolyai), 80
Applied Mathematics Panel, 173
Arabic/Islamic mathematics, 22–24
    combinatorics, 24
    decimal system, 22, 23
    geometry, 23
    numerical mathematics, 24
    trigonometry, 22, 24
Arab Journal of Mathematics and Mathematical Sciences, 34
Archimedes
    contributions, 25
    "Eureka", 90
    as the father of integral calculus, 90
    Greek mathematics and, 83
    siege of Syracuse and, 104
    volume of a sphere and, 90
Archytas of Terntum, 83
Arf invariant in algebraic topology, 33
Arf rings, 34
Arf semigroups, 34
Army Corps of Engineers, 55
Army Signal Corps, 134
Arnold, Benedict, 148
Arrow, Kenneth, 58, 72, 166
Arrow's Paradox, 58
artillery
    types of, 52
Aryabhata the Elder, 31
Aryabhatiya (Aryabhata), 31
Arzarchel, 119
Ascher, Marcia, 10, 129
Ascoli, Giulio, 83
Asia, Central and Northern (Russia), 24–26
Asia, eastern, 26–28
    China, 26
    educational philosophy, 26

Hong Kong, 27
Japan, 27
Mongolia, 27
North Korea, 27
number system, 26
South Korea, 27
Taiwan, 28
Asia, southeastern, 28–30
    Brunei, Myanmar, and the Philippines, 30
    Cambodia, Laos, and Vietnam, 30
    early history of, 28
    Indonesia, 30
    Singapore and Malaysia, 29
    Thailand, 29
Asia, southern, 30–32
    history of, 31
    mathematics education, 31
Asia, western, 32–35
    Babylon, 33
    Israel, 34
    Ottoman Empire and Turkey, 33
Asimov, Isaac, 138
    Foundation, 138
assets, 1, 2, 3, 40, 42
Association Mathématique Algérienne, 12
Assumption College (Bangkok), 29
Astronomia Nova (Kepler), 84
astronomy
    Chinese and, 49
    early mural sextant, 25
    Greeks and, 91
    Islamic Golden Age and, 11
    Mayans and, 99
Atanasoff-Berry Computer (ABC), 171
Atanasoff, John V., 171
Atiyah, Michael F., 82
atomic bomb (Manhattan Project), 35–37
    authorization of, 35
    Cold War and, 56, 57
    energy-mass equivalence, 36
    nuclear reactions, 35
    scientists and, 35
    World War II and, 35
Australia, 129
Australian Mathematical Society, 129
automobile manufacturers, 4
automobiles
    purchasing, 59, 60

autoregressive integrated moving average (ARIMA), 161
Avempace, 119
Averroës (Ibn Rushd), 119
Avicenna, 119
Azerbaijan Journal of Mathematics, 34
Aztec civilization, 127

Babbage, Charles
    mechanical computers and, 103
Babylonian mathematics, 37–39
    history of, 33
Bacon, Roger, 119
Ba-ila settlement, 10
Baker, Alan, 82
Baker, Garth A., 16
Bakhshali manuscript, 31
ballistic pendulum, 122
ballistics studies
    computers and, 172
    military research, 168, 172
    underwater, 173
ballot problem, 76
Banach algebra theory, 79
Banach, Stefan, 42, 80
BankAmericard, 63
bankruptcy, business, 39–41
Bankruptcy Code of 1978, 42
bankruptcy, personal, 41–43
Banzhaf, John, 72, 74
Banzhaf Power Index, 74
Barbilian, Dan, 80
bar codes, 43–44
barometers, 83
Barricelli, Nils, 67
base-10 system, 25
basketry, 10
Baudot-Murray code, 172
Bayesian decision theory, 85
Beaugris, Louis, 17
Bellaso, Giovan, 54
Bell Telephone Laboratories, 144
Benford, Frank, 2
Benford's Law, 2, 101
Bernard, Serge, 17
Bernoulli, Daniel
    Leonhard Euler and, 122
    professional life,, 78
    projectile trajectories and, 122
Bernoulli, Jacob (Jacques, or James)
    Law of Large Numbers and, 107

Berry, Clifford, 171
Bertrand, Joseph, 73
Bessarion, Cardinal, 83
Besson, Jacques, 147
Bhaskaracharya II
    Siddhanta Siromani, 31
Black, Duncan
    The Theory of Committees and Elections, 72
Blackett, Patrick M. S., 172
Black, Fisher, 66
Bliss, Gilbert, 122, 168
Bocaccio
    Decameron, 120
Boers, 13
Boethius, Anicius Manlius Severinus, 83, 118
    Consolation of Philosophy, 118
Boethius's arithmetic, 83
Bollettino dell'Unione Matematica Italiana (journal), 83
Bolyai, János
    Appendix, 80
Boma, A. N., 8
Bonferroni, Carlo Emilio, 83
Bonferroni inequalities theory, 83
book ciphering, 148
Book of Addition and Subtraction by the Indian Method (al-Khwarizmi), 25
Boolean algebra (Boolean logic), 81
Boole, George
    Investigation of the Laws of Thought, 81
    The Mathematical Analysis of Logic, 81
boosting, 68
bootstrap aggregation (bagging), 68
Borda election method, 73, 165
Borda, Jean-Charles de, 73
Borgia, Cesare, 83
Bosnian Mathematical Society, 83
Bougainville, Louis-Antoine de
    Traité du calcul–intégral, 149
Bourbaki, Nicolas (collective pseud.), 85
Bourgainof, Jean, 85
Bragg, Braxton, 54
Brahe, Tycho, 84
Brahmagupta
    negative numbers concept and, 125
Brink, Chris, 13
Brouwer, Luitzen Egbertus Jan "L. E. J.", 169

Brown, Earl, 16
Brownian motion, 125
Brown, Robert, 125
Brunelleschi, 146
Buckmire, Ron, 16
budgeting, 44–46
Buell, Don Carlos, 54
Bulgarian Competition in Mathematics and Informatics, 80
Buonarotti, Michelangelo, 146
Bureau of Labor Statistics, 159
Burundi, 10
Busa Xaba, Abraham, 13
Bush, George W., 109, 164, 166
business, economics, and marketing
    bankruptcy, 40, 41, 42
    budgeting, 44
    business-to-business marketing (B2B), 114
    business-to-consumer marketing (B2C), 114
    credit, 63, 64, 94
    debt, 94, 125
    economic order quantity (EOQ), 111
    European Economic Community (EEC), 74
    gross domestic product (GDP), 91, 93, 125, 126
    interest, 42, 94, 96, 112, 160
    inventory models, 111
    loans, 94, 95, 112
    market research and, 114
    mutual funds, 123
    stocks/stock market, 40, 123, 124, 126, 154
business-to-business marketing (B2B), 114
business-to-consumer marketing (B2C), 114
Butler, C. Allen, 16
buxiban (cram schools), 28
Byzantine culture, 119

Cadogan, Charles C., 16
Caesar, Julius, 53
Cahit Arf, 33
calculators in society
    early history of, 152
calendars
    ancient Mesoamerica and, 17
    farming, 143

    Julian and Gregorian, 151
    lunar, 33
    Mayan, 98
    menstrual, 13
    religious festivals and, 33
Campbell, Lucy Jean, 16
Campbell, Merville O'Neale, 16
Canada, mathematics in, 18
Canadian Mathematical Society (CMS), 19
cannons, 122
Canterbury Tales (Chaucer), 120
Cardano, Gerolamo
    cubic equations and, 147
Caribbean Journal of Mathematical and Computing Sciences, The, 16
Caritat, Marie Jean Antoine Nicolas de. Marquis de Condorcet, 74
Carleson, Lennart, 82
Carley, Kathleen, 109
Carolingian Renaissance, 118
car purchases, 60
Carraher, David, 60
Cartan, Henri, 86
Carte Blanche, 63
cartography, 104, 132
Catalan, Eugène, 73
catastrophe theory, 84
Catherine II (Catherine the Great), 78
Cauchy, Augustin Louis
    complex function theory and, 85
Cavalieri's Principle, 50
Cayley, Arthur, 103
census, 46–48
Census Act of 1800, 47
Center for Promotion of Mathematical Research (Thailand), 30
Central Limit Theorem (CLT), 107
Cetti, Francesco, 83
Ceva, Giovanni, 83
Chadwick, Edwin, 102
Chapter 7 bankruptcy (liquidation), 42
Chapter 13 bankruptcy (reorganization), 42
Charlemagne, 118
Chaucer, Geoffrey
    Canterbury Tales, 120
Chebyshev, Pafnuty, 103
Cheney, Dick, 110
Cheng Dawei
    Suanfa Tongzong (*General Source of Computational Methods*), 51

Chern, Shiing-Shen, 27
Chinese mathematics, 48–52
    East meets West period, 51
    foundation period, 49
    golden period, 50
    number system, 26
Chinese Remainder Theorem, 50, 51
Chitika, 5
Chokwe people, 8
Chouren Zhuan (*Biographies of Astronomers and Mathematicians*) (ed. Ruan Yuan), 51
Christian religious tenets, 118
chronometers, 104
Church, Albert, 54
Churchill, Winston, 108
Church of San Lorenzo, 146
ciphers, 52, 53
Citrabhanu, 31
Civil War, U.S., 52–55
    cryptography, 53
    impact on education, 54
classic mathematical problems
    ballot problem, 76
    Correlation Problem, 99
    coupon collector problem, 62
    river-crossing puzzle, 10
    three construction problems, 89
"Classification of Countable Torsion-Free Abelian Groups" (Campbell), 16
Closs, Michael, 127
Coca-Cola Company, 61
codebooks, 53, 170
code breaking, 53
Code of Hammurabi, 107
code talking, 171
coding and encryption
    Civil War and, 53
    code talking, 171
    Revolutionary War and, 148
    World War I and, 170
    World War II and, 134, 170
Cold War, 55–59
    arms race, 56, 57
    competition during, 19, 58, 78, 84
    intelligence/counterintelligence and, 108
Colojoara, Ion, 80
Columbus, Christopher, 21, 147
combinatorics, 24, 87
Commedia (Dante), 120

Commissions on Mathematics Education in Africa, 15
comparable transaction method, 40
comparison shopping, 59–61
Complete Mancala Games Book, The (Russ), 7, 10
compound interest, 42
computers
    hacking, 53
    steam powered, 103
    World War II and, 171
Comrade Deuch (Gaing Kek Ieu), 30
Condorcet criterion, 58
Condorcet winner, 58, 166
Connes, Alain, 85
Consolation of Philosophy (Boethius), 118
continuous quality improvement, 145
conversion rates, 5
Cooper, Lionel, 13
Copernicus, Nicolaus, 84
    heliocentric astronomy and, 147
Correlation Problem, 99
Corson, William, 163
Costley, Charles Gladstone, 16
cost per click (CPC), 5
cost per mile (CPM), 5
coupon collector problem, 62
coupons and rebates, 61–63
credit cards, 63–65
    credit bureaus and, 64
    data mining and, 6, 63
    fraud detection and, 63
credit ratings, 94
Critical Infrastructure Protection (Lewis), 109
Croatian Mathematical Society, 83
Crozet, Claudius, 54
Cuba, 16
currency exchange, 65–67
Current Population Survey (U.S. Census Bureau), 160
Custer, George Armstrong, 54
"Cylinder Measures" (Millington), 16
Cyprus Mathematical Olympiad, 34
Cyprus Mathematical Society, 34, 83
Czech-Polish-Slovak Match, 80

D'Ambrosio, Ubiritan, 21
dancing
    in West Africa, 14

Dante
    Commedia, 120
Dantzig, George, 58, 157
data mining, 67–69, 109
Davies, Charles, 54
Davis, Jefferson, 54
Day, Jeremiah
    An Introduction to Algebra, 54
debt-to-income ratios, 94
Decameron (Bocaccio), 120
decimal system, 22, 25, 49
de Gusmão, Bartolomeu, 21
Deligne, Pierre, 85
Demidovich, Boris Pavlovich, 79
Deming, W. Edwards
    Statistical Methods from the Viewpoint of Quality Control, 145
de Moivre, Abraham, 107
Dennison, William, 53
Descartes, René
    analytical geometry and, 84
    Discourse on the Method, 84
    Meditations on First Philosophy, 84
    as philosopher, 84
    Principles of Philosophy, 84
DeWitt, Simeon, 149
Dias, Bartholomeo, 147
dictatorships, 165, 166
Dieudonne, Jean, 85
Diners Club, 63
Diophantes, 91
Discourse on the Method (Descartes), 84
Discourses on Livy (Machiavelli), 146
Distance Measuring Equipment (DME) system, 173
Dodge, H. F., 144
Doll, Richard, 81
Domesday Book, 81
Donaldson, Simon, 82
Doppler, Christian, 134
Doubleday, Abner, 54
Dow Jones Industrial Average (DJIA), 124, 154
Dresden Codex, 99
Dzhumadildayev, Askar, 26

Eckert, John Presper, 172
École Polytechnique, 85, 103
economic order quantity (EOQ), 111
Eglash
    African Fractals, 8

Egyptian mathematics, 69–72
Einstein, Albert
    personal and proffessional life of, 151
    theory of relativity and, 168
Einstein Institute of Mathematics, 34
elections, 72–77
    ballot problem, 76
    exit polling, 75
    types of, 72, 73
    U.S. Electoral College, 75
    weighted voting, 74
Electrical Numerical Integrator and Calculator (ENIAC), 172–173
Elements (Euclid)
    Abraham Lincoln and, 54
    deductive logic and, 90
    terminology, 22
    translations, 11, 51
Enclosure Acts, 102
Engels, Friedrich, 102
Engle, Robert, 92
Enigma code, 171
Enron Corporation, 109
enterprise value (EV), 40
Epitome of Copernican Astronomy, The (Kepler), 84
Eratosthenes of Cyrene, 11, 91
Erdos, Paul, 80
Eritrea, 10
ethnomathematics, 14, 21, 88
Euclid of Alexandria
    Abraham Lincoln and, 54
    Arabic/Islamic mathematics and, 22
    axiomatic geometry and, 90
    as library leader, 90
Eulerian Graphs, 8
Euler, Leonhard
    functions and, 85
    at Saint Petersburg Academy, 78
    Seven Bridges of Konigsberg and, 85
"Eureka" (Archimedes), 90
European Economic Community (EEC), 74
Europe, eastern, 77–81
Europe, northern, 81–82
Europe, southern, 82–84
Europe, western, 84–86
expenditure method (GDP), 92

fair market value (FMV), 40
Faltings, Gerd, 85
Fannie Mae, 64
Farrell, Edward, 16
Federal Deposit Insurance Corporation (FDIC), 123
federal tax tables, 100
Fermat, Pierre de
    probability and, 84
Fermat's Last Theorem, 81, 84
Fermi, Enrico, 35
Feynman, Richard, 35
Fibonacci, Leonardo, 12, 23
Fibonacci sequence, 83
FICO score, 64, 94
Fields Medal, 27, 82, 85
Fiqh al-Hisab (Ibn Mun'im), 24
Fiss, Andrew, 55
fixed rate mortgages, 95
FLOW-MATIC, 132
Ford Motor Co., 4, 111
foreign exchange (FX) market, 65
formal concept analysis (FCA), 109
Foundation (Asimov), 138
Fourier, Joseph Baptiste Joseph
    Bonaparte, Napoleon and, 104
Fourier transforms, 79
fractals
    in African societies, 7
Franklin, Benjamin, 148, 151
fraud detection
    accounting and, 3
    credit card, 63
    data mining and, 68
    probability theory and, 3
    Social Security and, 137
    taxes, 101
Freddie Mac, 64
Friedman, William, 134, 168
futures market, 110
FX market, 65

"Gadget" (nuclear test bomb), 36
Gaing Kek Ieu "Comrade Deuch", 30
Galileo (Galileo Galilei)
    influence of, 83
    theory of gravity and, 83
Galois, Evariste, 81, 85
Gama, Vasco da, 147
game theory, 55
    Cold War and, 55, 164
    strategy and tactics in, 157, 158
Gaugin, Paul, 130

Gauss, Carl F.
    contributions of, 85
Gaussian distribution, 85
Geber (Jabir ibn Aflah), 119
GEE system, 173
Gelfand, Israel Moiseevich, 79
Gelfand representation, 79
General Electric Company (GE), 145
General Motors Company (GM), 4
General Theory of Employment, Interest and Money, The (Keynes), 160
genetic algorithms, 138
geometric art, 130
geometry and geometry education
    Arab/Islamic, 23
    Babylonian, 38
    early history of, 142
    plane and spherical, 23
    prehistorical, 142
Geometry From Africa (Gerdes), 7
Georgian National Mathematical Committee, 34
Gerdes, Paulus
    Geometry From Africa, 7
Gerry, Elbridge, 86
gerrymandering, 86–88
Gfarm Grid File System, 67
Gibbard, Allan, 167
Giotto di Bodone, 146
Gödel, Kurt
    Incompleteness theorem, 79, 85
Goldbach, Christian, 78
Goldstine, Herman, 173
Google, 5
Gore, Al, 164, 166, 167
Gorgas, Josiah, 54
Gougu theorem, 49
Gowers, Timothy, 82
Granger, Clive, 92
Grant, Ulysses S., 54
graphical user interfaces (GUIs), 67
graphs, 108
Greek mathematics, 88–91
    Archimedes, 83, 90
    astronomy and, 89
    deductive logic and, 90
    early mathematicians, 89
Gregorian calendar, 99
groma, 150
Gromov, Mikhail, 85
gross domestic product (GDP), 91–93, 125, 126

Grosseteste, Robert, 119
Grothendieck, Alexander, 163
group theory, 81
Guo Shoujing (Kuo Shou-ching), 50
Guthrie, Francis, 13

hacking, computer, 53
hagwons (academies), 28
Halmos, Paul Richard, 80
Hamilton-type circuits, 5
Hamilton, William Rowan, 81
Han dynasty, 49
Hardy, Godfrey (G. H.)
    contributions of, 32, 169
    Srinivasa Ramanujan and, 32
Hare system, 165
Harmonices Mundi (Kepler), 84
Hawking, Stephen, 81
Headley, Velmer, 16
heap (Egyptian mathematics), 71
Henry, Leighton, 16
Henry, Warren, 163
Hexagrams, 49
hieroglyphics, 96
Hilbert, David
    contributions of, 85, 168
Hilbert spaces, 80, 85
Hill, Austin Bradford, 81
Hilton, Conrad, 64
Hindu-Arabic numerals
    widespread usage, 21, 23, 24, 147
    zeros and, 98
Hironaka, Heisuke, 27
Hiroshima, Japan, 35
Hirzebruch, Friedrich, 86
Hispalensis, Isidorus, 83
Holland, John, 138
home buying, 93–96
Hood, John Bell, 54
Hooker, Joseph, 54
Hopper, Grace Murray, 132
Hormander, Lars, 82
Horner-Ruffini method, 50, 51
horsepower, 103
House of Wisdom, 33
house purchases, 60
hovercraft, 163
howitzers, 52
Hunt, Fern, 16
Hypatia of Alexandria, 11, 91, 151

Iacob, Caius, 80
IBM Corporation, 6, 43
Ibn al-Haytham (Alhazen), 12
Ibn Baija (Avempace), 119
Ibn Mun'im
    Fiqh al-Hisab, 24
Ibn Rushd (Averroës), 119
Ibn Turk, Abd Al-Hamid, 31
Ibn Yunus ibn Abd al-Rahman, 12
Identification Friend or Foe (IFF), 104, 105
Incan and Mayan mathematics, 96–99
    astronomy, 99
    base-10 system, 127
    calendars, 98
    Incan civilization, 20
    quipus, 97
    zero and, 98
income method (GDP), 91
income tax, 99–102
Incompleteness theorem, 79, 85
independence of irrelevant alternatives method, 166
Indian mathematics
    decimal system and, 22, 26
    negative numbers and, 31
    sine function and, 24
    zeros and, 31
industrial revolution, 102–103
Industrial Revolution
    accounting and, 1
    employment and, 102, 160
    mass production and, 91, 153
    mathematics and, 102
    steam engines and, 102, 103
infantry (aerial and ground movements), 104–106
infinity
    Euclid's fifth postulate and, 90
inflation, 126
information systems, 48
information theory, 141
Institute of Mathematics of the National Academy of Sciences (Republic of Armenia), 34
insurance, 106–108
integral calculus, 90
intelligence and counterintelligence, 108–111
intentional debt, 126
interest rate, 94, 95, 112

International Congress of Mathematicians, 163
International Mathematical Olympiad (IMO), 12, 13, 25
    Azerbaijan, 34
    China and, 27
    eastern Europe, 80
    Israel, 34
    Kuwait, 34
    Malaysia, 29
    Mongolia, 27
    northern Europe, 82
    Republic of Armenia, 34
    southern Asia, 32
    South Korea, 27
    Turkey, 34
    Vietnam, 30
    western Europe, 86
internet
    advertising, 5
Introduction to Computational Studies (*Suanxue Qimeng*) (Zhu Shijie), 51
inventory models, 111–112
Isaac, Earl, 64
Israel Journal of Mathematics, 34
Israel Mathematical Union, 34
Italian Mathematical Union, 83
iterative algorithms, 43
Ito, Kiyoshi, 27

Jabir ibn Aflah (Geber), 119
Jackson, Thomas "Stonewall", 54
James, Lancelot F., 16
Japanese Technology Board, 134
Jean-Michel, Jean-Michelet, 17
Jia Xian, 50
Jia Xian Triangle, 50, 51
Jigme Khesar Namgyel Wangchuck, King, 32
Jigu Suanjing (*Continuation of Ancient Mathematics*) (Wang Xiaotong), 50
Johnson, Neil, 139
Johnston, Albert Sidney, 54
Johnston, Joseph, 54
Jones, Vaughan, 129
juku schools, 27
Juran, Joseph M., 144
Jyesthadeva, 31

Kai Fang Shu, 49
Kaigun Ango-Sho D (JN-25B), 134

Kangaroo Mathematics Contest, 80
kappa test, 170
Karaji, 23
Karlin, Samuel, 111
Katyayana, 31
Kempe, Alfred, 103
kente cloth, 14
Kepler, Johannes
    Astronomia Nova, 84
    Harmonices Mundi, 84
    laws of planetary motion, 81, 84
    The Epitome of Copernican Astronomy, 84
Kepler's Laws, 81, 84
Kerala, 31
Keynes, John, 92, 160
    The General Theory of Employment, Interest and Money, 160
Key to Arithmetic, The (al-Kashi), 31
Khayyam, Omar, 23, 31
Kim Il-Sung University, 27
kinship systems, 129
Klein, Felix
    Aerodynamic Proving Ground and, 168
Knaster, Bronisław, 42
"Knee Deep in the Big Muddy" (Straw), 164
knowledge discovery in databases (KDD), 67
Kodaira, Kunihiko, 27
Kolmogorov, A. N., 79
Kuo Shou-ching (Guo Shoujing), 50

Lagrange, Joseph-Louis, 85
Lanchester, Frederick, 56, 168
Lanchester model of warfare, 56
Lanczos, Cornelius, 80
Langley Laboratory, 168
Lao People's Democratic Republic (Laos), 28
Laplace, Pierre de, 85, 107
Latini, Brunetto, 146
lattice theory, 108
law of exponential growth, 42
Law of Large Numbers (LLN), 107
laws of planetary motion, 81
Laws of Thought, The (Boole), 81
Lay, Kenneth, 109
Lebesgue, Henri Léon, 79
Lebesgue integral, 79
Lebombo bone, 12

Lecroix, Sylvestre-Francois, 103
Lee, Robert E., 54
Legendre, Adrien-Marie, 103
Leibniz, Gottfried Wilhelm
    calculus and, 81, 84
    mathematics education and, 78
Lenin, Vladimir, 55
Leonardo da Vinci
    paintings, 146
    siege machines and, 83
Leontief, Wassily, 59
Leray, Jean, 85
Leslie, Joshua, 16
Lewis, Ted
    Critical Infrastructure Protection, 109
Liber Abaci (Fibonacci), 23
Library of Alexandria, 11, 25, 90, 91
Li Chunfeng, 50
Lie groups, 79
Lincoln, Abraham, 53, 54
Lindenstrauss, Elon, 34
linear programming problems, 58, 156, 157
Linnaeus, Charles, 147
Lions, Pierre-Louis, 85
Littlewood, John, 122, 168
Liu Hui
    Sea Island Mathematical Manual (*Haidao Suan Jing*), 50
    Six Arts, 49
Li Zhi (Li Yeh), 50
    Sea Mirror of the Circle Measurements (*Ce Yuan Hai Jing*), 51
    Yi Gu Yan Duan (*New Steps in Computation*), 51
loans, 94, 95, 112–114
Lobachevsky, Nicolai Ivanovich, 78
locations systems, 123
Lockheed Corporation, 163
Long Count days, 98
Long Range Navigation (LORAN), 173
Longstreet, James, 54
lotteries, 120, 121
Louis XIV, King, 47
Lovász, László, 80
Lovelace, Ada, 103
Lovell, James, 148
Luoshu, 49

Machiavelli, Niccolò
    Discourses on Livy, 146
    Prince, 146

Maggi, Girolamo, 83
MAGIC code, 135
Magnus, Albertus, 119
Magnus Effect, 122
Magnus, Heinrich, 122
Maimonides, Moses, 41
Mair, Bernard, 16
Malaysian Mathematical Sciences Society, 29
Malthus, Thomas, 47
Mancala, 7, 9
Mandelbrot, Benoit, 80, 154
Mandelbrot set, 80
Mansur, Abu Nasr, 25
Maori culture, 130
Marar, K. M., 31
Marchetti, Alessandro, 83
marine navigation, 132
market research, 114–116
market-value-weighted stock indices, 155
Markopoulou, Athina, 139
Markowitz, Harry, 124
Marshall Islands, 131
Marx, Karl, 55
Maslow, Abraham, 141
Maslow's hierarchy, 141
Mathematical Analysis of Logic, The (Boole), 81
Mathematical Association of America (MAA), 19
Mathematical Correspondent, The (journal), 149
mathematical modeling
    in accounting, 3
    combat modeling, 106
    comparison shopping and, 61
    for economic order quantity (EOQ), 111
    of kinship systems, 129
    linear programming models, 58, 156, 157
    for predicting attacks, 138, 139
    for taxes, 101
"Mathematical Psychology of War, The" (Richardson), 169
Mathematical Society of Serbia, 83
Mathematico-Physical Journal, 79
mathematics: discovery or invention, 116–118
Mathematics in Africa (Int'l Mathematical Union), 8

Mathematics, Physics, and Astronomy Society of Slovenia, 83
Math Kangaroo, 80
Mattangs, 131
Mauchly, John, 172
McCalla, Clement, 16
McClellan, George, 54
McNamara, Frank, 63
McNamara, Robert, 164
Meade, George G., 54
Mecca, 33
Meddos, 131
Meditations on First Philosophy (Descartes), 84
Meet the Press (news program), 110
Mehmed-II, Sultan, 33
Mei Juecheng, 51
Melanchthon, Philip, 147
Melville, Herman
    Moby Dick, 130
Meril, Alex, 17
Merton, Robert C., 66
Michell, Keith, 16
Middle Ages, 118–120
Miftah al-Hisab (*Calculators' Key*) (al-Kashi), 24
military code
    code talking, 171
    Enigma code, 171
    Morse code, 53, 168
    superencypherment, 170, 172
    trench codes, 168
military draft, 120, 120–122
military research in mathematics, 134, 168
Millington, Hugh G. R., 16
Ming Antu, 27
Ming dynasty, 49
Minié ball ammunition, 52
Minié, Claude, 52
missiles, 122–123
Mitofsky, Warren, 73
Mittag, Leffler, Gosta, 169
Moby Dick (Melville), 130
Moisil, Grigore C., 80
Mo Jing, 49
Monge, Gaspard, 103
Mongkut, King, 29
monotonicity criterion, 166
Montenegro Mathematical Society, 83
moon
    lunar calendars, 33

Mori, Shigefumi, 27
Morse code, 53, 168
mortgages, 94
Moser, Jurgen, 86
Motorola, Inc., 145
Mouhe Fanggai (double vault), 50
Moulton, Forest Ray, 168
"Multiobjective and Large-Scale Linear Programming" (Osei-Bryson), 16
multivariate probability inequalities, 4, 5
Mumford, David, 82
Munk, Max, 134
mutual funds, 123–125
Mutually Assured Destruction (MAD), 56, 164
myriad-grouping system, 26

Nader, Ralph, 164, 166, 167
Nagasaki, Japan, 35
Napoleon Bonaparte (Napoleon III), 52, 104
Napoleon (howitzer), 52
Napoleon's Theorem, 104
NASDAQ, 124
Nash Equilibria, 56
Nash, John, 57
National Academy (French Indonesia), 30
National BankAmericard, Inc. (NBI), 63
National Council of Teachers of Mathematics (NCTM)
    No Child Left Behind (NCLB) Act and, 19
National Mathematical Olympiad (Malaysia), 29
National Popular Vote Compact, 75
National Security Agency (NSA), 108, 109
Native American mathematics, 127–129
    code talking and, 171
    minorities in mathematics and science, 48
Natural Magic (Porta), 147
navigation systems, 89, 123
Nazca lines, 20
negative numbers
    Chinese mathematics and, 51
    Indian mathematics and, 31
Neoplatonism, 147
Netflix, 6
Neugebauer, Otto, 37, 38

Neumann, John von, 35, 57, 58, 80
New and Complete System of Arithmetick, The (Pike), 149
Newcomen engine, 103
New Math, 163
Newton, Sir Isaac
    invention of calculus, 81
    Philosophiae Naturalis Principia Mathematica, 81
New York Stock Exchange, 124
New Zealand, 129, 130
New Zealand Mathematical Society, 129
Ngo Bao Chau, 85
Nielsen, Arthur C., 114
Nielsen ratings, 114
Nigrini, Mark, 101
Nine Chapters on the Mathematical Art (Chinese text), 49, 50, 151
No Child Left Behind (NCLB) Act, 19
nominal GDP, 92
North Atlantic Treaty Organization (NATO), 56
Nuclear fission, 35
number and operations
    Roman numerals, 147, 150
numbers and God
    infinity and, 90
Nunes, Terezinha, 60

Oceania, Australia and New Zealand, 129–130
Oceania, Pacific Islands, 131–132
Once Were Warriors (movie), 130
On graphs not containing independent circuits (Lovász), 80
Onicescu, Octav, 80
"On Operations on Abstract Sets and their Application to Integral Equations" (Banach), 80
Opana Point, 135
operations research (OR), 58, 172, 173
Oppenheim, Alexander, 29
Oppenheimer, J. Robert, 35
optical scanners, 43
Osei-Bryson, Kweku-Muata Agyei, 16

Pacioli, Luca
    Summa de Arithmetica, Geometria, Proportioni et Proportionalita, 1, 132
painting, 130, 146
Paisano, Edna Lee, 48

Panini, 31
paradoxical preferences, 140
Pareto condition, 166
Park, Bletchley, 171
Pascal, Blaise
    invention of Pascaline, 151
    probability theory and, 84
Pascal, Étienne, 151
Passarola, 21
patterns
    decorative, 7, 128, 130
    geometric, 10, 14
Paulos, John Allen
    A Mathematician Plays the Stock Market, 154
payroll, 132–134
Pearl Harbor, attack on, 134–135
Peart, Paul, 16
Peaucellier, Charles-Nicolas, 103
Penrose-Banzhaf Power Index, 73
Penrose, Lionel, 73
pensions, IRAs, and Social Security, 135–138
Perelman, Grigori "Grisha", 79
Perelman, Yakov Isidorovich, 79
Persons, Jan, 10
Péter, Rózsa, 80
Peter the Great, 78
Petrarca, Francesco, 146
Philadelphia Storage Battery Company (Philco), 43
philosophers
    Neoplatonists, 147
    René Descartes, 84
Philosophiae Naturalis Principia Mathematica (Newton), 81
Phoenix Mathematics, Inc., 108
Pi, 24
Pickett, George E., 54
Picone, Mauro, 168
Piero della Francesca, 146
Pike, Nicholas, 149
    The New and Complete System of Arithmetick, 149
Pincherle, Salvatore, 83
place-value structures, 143
Plankalkül, 172
Platonists, 116
Playfair, William, 125
plurality elections, 166
plutonium bombs, 35, 36
Poincaré conjecture, 79

Poincaré, Jules Henri, 79
Pol Pot (Saloth Sar), 30
Pólya, George, 79
Pompeiu, Dimitrie, 80
Ponzi schemes, 137
Porta, Giambattista della
    Natural Magic, 147
Portuguese Society of Mathematics, 83
Post Cereal, 61
Post, Charles, 61
power laws, 138
Precious Mirror of the Four Elements (*Si Yuan Yujian*) (Zhu Shijie), 51
predicting attacks, 138–139
    National Security Agency (NSA), 108
predicting preferences, 139–142
preference ballots, 165
prehistory, 142–144
premiums, insurance, 106, 107
Price, G. Baley, 173
primary sampling units (PSUs), 160
Prince (Machiavelli), 146
Principia Mathematica (Russell and Whitehead), 81, 169
Principles of Philosophy (Descartes), 84
private mortgage insurance (PMI), 94
probability
    fraud detection and, 3
    Native Americans and, 128
probability theory, 84, 107
product method (GDP), 92
Ptolemy, Claudius
    Almagest, The, 11
    Earth centered universe, 91
    table of chords, 24
Pythagoras of Samos, 83, 89
Pythagorean theorem
    Greek mathematics and, 89
Pythagorean tuning, 25

Qin Jiushao
    Shushu Jiuzhang (*Mathematical Treatise in Nine Sections*), 50
Qin Shi Huang (emperor), 49
quality control, 144–145
quantum mechanics, 36
quasi-empirism, 117
Quetelet, Adolphe, 102
queuing theory, 108
quipus, 20, 97, 98

Racine, Father, 31
radar, 104, 105
Radio Corporation of America (RCA), 43
Radó, Ferenc, 80
Raffles Institution, 29
Raffles, (Thomas) Stamford, 29
Rajagopal, C. T., 31
Ramanujan, Srinivasa
    mathematics education and, 32
    number theory and, 168
Ramelli, Agostino, 147
RAND Corporation, 56
Raphael
    School of Athens, 146
Rebbelibs, 131
redistricting, 86, 87
Rees, Mina, 173
reflexive theory, 108
Rejewski, Marian, 171
Renaissance, 146–147
Rényi, Alfréd, 80
Revenue Act (1926), 136
Revere, Paul, 148
Revolutionary War, U.S., 148–149
Rhind papyrus
    doubling/halving numbers and, 70
    recreational mathematics and, 72
Ricci, Matteo, 51
Richards, Donald St. P., 16
Richardson, Lewis Fry, 138, 169
Riesz, Frigyes, 79
risk pooling, 106, 108
risk transfer, 106, 107
river-crossing puzzle, 10
Robins, Benjamin, 122
Robinson, Karl, 16
Rockefeller, John D., 45
Romanian Master of Sciences, 80
Romanian National Olympiad, 80
roman mathematics, 149–151
roman numerals, 147, 150
Romig, H. G, 144
Roosevelt, Franklin D., 35, 134
rostro, 150
Roth, Klaus, 82
Royal Air Force (RAF) Coastal Command, 172
Royal Spanish Mathematical Society, 83
Rozycki, Zerzy, 171
R (software), 67
Ruan Yuan, 51

Rumsfeld, Donald, 110
Russell, Bertrand
    Principia Mathematica, 81, 169
Russ, Laurence
    The Complete Mancala Games Book, 7, 10

Saint Petersburg Academy, 78
sales tax and shipping fees, 151–153
Samarkand Observatory, 25
Samphan, Khieu, 30
Sampling Methods for Censuses and Surveys (U.S. Census Bureau), 46
Sarrus, Frederic, 103
Sar, Saloth (Pol Pot), 30
satellites
    military uses of, 106
Sato, Mikio, 27
Satterthwaite, Mark, 167
Saudi Association for Mathematical Sciences, 34
scheduling, 153–154
Schliemann, Analucia, 60
Schmidt, Stefan, 110
Schoenberg, Isaac Jacob, 80
Scholasticism, 119
Scholes, Myron, 66
School of Athens (Raphael), 146
School of Mathematical and Navigational Sciences (Moscow), 78
schools
    buxiban (cram schools), 28
    juku schools, 27
Schramm, Oded, 34
Schwartz, Laurent, 85
Schwarz, Stefan, 79
Scientific and Technological Research Council (Turkey), 34
Scotus, Duns, 119
Sea Island Mathematical Manual, The (*Haidao Suan Jing*) (Liu Hui), 50
Sea Mirror of the Circle Measurements (*Ce Yuan Hai Jing*) (Li Zhi), 51
seasonal extension (SARIMA), 161
secondary sampling units (SSUs), 160
Selberg, Atle, 82
Selective Service, 120
Sequential Pairwise elections, 74
Serre, Jean-Pierre, 85
Seven Bridges of Konigsberg, 85
severe acute respiratory syndrome (SARS), 110

sexagesimal notation, 37, 38
Shafarevich, Igor, 79
Shapely, Lloyd, 41
Shelah, Saharon, 34
Sherman, William Tecumseh, 54
Shewhart control charts, 144
Shewhart, Walter
    Statistical Methods from the Viewpoint of Quality Control, 145
Shewhart, Walter E., 144, 145
Shisima, 10
Shuli Jingyun (ed. Mei Juecheng), 51
Shushu Jiuzhang (*Mathematical Treatise in Nine Sections*) (Qin Jiushao), 50
SIAM News (journal), 93
Siddhanta Siromani (Bhaskaracharya II), 31
Siegel, Carl, 85
Sierpinski, Waclaw, 168
Sieve of Eratosthenes, 11
Signal Intelligence Service, 134
Silver, Bernard, 43
Singapore Mathematical Olympiad, 29
Singapore Mathematics Project Festival, 29
Singapore Math Method, 29
Sino-Korean number system, 28
Six Arts (Liu Yi), 49
Skewes number, 13
Skewes, Stanley, 13
Smarter Planet campaign, 6
Smith, Edmund Kirby, 54
Sobolev, Sergei Lvovich, 79
Social Choice and Individual Values (Arrow), 72
social choice theory, 109, 164
Social Constructivism, 117
socialism, 58
social network analysis (SNA), 108
social networks, 108
Sociedad Matemática Mexicana (Mexican Mathematical Society), 18
Société des Sciences Naturelles et Physiques du Maroc, 12
Soldo, Fabio, 139
"Some Results Related to the Generators of Cyclic Codes Over Zm" (Beaugris), 17
soroban, 26, 27
South African Mathematics Olympiad, 13
Southern Africa Mathematical Sciences Association, 13

Soviet Union, 24, 55, 56
S&P 500, 124
Spencer (rifle), 52
spherical geometry, 23
Sputnik, 19, 78
Stager, Anton, 53
Stancel, Valentin, 21
static budget, 44
Statistical Methods from the Viewpoint of Quality Control (Stewhart and Deming), 145
statistical process control (SPC), 144
steam engines, 91, 102, 103
Steinhaus, Hugo, 42
stick charts, 131
Stochastic calculus, 125
stock market indices, 154–156
stocks/stock market, 40, 123, 124, 154, 155
Stoilow, Simion, 80
Strategic Air Command (SAC), 56
strategic voting, 166
strategy and tactics, 156–159
Straw, Barry, 164
Stuart, J. E. B., 54
Studies in Conflict and Terrorism (journal), 109
Suanfa Tongzong (*General Source of Computational Methods*) (Cheng Dawei), 51
Summa de Arithmetica, Geometria, Proportioni et Proportionalita (Pacioli), 1, 132
superencypherment, 170, 172
support vector machines (SVMs), 138
Sutras, 162
Swerling, Peter, 122
Sylvester, James Joseph, 103
symbolism
    of "zero", 49, 97, 98

Taichi, 49
tally marks, 143
Talon, Jean, 47
Tan, Tony, 29
Tao, Terence, 129
Tartaglia, Niccolo, 122
telegraphy, 52, 54
Teller, Edward, 35
Ten Computational Canons (ed. Li Chunfeng), 50
terrorist cells, 109

Index   189

Thales of Miletus, 82, 89
"theaters of machines" (Besson and Ramelli), 147
The Institute of Mathematical Sciences (University of Malaya), 29
Theory of Committees and Elections, The (Black), 72
theory of linkages, 102
theory of relativity, 36, 168
Thomas Aquinas, Saint, 119
Thom, René, 84, 85
three construction problems, 89
Tian Yuan Shu (Method of Coefficient Array), 50, 51
tilings, 128
time series data, 161
Tirthaji, Sri Bharati Krsna
    Vedic Mathematics, 163
Titeica, Gheorghe, 80
Tits, Jacques, 85
Torricelli, Evangelista, 83
total quality management (TQM), 145
Traité du calcul-intégral (Bougainville), 149
Transvaal, 13
Treatise on the Circumference (al-Kashi), 31
Treatise on the Quadrilateral (al-Tusi), 24
trench codes, 168
Trends in International Mathematics and Science Study (TIMSS), 34
trigonometry
    Arabic/Islamic mathematics and, 22, 24
Trigrams, 49
"Trinity" (nuclear bomb test), 35
Tsoro Yematatu, 10
Tunisian Mathematical Society, 12
Turing, Alan, 108, 171
Turkish Mathematical Society, 34

UK census, 47
Ulam, Stanislaw, 35, 36
Ulughbek, Muhammad Taragay (Abd al-Latif), 25
unemployment, estimating, 159–161
United Nations Commission on Statistical Sampling (U.S. Census Bureau), 46
Universal Product Code (UPC), 43
University of Stellenbosch, 13
uranium bombs, 35, 37

U.S. Army Signal Corps, 52
USA Today (newspaper), 108
U.S. Census Bureau
    history of, 46
    unemployment estimates and, 160
U.S. Department of Commerce, 92
U.S. Department of the Navy, 44
U.S. Electoral College, 73, 75
U.S. Military, 120
U.S. Military Academy, 52, 104
U.S. Missile Defense Agency, 123
U.S. Naval Academy, 55
U.S. Supreme Court, 48

Vaidyanathaswamy, Ramaswamy, 32
Veblen, Oswald, 168
Vedic mathematics, 30, 161–163
Vedic Mathematics (Tirthaji), 163
Vedic scholars, 30
Velez-Rodriguez, Argelia, 17
VENONA, 171
Vidinli Hüseyin Tevfik Pasa, 33
Vietnamese Mathematical Society, 30
Vietnam War, 121, 163–164
Vigenère, Blaise de, 54
Vigenère cipher, 54
vigesimal and duodevigesimal number system, 96
Villani, Cedric, 85
Visa, 63
Volterra, Vito, 168
von Neumann, John, 122, 172
voting methods, 164–167
Vranceanu, Gheorghe, 80

Wall Street crash, 126
Wang Xiaotong, 50
    Jigu Suanjing (Continuation of Ancient Mathematics), 50
war contracts, 173
Warlpiri kinship system, 130
"war on terror", 108, 109
War Policy Committee (War Preparedness Committee), 134
Warring States period (China), 49
Warschawski, Stefan E., 79
Washington, George, 149
Watt, James, 103
Weaver, Warren, 173
weaving, 14
Web sites, 5, 61
Wedgwood, Josiah, 1

weighted voting, 165
Weil, André, 85, 86
WEKA, 67
Werner, Wendelin, 85
West African Examinations Council, 15
Whish, Charles, 31
Whitehead, Alfred North
    Principia Mathematica, 81, 169
Wiener, Norbert, 168
Wiles, Andrew, 81, 82, 84
William the Conqueror, 46
Winkler, Johann, 132
Wittgenstein, Ludwig, 117
Wohlstetter, Albert, 56
Wohlstetter, Roberta, 56
Wolf Prize in Mathematics, 27, 82, 85
women
    professional employment and, 173
Women in Mathematics in Africa, 15
wood carvings, 129, 130
Woodland, Norman J., 43
World War I, 167–170
World War II, 170–174
    atomic bomb (Manhattan Project) and, 35
    codes/cyphers and, 170, 171
    computers and, 171, 172
    intelligence/counterintelligence and, 108
    mathematicians and, 104, 105, 173
    navigation and, 173
    operations research (OR) and, 172

Xiangjie Jiuzhang Suanfa (Yang Hui), 51
Xu Guangqi, 51

Yahoo, 6
Yamamoto, Isoroku, 134
Yang Hui, 50
    Xiangjie Jiuzhang Suanfa, 51
Yates, Frank, 46
Yau, Shing-Tung, 27
Yi Gu Yan Duan (New Steps in Computation) (Li Zhi), 51
Yi Jing (I-ching or Book of Changes), 49
Ying Yang, 49
Yoccoz, Jean-Christophe, 85
yupana, 98
Yupaporn Kemprasit, 30
Yusuf ibn Ibrahim, 12

Z3 computer, 172

Zaslavsky, Claudia, 7
    Africa Counts, 7, 10
zero, 50, 96, 98
Zhangjiashan's tomb, 49
Zhou Bi Suan Jing (Anon.), 49
Zhou dynasty, 49

Zhui Shu (Zu Chongzhi), 50
Zhu Shijie, 50
    Introduction to Computational Studies, 51
    Precious Mirror of the Four Elements (*Si Yuan Yujian*), 51

Zimbabwe, 10
Zu Chongzhi, 50
Zu Geng, 50
Zuse, Konrad, 171
Zygalski, Henyrk, 171